CLINICAL DETECTIVE STORIES

A Problem-Based Approach to Clinical Cases
in Energy and Acid-Base Metabolism

CLINICAL DETECTIVE STORIES

A Problem-Based Approach to Clinical Cases in Energy and Acid-Base Metabolism

Mitchell L. Halperin
University of Toronto

Francis S. Rolleston
Medical Research Council
Ottawa

PORTLAND PRESS

London and Chapel Hill

Clinical Detective Stories: A Problem-Based Approach to Clinical Cases in Energy and Acid-Base Metabolism

Published by Portland Press Incorporated, PO Box 2191, Chapel Hill, North Carolina 27515-2191, USA
Portland Press Limited, 59 Portland Place, London WIN 3AJ, UK

Distributed in North and South America by Portland Press, Chapel Hill
Distributed elsewhere by Portland Press, London

ISBN 1 85578 999 X

Library of Congress Cataloguing-in-Publication Data applied for
British Library Cataloguing-in-Publication Data - A catalogue record for this book is available from the British Library.

Printed in Canada
on recycled paper
March 1993

Editor:	Anne Harmon
Production:	LetroMac Design & Printing, Inc.
	Lorne Jacobs
Cover Art:	Lisa Shoemaker

PORTLAND PRESS

London and Chapel Hill

ACKNOWLEDGEMENTS

We are grateful to the following people for their expert help in preparing this book: Marcel Blanchaer, Sean Brosnan, Surinder Cheema-Dhadli, Aileen, Brenda, Frank, and Ross Halperin, Bob Jungas, Jolly Mangat, Bob Murray, Luisa Raijman, and Frank Vella. Special thanks to Anne Harmon, who edited the text, and to Lorne Jacobs, who produced it.

DEDICATION

With sincere appreciation to Brenda and Susan.

LIST OF CHAPTERS

LIST OF CASES

PREFACE

Medical students lead fiercely crowded lives. So do nursing students. Their curriculum is heavy with detail. Huge amounts of information must be absorbed and kept in mind at least long enough to get through the exams.

Given these realities, most students of health science are likely to see their biochemisy course as a rite of passage, a ritual that has little to do with clinical practice. They slog through the molecular details with uninspired diligence; memory replaces understanding; and physiological processes essential to sound clinical work may go unlearned.

We and many of our peers have discovered that these same students take strong interest in biochemical events if they approach the subject within the context of clinical cases and can actively engage in efforts to solve problems that occur in the "real world".

Hence this book.

We are writing for students who want to apply fundamental concepts of energy and acid-base metabolism to clinical problems. During several years of class testing, we have found this problem-solving approach to be very effective in teaching metabolism to students of the health professions at all stages of their training. By basing this book on clinical problems in human nutrition, exercise, and veterinary medicine, we show the vivid relevance of biochemistry and physiology to everyday life. Although we emphasize principles of biochemistry and physiology, we also include aspects of endocrinology, gastroenterology, and respiratory and cardiovascular physiology, thus allowing a thorough sense of the clinical problems here presented. This problem-based strategy prompts students to propose various hypotheses in response to the information given in each case. They then put these hypotheses to quantitative tests, a process that generates a deeper understanding of clinical events than could ever be gained by listening to lectures or reading textbooks.

By focusing on concepts and avoiding unnecessary detail, we encourage a logical and conceptual approach to understanding disease and treatment. We challenge readers to think about the biochemistry and physiology underlying cases, leaving them better equipped to deal with a great diversity of situations when they enter clinical practice.

Clinical Detective Stories is divided into five sections:
1. Energy and Acid-Base Physiology
2. Clinical Applications
3. Metabolic Function and Control
4. Discussion of Questions
5. Structures and Pathways

In the first two sections, which form the bulk of the book, we focus on clinical medicine, including only material we regard as essential for understanding the concepts presented. Those needing further biochemical information will find in Section Three the interactions between control mechanisms and physiological functions. In Section Four we discuss all questions raised in the book; in Section Five we give selected chemical structures and metabolic pathways.

There are three chapters in Section One. We cover whole-body energy metabolism in Chapter 1, highlighting the metabolism of carbohydrates, fats, and proteins. In Chapter 2, we address acid-base metabolism and, in Chapter 3, concentrate on the unique energy demands of exercise. Each chapter is divided into a number of parts. Each part is introduced by two to four clinical cases, followed by the biochemical and physiological concepts needed to understand those cases. Each case is then discussed in the light of these concepts.

Throughout Section one, we focus on the roles of the five major metabolic tissues (the brain, liver, muscle, adipose tissue, and kidneys), the five major systems of energy metabolism (the carbohydrate system, fat system, protein system, ATP generation system, and pyruvate dehydrogenase system), the control mechanisms that integrate these systems, and the effects of different dietary states. To facilitate an understanding of metabolic functions, we present the pathways and controls of metabolism in broad physiological terms.

Section Two also contains three chapters. Here we continue the problem-based approach, discussing broad areas of clinical medicine. We provide clinical cases first, following with background information and discussions of these cases. We have included flow charts to help in differential diagnoses of conditions resulting from multiple causes, and we have enhanced our presentation of human physiological principles by including contrasting data from elsewhere in the animal kingdom.

There are six chapters in Section Three. In the first, Chapter 7, we give an outline of metabolic control mechanisms, the roles of membranes, and second-messenger systems. We next discuss, in Chapters 8-12, the five metabolic systems of energy metabolism, focusing on the mechanisms of control that cause the pathways and systems to function together in the body.

In Section Four we discuss the many questions distributed throughout the text. These questions and their discussions provide elaborations of clinical data and concepts.

In Section Five we include a brief digest of the major chemical structures and pathways of energy metabolism; this is the only section that contains chemical structures.

Clinical practice day by day is essentially a problem-solving enterprise. A good doctor or a good nurse is a good detective. So it makes sense to us, and to increasingly many of our peers, that we give our students ample exercise in confronting the mysteries of metabolism in a "natural" setting. That's what Clinical Detective Stories is meant to do.

Mitchell L. Halperin
Francis S. Rolleston
Toronto and Ottawa
March 1993

SECTION ONE

Energy and Acid-Base Physiology

CHAPTER 1
ENERGY METABOLISM: FUELS, ORGANS, PATHWAYS, AND CONTROLS

PART A
ENERGY FUELS AND STORES

Carbohydrates, triacylglycerols, and proteins are dietary fuels and stores of energy in the body. Triacylglycerols are the most efficient energy stores on a kilocalorie per weight basis.

PART B
ORGANS AND FUELS

The brain's need for glucose drives energy metabolism. The liver and adipose tissue maintain supplies of energy fuels for consumer tissues (the brain and muscle).

PART C
INSULIN AND COORDINATION OF ENERGY METABOLISM IN ORGANS

Insulin coordinates energy metabolism, directing fuels to stores during meals and orchestrating the controlled release of stored energy between meals. Absence of insulin leads to uncontrolled release of these energy stores.

PART D
NUTRITION, METABOLIC PATHWAYS, AND CONTROLS

Fuels are directed through metabolic pathways and processes. If you know the function of a metabolic process, you can deduce the controls.

PART E
CASES FOR REVIEW

PART F
SUMMARY OF MAIN POINTS

PART A
ENERGY FUELS AND STORES

Case 1·1
Weight Gain on a Low-Calorie Diet
(Case discussed on pages 6–7)

Elizabeth, an experienced nurse, underwent routine abdominal surgery. She was slightly overweight upon entering surgery and had eaten normally to that time. Because the surgeon nicked her bowel, she consumed nothing by mouth for seven days, receiving instead an intravenous infusion of glucose (400 *kilocalories* per day). She had no infection, and recovery was uneventful except for a large weight loss (4.5 kg, 10 lb), which pleased her greatly. After being discharged from the hospital on day seven, she deliberately ate a low-calorie diet high in proteins and carbohydrates but low in sodium chloride and fats. To her chagrin, she gained 0.25 kg (about 0.5 lb) per day. Despite her doctor's suspicions, she knew that she had not cheated on her diet. Please explain her weight gain.

Kilocalorie (kcal):
The classical unit of energy in human nutrition. A calorie (cal) is defined as the quantity of heat required to raise the temperature of 1 g of water from 14.5°C to 15.5°C. Although the modern unit of energy is the joule (J), defined as the amount of heat generated by a current of 1 A acting for 1 s against a resistance of 1 Ω, we shall refer to kilocalories in this book (1 J = 4.19 kcal).

Case 1·2
Weight Gain Without Exercise
(Case discussed on page 7)

Anne, a star middle distance runner, typically exercised vigorously every day, weighing herself afterwards under the same conditions. Because of exams, she did not exercise as usual for two days but kept the same diet and other habits. She noted a weight gain of 1.5 kg. What happened?

Questions to Consider

What are the main storage forms of energy?

How efficient are the various energy stores?

Is weight change due only to an increased or decreased intake of energy fuels?

What causes spectacular weight loss (more than 0.5 kg/day)?
Is it good for you?
Can you maintain this weight loss?

What will cause a rebound gain of weight after a diet designed to lose weight?

BACKGROUND

ENERGY AND ATP

Adenosine triphosphate (ATP):
The currency of energy in cells. The structure is shown on page 284.

Metabolism:
The generic word for the host of reactions that interconvert biomolecules in living cells. Metabolism involves the acquisition and use of free energy to perform mechanical work, active transport, and synthesis of biomolecules.

Anabolism:
The formation of body constituents (with an increase in mass).

Oxidation:
Loss of electrons, which are ultimately transferred to molecular oxygen in a process that regenerates ATP.

Catabolism:
The utilization or breakdown of body constituents (with a decrease in mass).

To maintain existence, the body must regenerate *adenosine triphosphate (ATP)* continuously—when and where needed—or life stops. The diet supplies fuels (consisting of an enormous range of organic molecules) that, when oxidized, provide energy for the body. This energy characterizes *metabolism,* the multitude of reactions required by living organisms for such bodily functions as biosynthesis and maintenance of cell structures (a process called *anabolism*), transmission of nerve impulses, movement, and reproduction. The body incorporates most fuels into its structures for a while, but in steady state it uses all fuels ingested for production of energy.

The *oxidation* of fuels (involving *catabolism,* or breakdown, of fuel molecules) conserves some of the available energy as ATP; the rest is lost directly as heat (Figure 1·1). Although very little energy in the body exists in the form of ATP at any one time, energy released from ATP drives virtually all bodily processes.

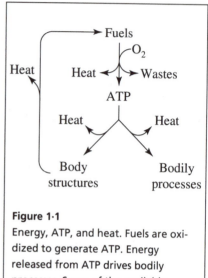

Figure 1·1
Energy, ATP, and heat. Fuels are oxidized to generate ATP. Energy released from ATP drives bodily processes. Some of the available energy from fuels is lost as heat during the generation of ATP.

NEED FOR ENERGY STORES

> The body needs energy stores because intake of food does not match the minute-to-minute need for ATP.

Meals provide fuels for the body but may be irregular and rarely coordinate with major demands for energy. Every organ requires energy all the time, but some have widely varying demands. Energy use by muscle, for example, can vary 20-fold from rest to strenuous exercise. Table 1·1 lists the daily energy requirements of humans and the average amount of energy expended for various activities.

A meal, which is usually absorbed within three hours, typically supplies more energy than the body needs at the time of the meal. The blood, on the other hand, has a very low supply of energy fuels—only a few minutes' worth. Because of the abundance of fuels at mealtime and the inability of the blood to contain large quantities of energy fuels, the body requires storage forms of these fuels (Figure 1·2).

Table 1·1
Energy requirements of humans

Description	Energy use (kcal/day)
Normal adult female	700 – 2000
Normal adult male	2400 – 2800
Bedridden patient	1300 – 1800
Newborn infant	350 – 450
Active teenage female	2400 – 2600
Active teenage male	3100 – 3600

Activity	Energy use (kcal/min)
Sitting at rest	0.7 – 2.0
Walking	2.0 – 6.0
Sprinting	15 or more
Long-distance running	10 or more
Competitive cycling	10 or more

The values are ranges for subjects in each group. A person keeping a constant weight burns all the energy consumed each day. Rates of energy use for various activities vary with speed and body weight.

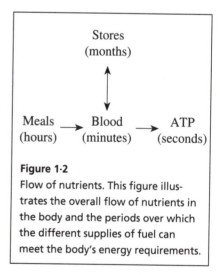

Figure 1·2
Flow of nutrients. This figure illustrates the overall flow of nutrients in the body and the periods over which the different supplies of fuel can meet the body's energy requirements.

ENERGY FUELS

> *Proteins and carbohydrates are "heavy" calories; triacylglycerols are "light" calories.*

Energy fuels for humans are *carbohydrates,* fats (hereafter referred to as *triacylglycerols,* the storage form of fats), and *proteins* (Table 1·2). Whereas *glycogen* (the major storage form of carbohydrates) and triacylglycerols function as energy reserves, proteins are structural and functional components of the body that can supply energy only at the expense of their structure and function.

Because molecules of carbohydrates and proteins attract water in the body (and in most foods), they have different contents of calories in wet and dry weights. In contrast to carbohydrates and proteins, which are "heavy" calories because they attract water, triacylglycerols have no affinity for water and are therefore "light" calories, with the same energy content in wet and dry weights (Table 1·3). Note from Table 1·3 that the marked differences in energy content per gram of fuel largely disappear when the energy released is related to the amount of oxygen consumed in burning these fuels.

Carbohydrates

> *Carbohydrates are the major fuels for the brain.*

Diet. Foods contain carbohydrates, mainly as polymers of *glucose* (starch in bread, flour, sweet corn, potatoes) but also as disaccharides (sucrose in table sugar, maltose in beer, lactose in milk). Because humans usually consume more carbohydrates than triacylglycerols or proteins, carbohydrates supply about 50% of dietary calories.

Storage Form. The body can store approximately 500 g of glycogen—about 100 g in the liver and 400 g in muscle. The liver, but not muscle, can convert its store of glycogen to glucose, releasing it to the blood as needed.

Dry carbohydrates yield 4 kcal/g when oxidized. However, because of the binding of 2–3 g of water for each gram of carbohydrates stored, each gram of glycogen (and associated water) in the body yields only 1–1.5 kcal.

Carbohydrates:
Fuels that constitute most of the dietary calories but, because of their water solubility, are not stored in large amounts.

Triacylglycerols:
The storage form of fats. Each triacylglycerol molecule contains three molecules of fatty acids and one molecule of glycerol. The structure is shown on page 286.

Proteins:
Nitrogen-containing compounds that have specialized functions (enzymes, structural elements, hormones, etc.) and are less important than carbohydrates or triacylglycerols as energy fuels.

Glycogen:
A branched polymer of glucose; the major storage form of carbohydrates. The structure is shown on page 280.

Glucose:
The fuel in the circulation derived from dietary carbohydrates or from glycogen in the liver. The primary fuel for the brain, glucose is also used by most organs after meals. See page 279 for the structure.

Plasma:
The fluid portion of the blood, excluding red blood cells.

Table 1·2
Daily consumption, body stores, and turnover of energy fuels

Fuel	Dry weight (grams)			Duration of energy supply in stores (days)
	Diet (per day)	Plasma (total)	Body (total)	
Carbohydrates	270	3	600	< 1
Triacylglycerols	100	5	15 000	111
Proteins	100	240	6 000	18

The values are approximations.

Table 1·3
Energy values of foods stored in the body

Fuel stored	Dry fuel energy content (kcal/g)	Dry weight of store (kg)	Water content of store (g/g dry)	Wet weight of store (kg)	Wet fuel energy content (kcal/g)	Energy released per mmol of O2 (kcal)	Total energy stored (kcal)
Carbohydrates	4	0.6	2–3	1.8	1–1.5	0.22	2 400
Triacylglycerols	9	15	0	15	9	0.21	135 000
Proteins	4	6	2–3	18	1–1.5	0.19	24 000

Triacylglycerols

> *The bulk of energy in the body is stored as fatty acids in triacylglycerols.*

Diet. Examples of foods rich in triacylglycerols are butter, vegetable oils, and meat fats.

Storage Form. Each triacylglycerol molecule contains one molecule of *glycerol* and three molecules of *fatty acids*. Although oxidation of fatty acids yields much of the ATP generated in the body, only the glycerol component of triacylglycerols can be converted into glucose. Fatty acids cannot be converted to glucose and therefore provide little fuel for the brain. However, the brain can use *ketoacids,* a fuel derived from fatty acids, during prolonged fasting.

The body can store triacylglycerols in virtually unlimited quantities. Because triacylglycerols do not mix with water and are therefore calorically dense, they are the body's most efficient energy store (Figure 1·3). Oxidation of stored triacylglycerols yields 9 kcal/g. Thus, triacylglycerols can store six to nine times more calories per unit weight than carbohydrates.

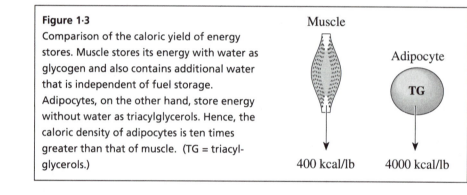

Figure 1·3
Comparison of the caloric yield of energy stores. Muscle stores its energy with water as glycogen and also contains additional water that is independent of fuel storage. Adipocytes, on the other hand, store energy without water as triacylglycerols. Hence, the caloric density of adipocytes is ten times greater than that of muscle. (TG = triacylglycerols.)

Muscle

Adipocyte

TG

400 kcal/lb   4000 kcal/lb

Proteins

> *The body can take proteins that are functioning as essential structural components and convert them into glucose for use as an energy fuel during fasting.*

Diet. Proteins, which consist of chains of *amino acids* joined together by peptide linkage, are usually a minor but essential part of the diet (close to 15% of energy intake). Meat, fish, poultry, eggs, and milk are some of the best sources of proteins.

Storage Form. Because proteins attract water, they provide 4 kcal/g dry weight but approximately 1–1.5 kcal/g wet weight. Unlike the other energy fuels, proteins have no storage forms independent of function (proteins function as enzymes, structural components, hormones, etc.). Because their use as a fuel involves damage to the body, they are a fuel of "last resort." Nevertheless, proteins form a large energy reserve since they constitute much of the structure of the average person.

Discussion of Case 1·1
Weight Gain on a Low-Calorie Diet
(Case presented on page 3)

Figure 1·4(a) examines the possible reasons for weight loss. Because Elizabeth lost a large amount of weight without excessive caloric expenditure, she must have lost *extra-*

Glycerol:
The material of carbohydrate origin that binds fatty acids together in triacylglycerols. Page 286 shows the structure of glycerol.

Fatty acids:
The major energy fuel. Fatty acids are released from triacylglycerols stored in adipose tissue. Fatty acids constitute 94% of the carbon in triacylglycerols. Although most organs oxidize fatty acids in preference to glucose, the brain does not oxidize appreciable amounts of fatty acids. See page 286 for the structure.

Ketoacids:
Fat-derived fuels that the brain uses during prolonged fasting. Refer to page 286 for the structure.

Amino acids:
The 20 building blocks of proteins. Amino acids differ from one another only in their side chains. Page 291 shows the structure of an amino acid.

cellular fluid or water associated with heavy calories. There was no evidence that she lost extracellular fluid, so loss of water associated with heavy calories is likely.

During her recovery, the intravenous infusion of glucose did not provide a sufficient quantity of glucose for her brain. Therefore, Elizabeth used her energy stores to make glucose. Her stores of glycogen lasted for about a day, after which she used her protein reserves. Stored fat supplied the remaining calories needed. Since the body stores glycogen and proteins with water, loss of intracellular water accounted for much of her weight loss.

After resuming normal activity and starting a low-calorie diet, Elizabeth's body began to replace the carbohydrate and protein reserves that she lost during the post-operative period. As outlined in Figure 1·4(b), Elizabeth experienced a gain of water from the accumulation of proteins and glycogen, materials that had a greater net weight (because of water) than the fat stores she burned to meet her overall expenditure of energy.

Discussion of Case 1·2
Weight Gain Without Exercise
(Case presented on page 3)

According to the principles outlined in Figure 1·4(b), if weight gain exceeds 0.25 kg in one day, there must be a gain of water, either related to nutrition (gain of proteins or glycogen) or independent of nutrition (gain of extracellular fluid or *intracellular fluid*). Since Anne is unlikely to have a condition that compromised the ability of her kidneys to excrete salt or water, it appears that she had an accumulation of proteins or glycogen. Her normal exercise would not have consumed 13 500 kcal, the equivalent of 1.5 kg of fat. Anne's vigorous exercise each day depleted her store of glycogen in muscle. Missing her workout for two days permitted glycogen and its associated water (1.2–1.5 kg wet weight) to remain in her muscle after meals.

Extracellular fluid (ECF):
Fluid outside cells. It's volume depends mainly on the amount of sodium in the body. Sodium in the ECF keeps some water outside cells (by osmosis). Excessive loss of sodium can occur in the urine (e.g., from diuretic therapy), skin (from sweating), or the gastrointestinal tract (from vomiting or diarrhea).

Note:
One-fourth of a kilogram of fatty acids will provide the body with 2250 kcal, which will enable the body to maintain a modest level of activity for 24 hours.

Intracellular fluid (ICF):
Water within cells. ICF constitutes two-thirds of the water in the body.

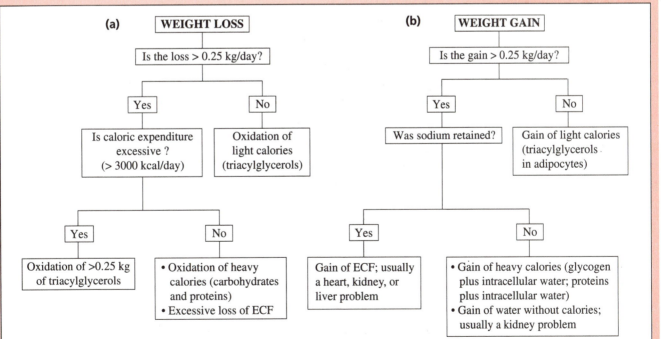

Figure 1·4
Consideration of excessive changes in weight. (a) Weight loss. (b) Weight gain. Energy and weight are not related directly because some forms of energy are stored with water (heavy calories) and thus have a larger bulk than those without water.

PART B
ORGANS AND FUELS

Case 1·3
Confusion in a Marathon Runner
(Case discussed on page 14)

Jon entered the stadium for the final lap of his marathon race. He was well ahead of his competitors. In the last few minutes, he became confused. In the stadium, he started running around the track in the wrong direction and then collapsed. What went wrong?

Case 1·4
Death of a Trauma Victim
(Case discussed on page 14)

Joe suffered multiple fractures and severe internal injuries in a motorcycle accident. Because he could not move or eat, he received an intravenous infusion of glucose (2 liters of 5% glucose in water per day). He lost weight rapidly and died 15 days later. The death certificate stated that he died of pneumonia (inflammation in the lungs from infection). What do you think may have happened from the perspective of energy metabolism?

Questions to Consider

Which organ must be kept fully supplied with energy all the time?
What fuels can it use?
What are the usual sources of these fuels?
Do alternative sources exist?

What are the dominant organs with respect to energy metabolism?

What are the demands and contributions of the various organs in relation to the functions of other organs?

BACKGROUND

MAJOR ORGANS IN ENERGY METABOLISM

> *A major factor controlling overall energy metabolism is the need to supply the brain with a water-soluble fuel.*

The dominant organs in energy metabolism are the brain, muscle, adipose tissue, the liver, and the kidneys. These five organs can be grouped into three categories, based on function.

1. *Consumer organs* consume energy for their own purposes.

2. *Maintainer organs* ensure that consumer organs have the fuels they need.

3. *Excretory organs* excrete metabolic wastes.

CONSUMER ORGANS

> *Consumer organs consume energy for their own purposes.*

Consumer organs (brain and muscle) use fuels only to regenerate the ATP consumed to perform biological work. Accordingly, their demands for ATP regulate the use of these fuels. When these organs require more energy, they oxidize fuels at a faster rate.

Brain

The brain, which uses 20–25% of the body's daily production of ATP when the body is at rest, requires a supply of fuel and oxygen from the circulation to meet its constant need for ATP. The brain always requires glucose, usually 5 g/h for 100% of its energy (Table 1·4). The brain's need for glucose drives energy metabolism. Paradoxically, the main component of the body's largest energy store (fatty acids from triacylglycerols) cannot be converted to glucose. However, the brain can use ketoacids, a fuel derived from fatty acids, for 80% of its energy if they are available in the blood (glucose must supply the remaining 20%). Unlike other organs, the brain has almost no storage forms of energy. Insulin (pages 15–21 and 99–120) has no significant effect on energy metabolism in the brain.

Muscle

Muscle can use ATP at a variable rate, consuming 20% (at rest) to 80% (during vigorous exercise) of the body's daily energy production. It oxidizes carbohydrates and fatty acids (Table 1·4) but must spare brain fuels (glucose, ketoacids), especially during severe and prolonged exercise. Muscle, a major reserve of proteins because it is collectively the largest organ, also contains significant carbohydrate reserves (glycogen) for its own use.

Consumer organs:
Organs that generate only enough ATP to meet their immediate needs for energy.

Maintainer organs:
Organs, such as the liver and adipose tissue, that primarily regulate the concentrations of fuels (in the blood) for the brain and other tissues.

Excretory organs:
Organs that excrete wastes; the lungs excrete CO_2, and the kidneys help eliminate acid indirectly by excreting NH_4^+ (ammonium ions).

> The brain has a high demand for turnover of ATP but cannot regenerate ATP from fatty acids.

> Muscle has an enormous capacity to utilize ATP.

Table 1·4
Hierarchy of fuels oxidized in consumer organs

Fuel	Brain	Muscle
Fatty acids	Not used	First priority, if available
Ketoacids	First priority, if available	Not used if fatty acids are available
Glucose	Used exclusively most of the time (ketoacids are rarely available)	Used only if fatty acids and ketoacids are not available

MAINTAINER ORGANS

> *Maintainer organs interconvert fuels to supply the brain with a steady source of fuels from the circulation.*

Maintainer organs (liver and adipose tissue) regulate the concentrations of fuels in the blood, disposing of excess fuels from the diet and providing fuels for other organs between meals (Table 1·5). They interconvert fuels so that the brain can always obtain enough glucose or ketoacids from the circulation.

Liver

The liver regulates blood levels of fuels (particularly for the brain), the most important of them being glucose. The liver not only stores glucose in the form of glycogen and releases glucose derived from glycogen but also makes glucose, as needed, from precursors (amino acids, glycerol, and *lactic acid* produced by other organs). When the concentration of glucose is low, the liver produces ketoacids for the brain. When glucose is excessive, the liver converts it to triacylglycerols for storage. The liver also converts NH_4^+ to *urea* (a nontoxic waste), functions as the "detox center," detoxifying drugs, by-products, etc., and is the initial site of oxidation of unique fuels (e.g., *ethanol*).

Adipose Tissue

Adipose tissue stores energy fuels as triacylglycerols and releases fatty acids and glycerol into the blood. The terms "adipose tissue" (the organ) and "adipocytes" (the cells within adipose tissue) are hereafter used interchangeably.

EXCRETORY ORGANS

Kidneys

The kidneys remove nonvolatile wastes, excreting nitrogen (as urea) and the excess acids formed during energy metabolism (primarily as ammonium salts). The kidneys may also act as maintainer organs, synthesizing glucose.

Lungs

The lungs remove CO_2 from the blood and add O_2 to the blood (see Chapter 2).

Note:

An acid is a compound that, when dissolved in water, releases a hydrogen ion and an anion. For example, lactic acid dissociates in water to form a lactate anion and a hydrogen ion.

Lactic acid:

The compound synthesized from the metabolism of glycogen in anaerobic muscle and glucose in red blood cells. Lactate anion, the circulating form of lactic acid, is a key intermediate in carbohydrate metabolism that can accumulate to very high levels in the extracellular fluid (see page 279 for the structure of lactate). One mmol of hydrogen ion accompanies each mmol of lactate anion that the body gains.

Urea:

A nontoxic, water-soluble waste product of protein metabolism that contains two nitrogen atoms (see page 291 for the structure). Urea is synthesized in the liver from NH_4^+, which are toxic.

Ethanol:

The alcohol in wine, spirits, and beer. Ethanol may be ingested in large amounts and must initially be metabolized in the liver. The structure of ethanol is shown on page 292.

Table 1·5

Major pathways for fuel metabolism in consumer organs

Metabolic event	Liver	Adipose tissue
Generation of ATP		
During meals	Uses amino acids	Uses glucose (small amount)
Between meals	Uses fatty acids	Uses fatty acids
Synthesis of storage compounds	Stores carbohydrates as glycogen (large amount) and triacylglycerols (small amount)	Stores energy as triacylglycerols (virtually unlimited amounts)
Breakdown of storage compounds		
To supply ATP for the brain	Converts glycogen to glucose, amino acids to glucose, fatty acids to ketoacids	–
To supply fatty acids for the liver	–	Converts triacylglycerols to fatty acids

OVERVIEW OF FUEL USE BY ORGANS

Figure 1·5 illustrates the volume of blood delivered to each organ and the quantity of oxygen that each organ consumes. Figure 1·6 summarizes the fuels that these organs use.

ENERGY FUELS IN THE BLOOD

Carbohydrates circulate as glucose, lactate, pyruvate, and glycerol.

Glucose may be derived from the diet or from glycogen or *gluconeogenesis* in the liver. The brain always requires glucose to generate at least some of its ATP. However, ketoacids, when available, can replace some of the brain's need for glucose. Although no other tissues with mitochondria require glucose specifically, all can use it to generate ATP. When glucose is in excess, the liver stores it as glycogen, releasing it, when needed, as glucose. Similarly, muscle stores glucose as glycogen, but, once stored, it can only be used in muscle and cannot be released as glucose.

Gluconeogenesis:
Synthesis of glucose from amino acids or glycerol.

Lactate plus a hydrogen ion (lactic acid) comes from the metabolism of glycogen in anaerobic muscle and glucose in red blood cells. Lactic acid can be oxidized in the heart and kidneys and converted to glucose in the liver and kidneys.

Pyruvate plus a hydrogen ion (pyruvic acid) is the important partner of lactic acid in the intracellular fluid. Pyruvic acid functions in the carbohydrate system as a key intermediate.

Pyruvate or pyruvic acid:
A key intracellular intermediate in carbohydrate metabolism. The structure is shown on page 279.

Glycerol derived from stored triacylglycerols can be converted to glucose or to triacylglycerols in the liver.

Question

(Discussion on page 247)

1·1 The brain consumes about 10% more glucose than it needs to regenerate ATP, converting the excess to lactate. Why might this consumption be advantageous?

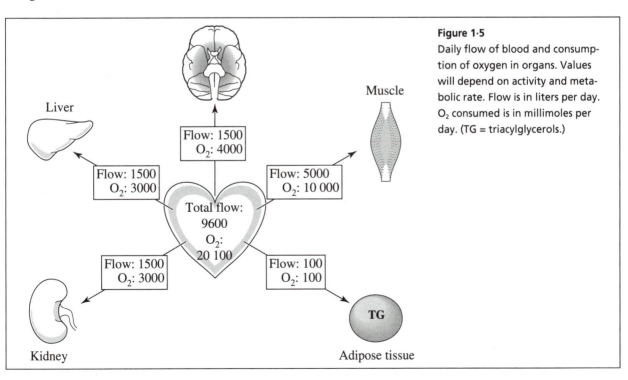

Figure 1·5
Daily flow of blood and consumption of oxygen in organs. Values will depend on activity and metabolic rate. Flow is in liters per day. O_2 consumed is in millimoles per day. (TG = triacylglycerols.)

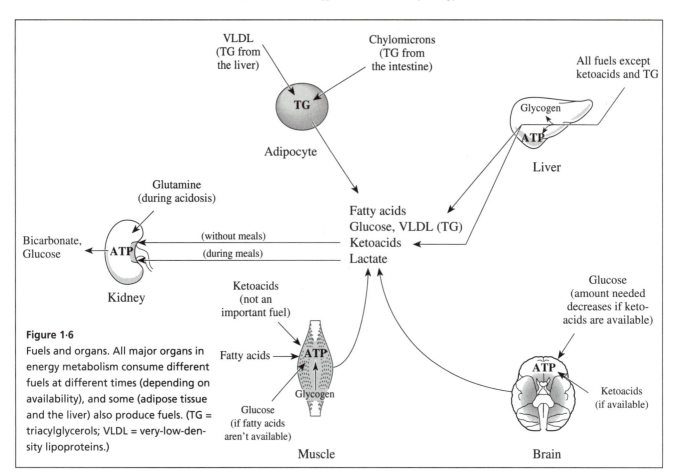

Figure 1·6

Fuels and organs. All major organs in energy metabolism consume different fuels at different times (depending on availability), and some (adipose tissue and the liver) also produce fuels. (TG = triacylglycerols; VLDL = very-low-density lipoproteins.)

Albumin:

The most abundant protein in plasma. Albumin carries water-insoluble fatty acids from adipose tissue to organs for oxidation. Albumin is also important in shifting extracellular fluid from the interstitial space into plasma via an osmotic force.

Plasma lipoproteins:

Compounds containing an insoluble fat core and a highly charged surface made up of phospholipids and proteins. Fats, which are insoluble in water, travel through plasma in this form.

Chylomicrons:

The largest plasma lipoproteins. They carry dietary lipids from the intestine to adipose tissue and muscle.

Very-low-density lipoproteins (VLDL):

The second-largest plasma lipoproteins. They carry triacylglycerols made in the liver to adipose tissue and muscle.

Fats circulate as fatty acids, ketoacids, and triacylglycerols.

Fatty acids (bound to *albumin* in plasma) are released with free glycerol from adipose tissue triacylglycerols. The liver, muscle, and kidneys can oxidize fatty acids to yield ATP.

Ketoacids, water-soluble derivatives of fatty acids, are formed by the liver during prolonged fasting. After a sufficient concentration accumulates in the blood, the brain and kidneys can use them to regenerate ATP. Muscle, however, uses them sparingly.

Triacylglycerols are typically transported in two types of "envelopes" composed mainly of *plasma lipoproteins—chylomicrons* and *very-low-density lipoproteins (VLDL)*. Whereas chylomicrons contain dietary fatty acids incorporated into triacylglycerols by intestinal mucosa, VLDL contain triacylglycerols that the liver produces (from fatty acids, glucose, or amino acids). The blood carries VLDL to adipose tissue for storage and to muscle for oxidation.

Proteins circulate primarily as amino acids.

Amino acids, the 20 building blocks of proteins, can be synthesized into proteins and released from proteins. The liver (and the kidneys in acidosis) can convert amino acids to glucose, an irreversible process.

Question

(Discussion on page 247)

1·2 Why should muscle use ketoacids sparingly?

FACTORS THAT INFLUENCE THE BODY'S CHOICE OF ENERGY FUELS

PRIORITIES FOR THE USE OF FUELS

What is the fuel to be used for?

Satisfy the immediate needs for ATP.

Replenish stores of glycogen and proteins if fuels are available in the diet.

Store excess fuels as triacylglycerols.

Which fuel is to be used?

Oxidize fatty acids for energy when their levels are high in the blood.

Proteins or carbohydrates will be oxidized when in excess from dietary intake because their abundance signals adipose tissue to retain fatty acids.

Spare body proteins whenever possible.

Ability of Organs to Oxidize Fuels

The major organs differ in the fuels that they are capable of oxidizing to regenerate ATP. For example, the brain oxidizes glucose readily, but the liver does not.

Availability of Fuels

The choice of fuels to oxidize depends on the type of diet and the status of energy stores. Different diets affect the balance between the three major fuels available after meals, and hormone levels (insulin, glucagon) affect the availability of fuels between meals.

Hierarchy of Fuel Utilization

The body has a hierarchy of fuel use, oxidizing fatty acids first if available, then ketoacids, and finally glucose if neither fatty acids nor ketoacids are available. The brain, however, is an exception. Because few fatty acids are able to cross the blood-brain barrier, the brain cannot derive an appreciable amount of ATP from the oxidation of fatty acids.

INTERCONVERSION OF FUELS

1. Proteins cannot be made from either carbohydrates or triacylglycerols but can be made into both.

    Proteins break down to amino acids, which the liver can convert to glucose or to triacylglycerols (when glucose is in excess). This irreversible conversion involves a permanent loss of nitrogen that renders the new product—glucose or triacylglycerols—unfit for synthesis into proteins.

2. Fatty acids, the major portion (more than 90%) of triacylglycerols, can be made from both carbohydrates and proteins but cannot be converted back to either. The glycerol portion (less than 10%) of triacylglycerols can be converted back to glucose.

    Carbohydrates and proteins are precursors of acetyl-CoA. Once converted to acetyl-CoA, they cannot be converted back to their original form. However, acetyl-CoA is easily converted to fatty acids, and vice versa. (Refer to Figure 1·16, page 25.)

3. Carbohydrates can be made from proteins and can be used to make triacylglycerols.

As a result of these limited interconversions, an *energy paradox* exists.

Energy paradox:
1. The brain needs 500 kcal of water-soluble fuels (usually glucose) per day.
2. Almost all energy is stored as fatty acids, not as glycogen.
3. Fatty acids cannot be converted to glucose.

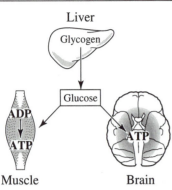

Figure 1·7
Marathon runner. During strenuous exercise, glycogen in the liver will normally replenish glucose in the blood, thereby satisfying all needs for glucose. Prolonged exertion, however, will deplete the supply of glycogen in both the liver and muscle, and, if continued, will deprive the brain of fuel to regenerate ATP.

Discussion of Case 1·3
Confusion in a Marathon Runner
(Case presented on page 8)

During strenuous exercise, the brain and muscle compete for glucose (Figure 1·7). Although the brain can use either glucose or ketoacids for its energy needs, ketoacids were not available because Jon had not been deficient in carbohydrates long enough. Hence, his brain could use only glucose. Muscle, on the other hand, can use fatty acids, muscle glycogen, or circulating glucose for energy. Because fatty acids are insoluble in water, they cannot be delivered to muscle at a sufficiently rapid rate. Thus, muscle will use glycogen and glucose when the demand for fuels is high. (Chapter 3 examines the use of energy in muscle.)

The marathon run consumed glycogen, the reserve of glucose in Jon's body. Pushing himself to the limit at the end of the race caused a sudden depletion of glucose in his blood, starving his brain of its only source of energy. Without fuel, his brain malfunctioned.

Discussion of Case 1·4
Death of a Trauma Victim
(Case presented on page 8)

For unknown reasons, some patients, like Joe, do not release (and oxidize) enough fatty acids from adipose tissue. After 24 hours, Joe depleted the reserves of glycogen in his liver. The liver makes glucose primarily from amino acids derived from proteins in muscle. Hence, breakdown of proteins, which occurred rapidly, supplied Joe with much of the ATP that he needed (Figure 1·8). The brain consumes the equivalent of 1 kg of muscle to meet its daily need for ATP.

- 1 kg of muscle   = 200 g of proteins (80% water; assume proteins account for the rest)
- 200 g of proteins = 120 g of glucose (not all amino acids nor all their carbons can be converted to glucose)
- 120 g of glucose = the brain's need for fuel in a day

Severe catabolism of proteins results from multiple injuries and possibly infection. After Joe consumed about half the proteins in his muscle, he became too weak to cough up secretions. These secretions lingered in his lungs, became infected, and Joe died. A diagnosis of bronchopneumonia was made on post-mortem examination.

Question

(Discussion on pages 247–48)

1·3 What nutritional therapy could have prevented Joe's death (Case 1·4)?

Figure 1·8
Trauma victim. The problem is insufficient fatty acids released from adipocytes or not enough ketoacids formed in the liver. As a result, proteins in muscle supply the brain with glucose. (β-HB = β-hydroxybutyrate, the most abundant so-called ketoacid.)

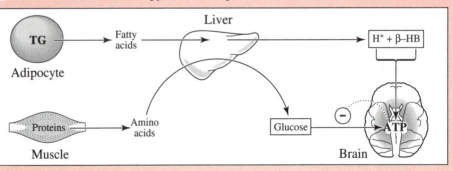

PART C
INSULIN AND COORDINATION OF ENERGY METABOLISM IN ORGANS

Case 1·5
Weight Loss with an Excessive Intake of Food
(Case discussed on pages 18–19)

Mary, 10 years old, started to lose weight rapidly, to pass large volumes of urine (polyuria), to eat excessively (polyphagia), and to drink excessively (polydipsia). What happened?

Case 1·6
Suicide Attempt by Insulin Overdose
(Case discussed on pages 20–21)

Algernon arrived at the emergency room with a convulsion. He had a vial of *insulin* and a syringe in his pocket. Upon receiving an infusion of glucose, he promptly became alert. Why? What treatment should he receive over the next 24 hours?

Insulin:
The hormone that signals storage of energy. The pancreas secretes insulin in response to an abundance of fuels (primarily glucose) in the blood.

Questions to Consider

What would happen if there were no controls over energy metabolism?

Should controls be inherently stable or unstable?

What kinds of controls do you expect to be self-limiting?

Given the need to preserve glucose and proteins (see previous sections), what kinds of controls would you expect?

BACKGROUND

CONTROLS

> *If you know a function in energy metabolism, you can deduce the likely controls.*

The rest of this book focuses on the relation between functions and controls in energy metabolism. Controls over metabolism, which are necessary to regulate the rate of formation of ATP, allow net synthesis or utilization of energy stores as needed. Controls select the fuels to be used, determine the rates at which organs use these fuels, and regulate the interconversions between fuels and stores. There are two overall types of control mechanisms.

Substrate activation:
When the substrate of a metabolic sequence exerts control over that sequence (positive feedforward).

Product inhibition:
When the product of a metabolic sequence exerts control over that sequence (negative feedback).

1. *Substrate activation,* or positive feedforward, occurs when a substance stimulates its own use. Use of a substrate (the substance upon which an enzyme acts) increases when its concentration rises; use of a substrate decreases when its concentration falls.

2. *Product inhibition,* or negative feedback, occurs when the product of a reaction prevents the reaction from continuing. Formation of the product decreases when the concentration of the product rises; likewise, formation of the product increases when its concentration falls.

We use the following notation to indicate the effects of controls on reactions. Reactions or pathways proceed from substrate (S) to product (P).

$$S \longrightarrow P$$

A broken line with a "+" for activation (speeding up) and a "–" for inhibition (slowing down) shows the effect of a controlling influence on an enzyme. Note: These "control lines" represent the overall influence in general terms. They do not imply a direct mechanism.

$$S \xrightarrow{\ \ \oplus\ \ } P \qquad\qquad S \xrightarrow{\ \ \ominus\ \ } P$$

This notation can also demonstrate the effects of a third metabolite on the reaction. An increase of X will inhibit the conversion of S to P, either indirectly (because X influences other reactions) or directly. An increase of Y will activate the conversion of S to P, either indirectly or directly.

$$\begin{array}{cc} X & Y \\ \ominus & \oplus \end{array}$$
$$S \longrightarrow P$$

Homeostatic Controls

Homeostasis:
Maintaining a stable internal environment for the body.

Although the external environment may fluctuate greatly, and the body's energy requirements constantly change, the body maintains internal stability (*homeostasis*). Because of homeostatic controls, which are usually self-limiting, the body has relatively constant levels of glucose and amino acids in the blood (despite large intermittent dietary loads) and a relatively constant level of ATP in cells (despite up to 20-fold changes in the rates at which muscle uses ATP).

Signals

Changes in the concentration of a metabolite in the body signal control mechanisms, which alter the rates of metabolic processes. Two major types of signals are:

1. metabolites that may or may not be components of the metabolic pathway;
2. hormones via intracellular signals.

Short-Term vs Long-Term Control

Short-term control, which occurs before the body can synthesize or degrade enzyme molecules, is mediated by a change in activity of existing enzyme molecules. Long-term control, on the other hand, occurs when time permits the synthesis or breakdown of enzyme molecules.

INSULIN AND CONTROL OF ENERGY METABOLISM

> *Insulin has five main actions in energy metabolism:*
> *1. It prevents the release of fatty acids from adipose tissue.*
> *2. It promotes the synthesis of glycogen.*
> *3. It accelerates the transport of glucose into muscle.*
> *4. It accelerates the synthesis of triacylglycerols.*
> *5. It inhibits the synthesis of glucose in the liver.*

Insulin is secreted from the endocrine pancreas (the β cells of the islets of Langerhans) primarily in response to an elevated concentration of glucose in the blood. The increased concentration of insulin then causes the body to convert energy fuels to storage forms (Figure 1·9).

1. The liver stores glucose as glycogen until the stores are full. The liver converts excess glucose to triacylglycerols, which the blood transports as VLDL to adipose tissue for storage.
2. The body stores amino acids as proteins (primarily in the liver and muscle) until the full complement is present. The liver converts excess amino acids to glucose, glycogen, or triacylglycerols (minor).
3. Circulating triacylglycerols are stored in adipose tissue.

High levels of insulin encourage the oxidation of glucose (Figure 1·10). The most important direct effect of increased insulin is an inhibited release of fatty acids from adipose tissue. Without competition from fatty acids, a fuel that the body oxidizes more readily than glucose, more oxidation of glucose can occur. Insulin also aids the oxidation of glucose by permitting glucose to enter muscle cells. Of lesser importance, insulin promotes the synthesis of triacylglycerols in adipose tissue.

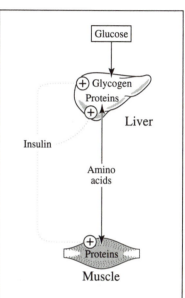

Figure 1·9
Insulin, glycogen, and glucose. Increased levels of insulin stimulate the synthesis of glycogen (in the liver) and proteins (in muscle and the liver).

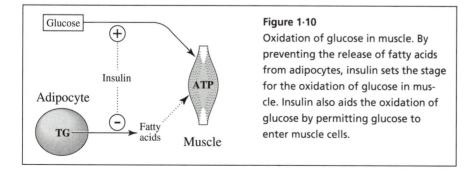

Figure 1·10
Oxidation of glucose in muscle. By preventing the release of fatty acids from adipocytes, insulin sets the stage for the oxidation of glucose in muscle. Insulin also aids the oxidation of glucose by permitting glucose to enter muscle cells.

GLUCAGON AND CONTROL OF ENERGY METABOLISM

> *Glucagon promotes the release of glucose from the liver.*

Glucagon:
The major hormonal signal for catabolism. The pancreas secretes glucagon in response to a deficiency of fuels in the blood.

Whereas insulin stimulates the use and storage of glucose, *glucagon*, a hormone secreted from the α cells of the islets of Langerhans, promotes the release of glucose. Although these hormones are complementary—a high level of insulin occurs in conjunction with a low level of glucagon, and vice versa—both operate under a product-inhibition system that is regulated by the concentration of glucose in the blood.

Discussion of Case 1·5
Weight Loss with an Excessive Intake of Food
(Case presented on page 15)

Diabetes mellitus:
The disease process that results from a deficiency of insulin and is characterized by excessive hunger, thirst, urination, and weight loss.

In a normal person, secretion of insulin is matched, moment-to-moment, to the concentrations of glucose in the blood. *Diabetes mellitus* results when the pancreas does not secrete sufficient insulin in response to an increased concentration of glucose in the blood. With destruction of the β cells of the pancreas, as in insulin-dependent (Type I) diabetes mellitus, less insulin is secreted than the body needs. By the time symptoms appeared in Mary, she lacked much of her endocrine pancreatic function.

Without insulin, the deposition of fuels in stores decreases markedly, and the release of fuels from stores increases. The rapid breakdown of triacylglycerols in adipose tissue releases fatty acids into the blood (Figure 1·11). As more fatty acids enter the blood, all tissues that can use them as fuels (the liver, kidneys, and muscle) oxidize a greater proportion of them. The liver also releases ketoacids, which the brain uses in preference to glucose. The presence of fat-derived fuels inhibits the oxidation of glucose. Consequently, a lack of insulin triggers the release of fat-derived fuels for ATP and ultimately inhibits the body's use of glucose.

Figure 1·11
Use of triacylglycerols when insulin is absent. The oxidation of fatty acids provides all the ATP in most organs but requires partial oxidation in the liver to provide the fuel (ketoacids, β-HB) needed to regenerate ATP in the brain.

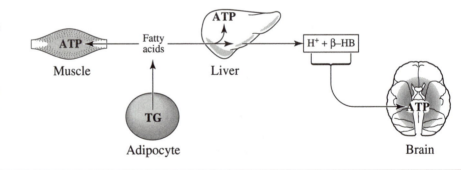

The deficiency of circulating insulin also causes an increased rate of formation of glucose (Figure 1·12). Proteins (mainly from muscle) break down rapidly, releasing amino acids, which the liver converts to glucose. The liver also converts the glycerol released from the breakdown of triacylglycerols to glucose. Without insulin, the body cannot store this increased amount of circulating glucose as glycogen. Worse yet, it uses some of its existing stores of glycogen as an added source of glucose. The liver breaks down its glycogen to glucose, but glycogen in muscle is usually reserved to meet needs during exercise (lactate released at that time, however, can be converted to glucose). If Mary were to exercise strenuously, conversion of glucose to lactic acid in muscle could be rapid, but this process would not remove carbohydrates because of the Cori cycle (shown in Figure 1·19 on page 27 and defined as glucopaleogenesis in Chapter 8).

Hyperglycemia and hyperketoacidemia result from excess glucose, fatty acids, and ketoacids in the blood. Because the kidneys cannot reabsorb a sufficient quantity of glucose and ketoacid anions from the glomerular filtrate (Chapter 2), these fuels are excreted with their associated water and electrolytes, causing thirst and a low circulating volume.

The rapid weight loss results from:

1. loss of proteins (heavy calories);
2. loss of water associated with glycogen;
3. loss of adipose tissue triacylglycerols (minor weight loss);
4. excretion of glucose and ketoacid anions in the urine (minor weight loss);
5. excretion of extracellular fluid (water and sodium) in the urine.

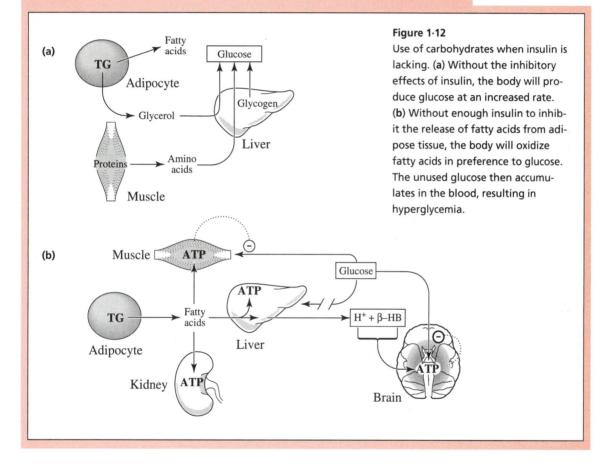

Figure 1·12

Use of carbohydrates when insulin is lacking. (a) Without the inhibitory effects of insulin, the body will produce glucose at an increased rate. (b) Without enough insulin to inhibit the release of fatty acids from adipose tissue, the body will oxidize fatty acids in preference to glucose. The unused glucose then accumulates in the blood, resulting in hyperglycemia.

Discussion of Case 1·6
Suicide Attempt by Insulin Overdose
(Case presented on page 15)

Hypoglycemia:

A condition in which the concentration of glucose in the blood is lower than normal. When the level is very low, the regeneration of ATP in the brain may be limited, resulting in loss of consciousness.

Algernon's rapid recovery upon receiving an infusion of glucose indicates that he had *hypoglycemia*. The syringe and vial suggest that he suffered from an overdose of insulin. Excess insulin causes:

1. conversion of glucose to glycogen (by the liver and muscle) and to triacylglycerols (by the liver);

2. use of glucose to regenerate ATP (in muscle);

3. uptake of amino acids by the liver and muscle, followed by the conversion of amino acids to proteins;

4. inhibited synthesis of glucose by the liver because few amino acids are available (there is less net breakdown of proteins) and little glycerol is available (release from adipose tissue is inhibited);

5. net uptake of triacylglycerols by adipose tissue (through deposition from circulating triacylglycerols and inhibited release of fatty acids).

Thus, Algernon had insufficient circulating fuels (Figure 1·13). The absence of glucose caused his convulsions.

When planning a course of treatment for a patient with excess insulin, first consider the brain, which requires 120 g of glucose per day (levels of ketoacids are low because of the inhibited release of fatty acids from adipose tissue and the decreased oxidation of fatty acids in the liver). The rest of the body also requires glucose, since fat-derived fuels are scarce. In addition, the liver and muscle will convert glucose to glycogen if their stores are not full. Hence, therapy is to provide glucose, orally or intravenously.

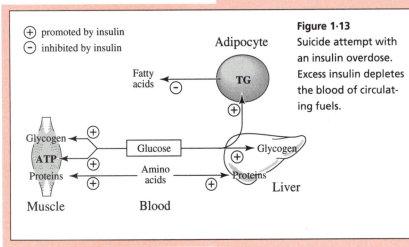

Figure 1·13
Suicide attempt with an insulin overdose. Excess insulin depletes the blood of circulating fuels.

⊕ promoted by insulin
⊖ inhibited by insulin

PRIORITIES OF MANAGEMENT

1. Provide the brain with 20 kcal/h.

   The fuels for the brain are glucose and ketoacids. Since ketoacids are less likely to be available, and they must be given with a load of H^+ or Na^+, give Algernon 5 g of glucose per hour. A difficulty is that other organs will oxidize or store some of this glucose.

2. Some glucose will be used to replenish stores.

   The major stores of carbohydrates are glycogen in the liver (about 100 g) and muscle (about 400 g). The amount of extra glucose needed depends on how full these stores were prior to therapy. The conversion of glucose to triacylglycerols is a slow process and will not use much glucose in 24 hours, but this rate will increase with time in this setting.

3. Provide glucose to regenerate ATP in other organs.

   The excess insulin means that fatty acids are unlikely to be released from adipose tissue stores. Therefore, other organs must utilize glucose, which they can oxidize at a rate of 15–20 g/h (60–80 kcal/h).

4. Caution: Do not give too much glucose.

   If more glucose is infused than can be oxidized or stored, its concentration will rise in the blood, and glucose will be excreted in the urine, dragging out useful ions (sodium, potassium, chloride) and water (see pages 115–16). When administered intravenously, glucose is usually given as 50 g (278 mmol) of glucose per liter. Providing 25 g of glucose per hour to meet the body's total energy needs for 24 hours requires an infusion—12 liters of water—that the kidneys cannot readily excrete. Therefore, consider the following options:

   a. Give a more concentrated solution of glucose intravenously (there is a limit to this concentration), or feed the patient glucose with a little water by mouth.

   b. Give ketoacids (β-HB), since they are an excellent fuel for the brain. Algernon's liver is not making β-HB because of the presence of insulin. However, his liver can still synthesize ketoacids if it can utilize another pathway for their formation. The switch to an alternate pathway can be accomplished by giving 1,3-butanediol, which is a precursor of β-HB that does not use the regular pathway of ketogenesis in the liver (instead, it uses alcohol and aldehyde dehydrogenases). Since this compound is an alcohol, it can be given as a very concentrated solution that does not contain sodium or potassium. Thus, it will not create salt, acid-base, or water problems. This form of therapy is superior to the administration of the sodium salt of β-HB.

Question

(Discussion on page 248)

1·4 What changes will occur in the types of fuels oxidized in a patient who takes too much insulin?

   How long will the patient remain conscious?

Case 1·7
Fulminant Hepatitis
(Case discussed on page 23)

Vic, a surgeon, pierced his finger with a needle while performing an operation on a patient with infectious hepatitis (viral inflammation of the liver, a disease that is readily transmitted by needles). Over the next week, Vic became severely ill with acute and massive destruction of most of his liver. When the case was discussed at clinical rounds, it was decided that Vic might survive if the hospital's nutrition team could undertake the functions of a normal liver and maintain the supply of ATP in other organs for the next seven days. Mortality from failure of the liver with normal conservative management is close to 100%. What advice would you give? (Reason from first principles.)

BACKGROUND

HYPOGLYCEMIA RESULTING FROM IMPAIRED LIVER FUNCTION

A catastrophic disease such as fulminant hepatitis will prevent the liver from functioning properly (Figure 1·14). Since the liver cannot provide glucose, hypoglycemia will develop. Hypoglycemia causes low blood insulin, which triggers rapid release of two types of fuels—fatty acids from adipose tissue and amino acids from muscle. The adrenergic response to hypoglycemia also results in the release of fatty acids from adipose tissue. Breakdown of proteins in muscle further stresses the patient who does not have a normally functioning liver and leads to the accumulation of lactic acid and NH_4^+.

Normally the liver maintains a concentration of glucose in the blood that is sufficient for the brain; the liver also produces fuels for the brain when needed (glucose from glycogen or proteins, or ketoacids from fatty acids during fasting). Because fulminant hepatitis impairs these functions and renders the patient too sick to eat, the brain must use endogenous and intravenously infused fuels.

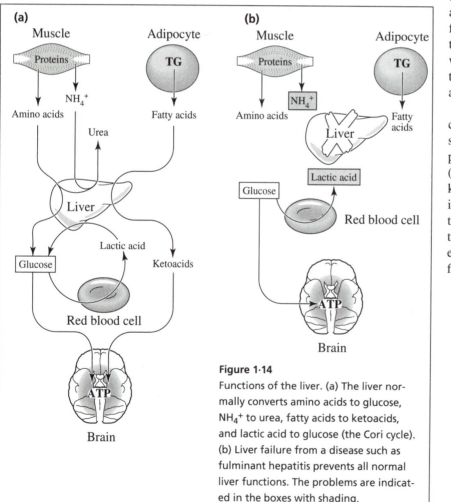

Figure 1·14

Functions of the liver. (a) The liver normally converts amino acids to glucose, NH_4^+ to urea, fatty acids to ketoacids, and lactic acid to glucose (the Cori cycle). (b) Liver failure from a disease such as fulminant hepatitis prevents all normal liver functions. The problems are indicated in the boxes with shading.

Discussion of Case 1·7
Fulminant Hepatitis
(Case presented on page 22)

POSSIBLE TREATMENT

1. Provide fuels for the brain—glucose or lactate.

   Since the problem is lack of fuel for the brain (there are no ketoacids and little or no glucose), the simple response is to give glucose, keeping its level at about 90 mg/dl (5 mmol/l). This conservative treatment may cause complications because the use of glucose involves production of lactic acid, which is formed in red blood cells, muscle, and other tissues. Because the sick liver cannot reconvert lactic acid to glucose (Figure 1·19, page 27), lactic acid will accumulate in the blood and can become lethal when above 20 mmol/l; this toxicity may be a major factor in the high mortality observed with conservative management.

   The brain, however, can oxidize lactic acid if its concentration in the blood is high enough (5–10 mmol/l) and glucose is low (1–2 mmol/l), as shown in Figure 1·15(a). Therefore, possible therapy is to provide glucose by the intravenous route to keep the concentration of lactate anion in blood in the 5–10 mmol/l range; give more glucose if the concentration of lactate anion falls, less if it rises. Use the anion gap in plasma (see pages 56–57) as a quick indicator of the concentration of lactate anion in plasma.

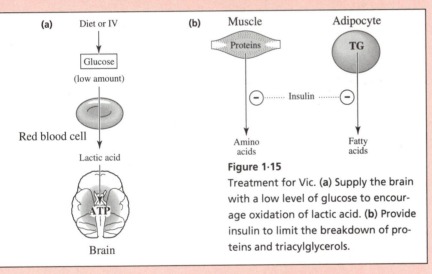

Figure 1·15
Treatment for Vic. **(a)** Supply the brain with a low level of glucose to encourage oxidation of lactic acid. **(b)** Provide insulin to limit the breakdown of proteins and triacylglycerols.

2. Provide insulin.

   When levels of insulin are low, NH_4^+ will accumulate from the catabolism of amino acids released from proteins. As Figure 1·15(b) illustrates, insulin limits the breakdown of proteins and also inhibits the release of fatty acids from adipose tissue. Because insulin promotes the oxidation of glucose and lactic acid to regenerate ATP, a possible treatment for Vic is to provide him with insulin in amounts that maintain low levels of amino acids in his blood. The drawback is that insulin will limit the breakdown of triacylglycerols; as a result, more glucose must be supplied to compensate for the lack of fatty acids to oxidize.

   Note: The release of NH_4^+ may be reduced further by infusing the ketoacid analogs of essential amino acids and by infusing insulin with the hope of driving the synthesis of proteins from amino acids.

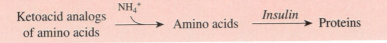

PART D
NUTRITION, METABOLIC PATHWAYS, AND CONTROLS

Case 1·8
Normal Diet in Control of Body Weight
(Case discussed on pages 27–29)

Henry, a yuppie but not an athlete, controls his weight strictly through diet. He knows how hard it is to lose weight since he was rather pudgy before he realized that he had to correlate his career goals with his personal appearance (see Case 1·9). He eats three regular, balanced meals a day, with no snacking. How do the organs of his body work together to use the energy fuels provided in a normal diet?

Case 1·9
Weight Loss in Fasting
(Case discussed on pages 29–31)

At an earlier period in his life, Henry (Case 1·8) realized that he did not have the lifestyle he wanted because he was overweight. In desperation, he decided to stop eating and only consumed vitamins, minerals, and water. What changes in his weight do you expect to occur in the first week? What pathways and controls do you expect to operate?

Questions to Consider

New

What would you expect to happen in energy metabolism between the start of a large meal and that of the next meal many hours later?

What would you expect to happen during prolonged starvation?

Why is weight loss usually greatest in the first few days of fasting?

Review

What are the main functions of the major organs in regard to energy fuels?

What interconversions are possible between energy fuels?

What are the primary controls over energy metabolism?

What are the circulating forms of energy fuels?

BACKGROUND

METABOLIC PATHWAYS

> *Metabolic pathways are specific sequences of reactions that interconvert fuels.*

Metabolic stores and their corresponding circulating fuels can be interconverted, but different enzymatic pathways are used for synthesis and degradation. All circulating fuels—except *essential amino acids*—can be both synthesized and degraded. The pyruvate dehydrogenase (PDH) system and the ATP generation system are one-direction pathways: the PDH system converts pyruvate to *acetyl-CoA*, and the ATP generation system oxidizes acetyl-CoA to generate ATP. All fuels can produce acetyl-CoA and can therefore generate ATP.

Energy Balances

All metabolic pathways involve exchange of energy. Some processes require energy (e.g., making large molecules out of smaller ones; all arrows pointing up in Figure 1·16 consume ATP). Other processes, most notably the ATP generation system, provide energy by yielding ATP. All pathways from circulating fuels to tissue intermediates involve oxidation, thereby releasing hydrogen, which binds to nicotinamide adenine dinucleotide (NAD+). This NADH (Chapter 10), which is used in the ATP generation system, yields water and ATP. Outside the ATP generation system, only conversion of glucose to lactic acid releases some ATP (but this flux can be very rapid indeed; see Chapters 3, 6, and 8).

METABOLIC SYSTEMS

> *In metabolic systems, functions and controls are linked.*

Consider the components of Figure 1·16 as systems centered on types of fuels and pathways. If you know the functions of a system, you can deduce the controls (Table 1·6).

METABOLIC PROCESSES

> *To define a metabolic process, examine the starting point and the final products of a series of metabolic pathways.*

Metabolic processes, which are defined by specific metabolic functions, begin with dietary or stored fuels and end with stored fuels or ATP, respectively. We identify three groups of metabolic processes; each involves more than one organ.

Essential amino acids:
Amino acids that cannot be synthesized by the body.

Acetyl-CoA:
The final common intermediate in the oxidation of glucose, amino acids, and fatty acids. It is used for the regeneration of ATP.

NAD+:
A complex molecule that accepts hydrogen atoms from reactions in metabolism and transfers them to other molecules. Reduced NAD+ (NADH) is a major intermediate in the regeneration of ATP using O_2 (see Chapter 10).

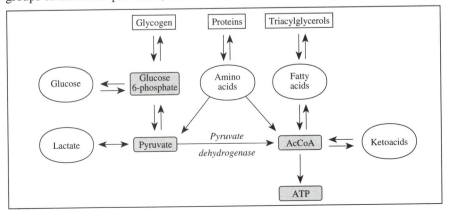

Figure 1·16
Pathways and systems in energy metabolism. Energy stores are shown as rectangles, circulating fuels as ovals, and the key intracellular intermediates as shaded rectangles. Notice that all fuels feed into a single ATP generating system, only some fuels can be interconverted, and control of PDH is the key to conserving glucose and proteins. (AcCoA = acetyl-CoA.)

25

Table 1·6
Metabolic systems

System	Functions	Controls
Carbohydrate	Formation of ATP from glucose or glycogen	Negative feedback by ATP; low flux if glucose is insufficient
	Conversion of glucose to stores: Glycogen	Stimulated by surplus glucose;* stops when store is full
	Triacylglycerols	Stimulated by surplus glucose;* no limits, apparently
	Conversion of precursors of glucose (lactate, amino acids, glycerol) to glucose	Supply of these precursors to the liver
Fat	Formation of ATP from fatty acids	Stimulated by insufficient glucose*
	Conversion of fatty acids to ketoacids (in the liver only)	Stimulated by insufficient glucose;* limited by the liver's need for ATP
	Transport of dietary triacyl-glycerols to stores	Stimulated by increased circulating insulin
Protein	Formation of ATP from proteins	Stimulated by surplus amino acids and by insufficient glucose*
	Conversion of amino acids to proteins	Stimulated by surplus glucose* and/or amino acids
PDH	Formation of ATP from glucose	Negative feedback by ATP
	Conversion of glucose to triacyl-glycerols (in the liver only)	Stimulated by surplus glucose*
ATP generation	This final common path leads from acetyl-CoA to the formation of ATP	Negative feedback by ATP and related compounds

*Surplus glucose implies surplus insulin; insufficient glucose implies insufficient insulin.

1. From Dietary or Stored Fuels to ATP

Fuels stored in one tissue can be oxidized to generate ATP in another. For example, the brain can generate ATP by using ketoacids, a "fatty acid product" derived from triacylglycerols stored in adipose tissue (Figure 1·17).

Figure 1·17

Ketoacids as a source of ATP for the brain. The liver converts fatty acids released from triacylglycerols in adipose tissue into ketoacids, which the brain can use to regenerate ATP.

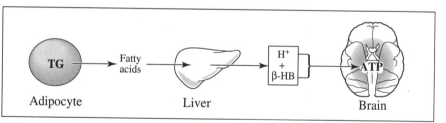

2. From Dietary Fuels to Energy Stores

After generating ATP in tissues, the second priority is to replace proteins and glycogen lost in the period between meals. The third priority is to store excess fuels as triacylglycerols in adipose tissue (Figure 1·18).

Figure 1·18

The storage of excess dietary glucose as triacylglycerols in adipose tissue. The liver converts excess glucose into very-low-density lipoproteins (VLDL), which transport triacylglycerols to adipocytes for storage.

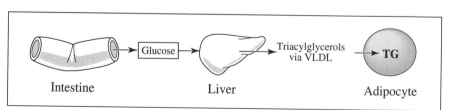

3. Energy Transfer Processes

In energy transfer processes, an organ converts one fuel to another so that a second organ can use it in the generation of ATP (one organ uses energy in order to provide a usable form of energy for another). An example is the Cori cycle (Figure 1·19).

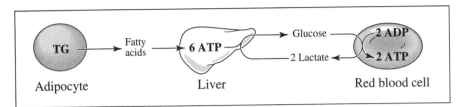

Figure 1·19

ATP "shuttle" from the liver to red blood cells (the Cori cycle). Red blood cells obtain ATP (two per glucose) from the conversion of glucose to lactate. The liver resynthesizes glucose from this lactate using six ATP for every molecule of glucose made. The ATP used in this process comes from the oxidation of fatty acids. Thus, through the Cori cycle, red blood cells may indirectly regenerate ATP via the oxidation of fatty acids.

Discussion of Case 1·8
Normal Diet in Control of Body Weight
(Case presented on page 24)

When Henry sits down to eat, he will have been without food since he finished absorbing his last meal, several hours before. His stores of glycogen in the liver are therefore partly depleted. The priorities for disposal of dietary fuels are shown in Figure 1·20. First, provide fuels for the brain and other tissues. Second, replenish stores of glucose (glycogen in the liver and muscle) and stores of proteins catabolized since the last meal. Third, store excess energy as triacylglycerols in adipose tissue.

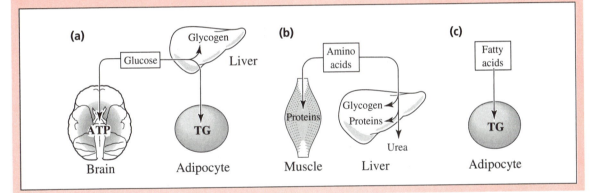

Figure 1·20

Priorities for disposal of dietary fuels. (a) Glucose travels to the brain for immediate use and then to the liver for conversion to storage forms. The liver fills its store of glycogen first and then converts glucose to triacylglycerols for storage in adipose tissue. (b) Amino acids are converted to proteins in muscle and in the liver. The liver converts amino acids to glucose for storage as glycogen. The liver also excretes urea, which results from the catabolism of amino acids. (c) Fatty acids are converted to triacylglycerols for storage in adipose tissue.

Table 1·7

Fuels and wastes from a balanced diet

Fuels	Moles	Weight (g)	Energy (kcal)
Glucose	1.5	270	1080
Amino Acids	1.0	102	408
Fatty Acids	0.36	104	936
Oxygen	21.6	690	–
Total fuels	–	1166	2424
Wastes			
CO_2	18.7	821	–
Water	17.8	320	–
Urea	0.5	30	–
Ammonium	0.04	0.7	–

During meals:

The time when energy needs are met by ingested dietary fuels.

Between meals:

The time when energy needs are temporarily met by stored fuels.

Without meals:

The time when energy needs are not met by dietary fuels. This time is subdivided further on the basis of the source of fuel for the brain.

1. Carbohydrate phase of fasting: Fuel for the brain must come from glycogen in the liver.

2. Protein phase of fasting: Fuel for the brain comes from proteins via gluconeogenesis (creation of glucose either by synthesis from amino acids or glycerol or by conversion—via lactate—from glycogen in muscle).

3. Ketotic phase of fasting: Fuel for the brain comes from ketoacids derived from fatty acids.

4. Terminal phase of fasting: The body has run out of triacylglycerols and must use proteins to regenerate ATP.

Figure 1·21

Use of carbohydrates from one meal. Unlike other organs, the kidneys cannot use glucose directly. Instead, they regenerate ATP from lactate. Most lactate comes from partial oxidation of glucose in consumer organs.

The daily diet usually contains more than twice the amount of glucose that is needed by the brain (Table 1·7). *During meals*, levels of glucose in the blood rise from the absorption of dietary carbohydrates (e.g., from 70–150 mg/dl, or 4–8 mmol/l), an action that stimulates the release of insulin. A high circulating level of insulin directs dietary fuels to stores; a low level leads to accelerated (but controlled) release of these stores. Increased levels of insulin in the blood signal:

1. deposition of glucose as glycogen in the liver and muscle;

2. deposition of amino acids as proteins in the liver and muscle;

3. deposition of triacylglycerols in adipose tissue;

4. conversion of excess energy fuels (glucose or amino acids) to triacylglycerols in the liver and their deposition in adipose tissue;

5. inhibited release of fatty acids from triacylglycerols in adipose tissue.

These signals by insulin limit the increase in concentration of glucose, amino acids, and fatty acids in the blood caused by the intake of food. The decreased levels of fatty acids in the blood reduce the rate of oxidation of fatty acids, thus favoring the use of glucose.

Intake of food provides fuels for about eight hours; absorption takes approximately two to three hours. Thus, the body must store enough fuel to last five or six hours (Figures 1·21 and 1·22). By the time the ingested food is absorbed, the concentration of glucose in the blood returns to normal, as does the concentration of insulin. The relatively low circulating level of insulin *between meals* allows for a net release of fuels from stores, at controlled rates. The control of most importance to the brain is the release of glucose from glycogen in the liver—to match the use of glucose by the brain. The concentrations of ketoacids remain low (they were kept low by the high levels of insulin during the meal and take several days to rise *without meals*). Therefore, the brain must rely totally on glycogen in the liver between meals. To conserve glucose for the brain, the rest of the body oxidizes fatty acids to supply as much energy as possible. The lower level of circulating insulin permits the release of fatty acids from adipose tissue to the blood.

At the cellular level, the oxidation of fatty acids to generate ATP prevents organs from oxidizing carbohydrates, since oxidation of fatty acids leads to inhibition of oxidation of carbohydrates at the pyruvate dehydrogenase (PDH) reaction (Chapter 9). This inhibition is the basis for the hierarchy of fuels (fatty acids are used first if they are available).

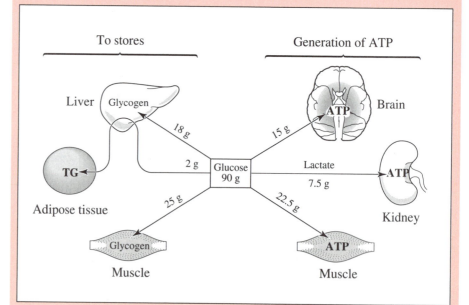

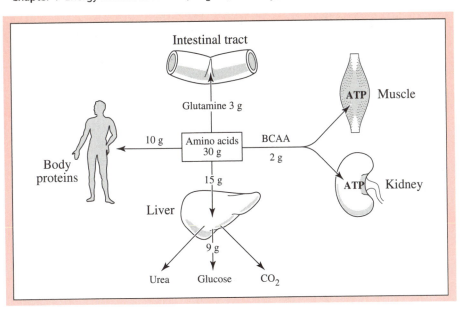

Figure 1·22
Use of proteins from one meal. If the liver were to metabolize all of the amino acids from the diet, it would have an oversupply of ATP. Accordingly, other organs must also oxidize amino acids. Both muscle and the kidneys use branched-chain amino acids (BCAA) to regenerate ATP, and the intestinal tract uses the dicarboxylic amino acids and their amide derivatives (represented in the figure as glutamine) as its primary fuels.

Discussion of Case 1·9
Loss of Weight in Fasting
(Case presented on page 24)

The problem is that Henry's brain constantly requires glucose (5 g/h unless the level of ketoacids is high), but his stores of glucose are limited; his only real store of glucose is glycogen in his liver, about 100 g (which would last for about 20 hours if used only by his brain). For Henry, fatty acids from adipose tissue are the major source of energy. Although fatty acids cannot be converted into glucose, the glycerol backbone of triacylglycerols (6% by weight of triacylglycerol carbon) can; this feature is important in starvation (page 30).

When his liver glycogen has been used up (in less than one day), his body must provide more fuels for his brain (Figure 1·23). As a short-term solution, his liver makes glucose from amino acids released from muscle proteins. As a long-term solution, his liver provides ketoacids from fatty acids released from adipose tissue. When the concentration of ketoacids rises to the 3–7 mmol/l range, these fuels can supply the caloric equivalent of about 80% of the glucose usually used by his brain.

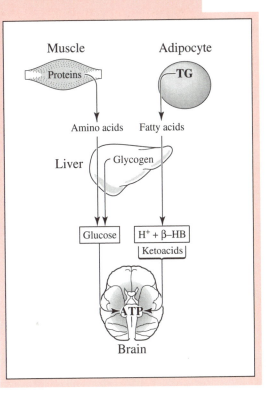

Figure 1·23
Fuels for the brain. Since fatty acids cannot cross the blood-brain barrier, they are not an important fuel for the brain. Ketoacids, if available, are the preferred fuel, but the brain always oxidizes some glucose. In the absence of ketoacids, the brain usually oxidizes glucose exclusively to regenerate its ATP. When dietary glucose is not available and glycogen in the liver has been depleted, proteins become the primary source of glucose.

Table 1·8
Concentrations of major fuels in the blood without meals

Duration of fast (days)	Concentration of fuel (mmol/l)		
	Glucose	Fatty acids	Ketoacids
0	5.0	0.5	0.01
1	4.0	0.8	0.05
2	3.6	1.2	1.2
3	3.3	1.2	3.0
21	4.2	1.2	6.0

THE THREE PHASES OF NORMAL FASTING

> *In fasting, the focus is on the source of fuel for the brain: glycogen from the liver, glucose derived from proteins, and ketoacids derived from fatty acids.*

1. **The carbohydrate phase occurs between meals and lasts up to 20 hours.**

   Assume that Henry's normal energy use is 2200 kcal/day. His brain uses glucose obtained from liver glycogen until available stores are exhausted—up to 20 h at 5 g/h. Triacylglycerol stores in adipose tissue provide the energy that the rest of his body needs.

2. **The protein phase is a harmful interim phase (see Cases 1·4, pages 8 and 14, and 1·10, pages 32 and 33) that lasts until the ketotic phase of fasting begins.**

   Henry's energy use drops to about 1750 kcal because he is less active and is no longer using energy to interconvert dietary fuels. Because his liver has not yet produced enough ketoacids, his brain can use only glucose. Most of this glucose must come from proteins (120 g of glucose come from 200 g of proteins), but some comes from the glycerol released from adipose tissue. Henry's energy use continues to drop as he becomes more and more listless. However, ketoacids start to accumulate in his blood (Table 1·8).

3. **The ketotic phase is the stable, final phase of fasting.**

   Henry now has built up sufficient levels of ketoacids for replacement of up to about 80% of his brain's need for glucose (Table 1·9). Although triacylglycerol stores supply his brain with some glucose directly (from the glycerol component), they primarily provide energy for the rest of his body, thereby meeting most of his needs.

Table 1·9

The flow of blood to the brain is close to 1 l/min, or 1500 l/day. The fuel menu for the brain is the same in the carbohydrate and protein phases of fasting, but it changes markedly in the ketotic phase of fasting. Negative values for A-V differences indicate net output rather than net uptake of metabolites.

Table 1·9
Quantitative description of energy metabolism in the brain

Fuels used	Concentration in blood (mmol/l)		Consumption (or production) (mmol)	A-V difference (mmol/l)	Source of ATP (%)
	Artery	Vein			
During and between meals					
Glucose	5.0	4.35	650–700	0.65	100
Lactate	1.0	1.4	(600)*	−0.4	0
Ketoacids	0.05	0.05	0	0	0
Oxygen	8.0	5.3	4000	2.7	100
Without meals: ketotic phase of fasting					
Glucose	3.5	3.35	240	0.15	20
Lactate	0.7	0.9	(300)*	−0.2	0
Ketoacids	5.0	4.5	750	0.5	80
Oxygen	8.0	5.3	4000	2.7	100

* Subtract half the output of lactate to deduce how much glucose was converted to lactate. Assume that half the remainder was oxidized completely to CO_2 + ATP.

WEIGHT LOSS

Table 1·10 outlines the theoretical weight loss in grams per day during prolonged fasting or starvation. The large weight loss in days one and two is due to the use of heavy calories—primarily glycogen and proteins. Proteins in the body are utilized more rapidly in the period immediately after depletion of glycogen in the liver (before ketoacid production has built up). In this period, the most damaging part of fasting, the rate of weight loss declines progressively between days two and five. The small weight loss in the ketotic phase of fasting reflects the use of the most efficient energy store—triacylglycerols, or light calories. The only good weight loss is loss of triacylglycerols!

Questions

(Discussions on pages 248–49)

1·5 What would you recommend that Henry (Case 1·9) add to his diet to minimize wasting of his muscles?

1·6 What will be the theoretical loss of carbohydrates, proteins, triacylglycerols, and weight if Henry (Case 1·9) exercises regularly, expending an extra 1500 kcal/day (assume it all comes from triacylglycerols)?

Table 1·10
Daily weight loss during prolonged fasting*

Phase	Energy used (kcal/day)	Glycogen dry	Glycogen wet	Proteins dry	Proteins wet	Triacylglycerols dry	Triacylglycerols wet	Net weight loss (g/day)
Carbohydrate, day 1	2200	100	400	50	200	215	215	815
Protein, day 2	1750	–	–	180	720	114	114	834
Protein, days 3 and 4	1500	–	–	15–150	60–600	140–150	140–150	200–750
Ketotic, days 5–50	1245	–	–	6	24	150	150	174
Terminal, day 50 or later	1000	–	–	250	1000	–	–	1000

* Calculations: The daily caloric consumption minus the daily use of glucose by the brain yields the quantity of triacylglycerols oxidized. Oxidation of triacylglycerols provides glucose (from glycerol—10% of the weight of triacylglycerols). The rest of the glucose must come from glycogen (day one only) or proteins, of which only 60% yields glucose (the rest yields CO_2 and urea). Simultaneous equations for the brain's requirements for glucose (g/day = 0.6 × amount of protein + 0.1 × amount of fat) and total calorie requirements (kcal/day = 4 × amount of protein + 9 × amount of fat) provide the figures in the table.

PART E
CASES FOR REVIEW

Cases 1·10 to 1·13 are provided for review but also introduce some new concepts.

Case 1·10
Cancer Cachexia: Rapid Weight Loss
(Case discussed on page 33)

Margaret had a fast-growing malignant tumor. Her appetite was somewhat decreased, and she led a sedentary life. Despite a modest decrease in caloric intake, she lost 7 kg (15 lb) in three weeks. She did not have a fever or excessive sweating, nor did she give a history of diuretic use, drug ingestion, or excessive activity. Her urine did not contain sugar or amino acids. What was her biochemical lesion (why did the cancer increase utilization of heavy calories)?

Case 1·11
Dumping Syndrome
(Case discussed on pages 34–35)

Eric had a partial gastrectomy as treatment for an ulcer. Once he began to eat again, he had unpleasant symptoms after meals, becoming dizzy, nauseated, light-headed, and anxious, and later even blacking out at times. What could account for these symptoms?

Case 1·12
A Panic Attack
(Case discussed on pages 35–36)

Sandy volunteered to be a subject in a project examining the effects of fasting on nitrogen wastes. Coming in to work on the subway after about 36 hours without food, she experienced a "panic attack" (classic signs of the effects of hypoglycemia on the brain). The symptoms diminished when her journey ended, though she did not eat. At the lab, about four hours after the attack began, the concentration of glucose in her blood was 4.5 mmol/l, but it had been 3.9 mmol/l the evening before. How do you explain this increase?

Case 1·13
Phil Is Grossly Overweight
(Case discussed on page 37)

Phil is grossly overweight but has been unable to lose weight despite a self-imposed diet. He read an advertisement guaranteeing that obese people could lose 30 lb in a month by adhering to a strict regimen. Phil comes to you for advice. Would you recommend the program? Explain your answer to Phil.

Discussion of Case 1·10
Cancer Cachexia: Rapid Loss of Weight
(Case presented on page 32)

Margaret was physically inactive and hence was not using much energy above her basal requirements for living. Her weight loss exceeds any reasonable expectation from simply oxidizing fatty acids, and her urine contained no lost nutrients. Thus, she must have been using heavy calories, signifying a breakdown of proteins (Figure 1·4(a), page 7, and Figure 1·24).

A fast-growing tumor needs a considerable amount of energy, but it cannot oxidize fatty acids or ketoacids because it lacks some of the necessary enzymes. Thus, it ultimately consumes carbohydrates or carbohydrate precursors (proteins). Administering an inhibitor of PDH that can enter tumor cells will diminish the regeneration of ATP in the tumor cells only. To save the other cells, encourage the oxidation of fatty acids in the rest of the body. (During the oxidation of fatty acids, high levels of acetyl-CoA and NADH normally inhibit PDH.) A carbohydrate-free diet supplemented with 1,3-butanediol, a precursor of ketoacids, may produce the desired results.

> **Note:**
> Pyruvate dehydrogenase (PDH) is discussed in more detail in Chapter 9, pages 211–13.

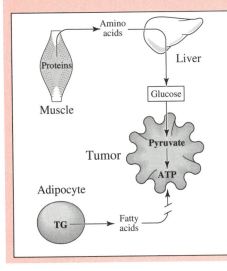

Figure 1·24
Cancer cachexia. The tumor, which cannot obtain energy from fatty acids or ketoacids, requires glucose for the production of ATP. Low levels of glucose in the blood are likely between meals, as the tumor oxidizes glucose. Eating causes intermittently high levels of insulin, so that ketoacids do not accumulate in Margaret's blood. Hence, to supply energy for herself as well as the tumor, Margaret uses liver glycogen rapidly and relies on protein stores between meals.

Questions

(Discussions on page 250)

1·7 Between meals, the body conserves glucose for use by the brain and oxidizes fatty acids to provide as much energy as possible for other organs. The oxidation of fatty acids diminishes the activity of PDH, a necessary action between meals because PDH destroys carbohydrates or their precursors by catalyzing an irreversible reaction in the body. Control by product inhibition also occurs: PDH produces acetyl-CoA and NADH, both precursors for the regeneration of ATP; as discussed in Chapter 9, ATP, NADH, and acetyl-CoA are strong inhibitors of PDH. However, high levels of insulin in the liver can override this inhibition.

In what metabolic situations and in what tissues would active PDH be required a) all of the time; b) some of the time; c) never?

1·8 If carbon from glucose must be conserved, especially in starvation, which is the most important point for control?

1·9 What types of controls would you expect over PDH in the liver?

1·10 Look back over the previous cases. What were the roles of PDH in the various tissues in each case?

Discussion of Case 1·11
Dumping Syndrome

(Case presented on page 32)

The stomach normally acts as a short-term reservoir of ingested food, releasing small amounts to the intestine. Because the stomach delivers a concentrated mix of foods to the intestine at a slow rate, it prevents the rapid movement of water from the blood into the lumen of the intestines. In contrast, in the dumping syndrome, excess loss of water causes a low circulating volume. The stomach reservoir also limits the rate of uptake of fuels and other nutrients into the blood. (The small intestine can absorb digested food products very quickly.) During digestion, ingested foods empty from the stomach to the intestine at approximately the same rate as the use of fuels by the body at rest (2.5 kcal of glucose per minute).

Some patients who undergo stomach surgery lose this short-term reservoir function of the stomach; the stomach therefore "dumps" its contents into the absorbing area (small intestine) more quickly than normal (Figure 1·25). Too rapid absorption of amino acids can affect the brain, since some amino acids are neurotransmitters or their precursors. Too rapid absorption of glucose causes hyperglycemia and may cause glycosuria. Rapid development of hyperglycemia also results in excessive secretion of insulin, which decreases the availability of fatty acids and increases the uptake of glucose by muscle. These reactions, in turn, cause hypoglycemia and lack of fuels for the brain.

Figure 1·25
The dumping syndrome.

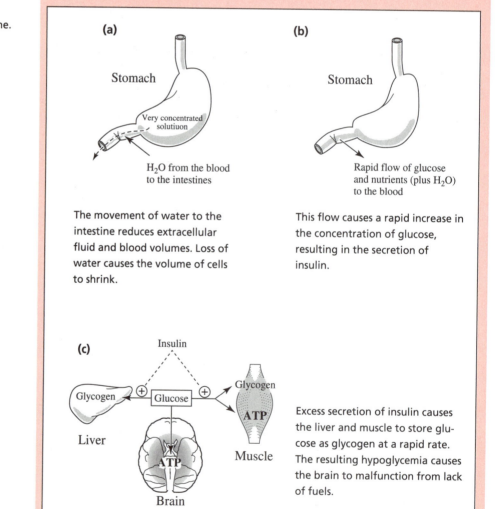

(a)

Stomach

Very concentrated solutiuon

H_2O from the blood to the intestines

The movement of water to the intestine reduces extracellular fluid and blood volumes. Loss of water causes the volume of cells to shrink.

(b)

Stomach

Rapid flow of glucose and nutrients (plus H_2O) to the blood

This flow causes a rapid increase in the concentration of glucose, resulting in the secretion of insulin.

(c)

Insulin

Glycogen   (+)   Glucose   (+)   Glycogen

Liver   ATP

ATP   Muscle

Brain

Excess secretion of insulin causes the liver and muscle to store glucose as glycogen at a rapid rate. The resulting hypoglycemia causes the brain to malfunction from lack of fuels.

THE SYMPTOMS

Patients suffer initially from symptoms of hypovolemia (contraction of extracellular fluid volume); loss of water into the intestinal lumen results in dizziness, nausea, sweating, and light-headedness. Secondary symptoms (similar to those in severe hyperglycemia) cause excess secretion of insulin and subsequent hypoglycemia. This hypoglycemia then results in symptoms from excessive stimulation of the sympathetic system (leading to elevated levels of adrenaline) and possibly from insufficient ATP in brain cells (Chapter 5).

Question

(Discussion on pages 250–51)

1·11   A super-marathon runner (50–100 miles) wants to eat enough fuel while running to compensate for energy used in a race (Chapter 3). Is this idea practical?

Discussion of Case 1·12
A Panic Attack

(Case presented on page 32)

Of a number of volunteers who underwent a 40-hour fast, Sandy was the only one who experienced a panic attack, which occurred 36 hours after commencing the fast. Let us assume that an inadequate delivery of fuels (glucose, ketoacids) to Sandy's brain caused the reaction.

At 24 hours, Sandy had a lower concentration of glucose in her blood than the other volunteers who fasted (3.9 vs 4.9 mmol/l), as shown in Table 1·11. Although the level at 24 hours was not yet low enough to cause symptoms of hypoglycemia, the concentrations of glucose fell as the fast continued for another 12 hours. Why did the concentration of glucose in Sandy's blood fall so sharply between 24 and 36 hours? Figure 1·26 offers a possible explanation. Note: At 24 hours, the concentration of ketoacids in Sandy's blood was relatively high (compared with the levels of the other volunteers), but not high enough to permit her brain to regenerate a sufficient quantity of ATP.

Table 1·11
Concentrations in blood

| | Concentrations in blood (mmol/l) | | | |
| | Sandy | | Others | |
Fuel	24 h	40 h	24 h	40 h
Glucose	3.9	4.5	4.9	4.2
Ketoacids	0.6	4.0	0.15	0.8
Fatty acids	0.6	1.67	0.6	1.35

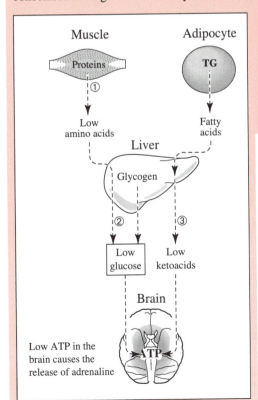

Figure 1·26

Fuel flow between tissues during a panic attack. Timing is everything! In the early stages of the fast, Sandy's brain probably used glucose at a faster rate than it was produced. If precursors of glucose (e.g., amino acids) do not reach the liver quickly enough (1), if the liver does not produce glucose from precursors such as glycogen or amino acids at a fast enough rate (2), or if the level of ketoacids is unduly low (3), then symptoms of hypoglycemia will develop.

Adrenaline:
A hormone released from the adrenal medulla that mediates the "fight or flight" response.

Lab analysis of a blood sample taken at 40 hours revealed that her concentration of glucose was higher than at 24 hours. The other volunteers did not experience such an increase; why did Sandy? Figure 1·27 offers a possible explanation: In response to a deficiency of fuels needed to generate ATP, the brain initiates a massive release of *adrenaline*, the "fight or flight" hormone (see Case 5·1 for details, page 126). Adrenaline stimulates the release of fatty acids from adipocytes, thereby stimulating the synthesis of ketoacids in the liver. Adrenaline also promotes the breakdown of glycogen in muscle, resulting in the synthesis of more lactate than is needed for the regeneration of ATP in muscle. The liver converts this excess lactate to glucose. Thus, the two primary actions of adrenaline—the production of ketoacids and glucose—temporarily abort the symptoms of hypoglycemia.

Questions

(Discussions on page 251)

1·12 Could Sandy (Case 1·12) have had a high level of ketoacids in her blood before the panic attack?

1·13 Why did Sandy's normal concentration of glucose in blood not cause a decrease in her level of ketoacids?

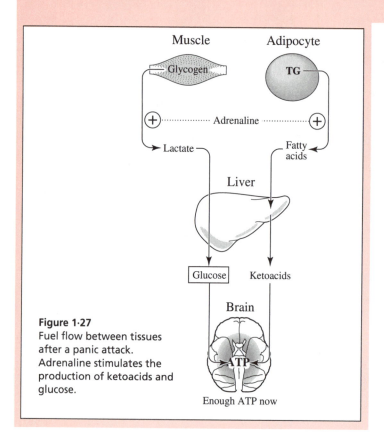

Figure 1·27
Fuel flow between tissues after a panic attack. Adrenaline stimulates the production of ketoacids and glucose.

Discussion of Case 1·13
Phil Is Grossly Overweight

(Case presented on page 32)

An obese person who wants to lose weight must lose adipose tissue (triacylglycerols), not proteins or carbohydrates (the heavy calories) or valuable body water. If Phil does not have a large excess of ECF volume (a disease state such as heart failure or cirrhosis of the liver could result in edema), his weight loss must depend on a negative caloric balance. If he reduced his caloric intake to zero, you could calculate his expected weight loss if fatty acids were the only fuels he oxidized. Since fatty acids have a high yield of kilocalories per unit weight (9 kcal/g) and Phil probably expends 2000–2400 kcal in his daily routine, the maximum amount of fat he could oxidize is only 220–270 g (close to 0.5 lb) per day. Any weight loss from fat in excess of this amount would require a very high metabolic rate (such as that achieved by running a marathon every day). Alternatively, Phil could lose some of his valuable lean body mass, or the diet plan could employ some trick to make him burn more calories (uncouple oxidative phosphorylation; see Chapter 10). All such strategies might harm the body and are therefore undesirable.

Question

(Discussion on pages 252–53)

1·14 Suppose Phil (Case 1·13) achieved rapid weight loss after taking diet pills. How would you determine the source of the weight he lost? If you obtained one of these pills, and had a research lab, animals, cells, mitochondria, etc., what tests would you perform to establish its biological effect(s)?

PART F
SUMMARY OF MAIN POINTS

- Chapter 1 enables the reader to recognize the general metabolic problems in clinical situations where there is an abnormality of energy metabolism.

- Humans derive energy mainly from dietary carbohydrates and triacylglycerols (much less from proteins). Energy metabolism converts dietary fuels to storage forms of energy in the body, using them as needed for generation of ATP, the currency of metabolic energy.

- Energy storage as carbohydrates and proteins is much less efficient than storage as triacylglycerols. Carbohydrates and proteins have a lower content of energy per gram dry weight, and, more importantly, the body stores them with water. Triacylglycerols, on the other hand, are stored as calorically dense particles. If weight loss during starvation exceeds 200 g/day, the body is probably using its reserve of proteins and/or carbohydrates and is losing water in the process.

- The body has large quantities of fats and proteins, but limited stores of carbohydrates. Proteins, though a source of energy, have specific functions in the body and therefore cannot be used without damaging the body somewhat.

- The major organs to consider in energy metabolism are:
  1. the brain and muscle, which consume energy fuels mainly to regenerate the ATP they need;
  2. adipose tissue, which serves as the major store of triacylglycerols;
  3. the liver, which converts fuels from storage forms to usable sources of energy (primarily for the brain), and vice versa;
  4. the kidneys, which help eliminate H^+ formed in metabolic reactions.

- The body has a hierarchy of fuel use, oxidizing fatty acids first if available, then ketoacids, and finally glucose if neither fatty acids nor ketoacids are available. The brain, however, is an exception. Because few fatty acid molecules are able to cross the blood-brain barrier, the brain cannot derive an appreciable amount of ATP from the oxidation of fatty acids. However, when they are available, ketoacids, which are derived from fatty acids, can supply most of the ATP needed by the brain.

- Insulin, the main hormone involved in controlling the distribution of energy fuels throughout the body, signals the body to store energy by stimulating the uptake of glucose and by decreasing the release of fatty acids and glycerol from triacylglycerols stored in adipose tissue. The concentration of insulin in the blood typically increases when the level of glucose increases, and it decreases when the concentration of glucose is low.

- Energy metabolism can be subdivided into five systems. There is one system for each of the three classes of fuels and one for the formation of ATP from acetyl-CoA, a common intermediate (one compound made from all three classes of fuels); the fifth system, pyruvate dehydrogenase (PDH), converts carbohydrates to triacylglycerols and hence must be strictly controlled (excess conversion might deprive the brain of energy).

CHAPTER 2
ACID-BASE METABOLISM: PRODUCTION AND EXCRETION OF WASTES

PART A
BUFFERING OF H+ AND RESPIRATORY ACID-BASE DISORDERS

The body contains a tiny amount of hydrogen ions (H^+)—approximately 0.000040 mmol/l, or 40 nmol/l—but each day it forms and uses huge quantities (5000 mmol or more). Minor changes in the concentration of H^+ ($[H^+]$)* can be life-threatening, so very efficient buffer systems are required. The mechanism for excreting CO_2 provides an effective way to adjust the $[H^+]$; it involves the bicarbonate buffer system.

*Square brackets denote concentration.

PART B
FORMATION AND CONSUMPTION OF H+ IN NORMAL ENERGY METABOLISM AND IN METABOLIC ACIDOSIS

The oxidation of sulfur-containing or cationic amino acids results in the production of H^+. People on a normal diet excrete approximately 1 mmol of H^+ per kilogram of body weight. The body forms and retains H^+ if the oxidation of carbohydrates is incomplete (as in hypoxia) or if the oxidation of fatty acids is incomplete (i.e., with a lack of insulin). Treatment of disorders involving metabolic acidosis requires a means of diagnosing the cause of the acidosis.

PART C
EXCRETION OF ACID AND NITROGENOUS WASTES: THE ROLE OF THE KIDNEYS

Only the kidneys can excrete H^+. The major route is through formation of ammonium ions (NH_4^+) and bicarbonate ions (HCO_3^-) from neutral amino acids. The NH_4^+ are excreted (with chloride or an organic anion) and the HCO_3^- are returned to the circulation, replacing those consumed by the H^+ that were produced from metabolic acids.

PART D
CASES FOR REVIEW

PART E
SUMMARY OF MAIN POINTS

PART A
BUFFERING OF H⁺ AND RESPIRATORY ACID-BASE DISORDERS

Case 2·1
Ursula Prepares for a Sprint
(Case discussed on pages 50–51)

Ursula, a world-class sprinter, always prepares for a race in the same way; just before the race, she breathes excessively in an attempt to get more O_2 into her lungs. Her trainer told her that her blood was already saturated with O_2 and that she could not derive further benefit from *hyperventilation*, but Ursula knows that her performance improves. Does Ursula benefit from hyperventilation? If so, why?

Hyperventilation:
A higher rate of breathing that causes the P_{CO_2} of blood to be lower than needed.

Case 2·2
Claude Has Chronic Lung Disease
(Case discussed on page 51)

Claude has smoked two packs of cigarettes a day for 30 years. He now has chronic obstructive pulmonary disease (COPD—emphysema and chronic bronchitis). Although the P_{CO_2} in his arterial blood is 55 mm Hg (Table 2·1), the [H⁺] in his blood is only slightly higher than normal. Last week, he developed an acute lung infection, which led to a further elevation in the P_{CO_2} of his arterial blood, accompanied by a large increase in the [H⁺]. The figures have remained constant for the past 24 hours. Why did this recent rise in P_{CO_2} lead to a large increase in the [H⁺]?

P_{CO_2}:
The pressure of CO_2 in the gaseous phase that is needed to maintain a certain concentration of CO_2 in solution.

Note:
The bicarbonate buffer system is discussed on pages 44–46 and 49.

Table 2·1
Claude's bicarbonate buffer system

		Normal	Case 2·2		Case 2·3
			Chronic infection	Acute infection	
[H⁺]	nmol/l	40	45	60	80
P_{CO_2}	mm Hg	40	55	70	90
[HCO_3^-]	mmol/l	25	30	29	28

Case 2·3
Nutrition in Very Serious Lung Disease
(Case discussed on page 52)

Smoking has further destroyed Claude's lungs, leaving no pulmonary reserve. He needs to have a much higher [CO_2] in his blood to excrete all the CO_2 formed by metabolism (Table 2·1). Oxidizing which fuel—carbohydrates or fatty acids—would benefit Claude more?

Hemoglobin:

The compound in red blood cells that binds O_2.

Buffer:

A compound that minimizes the change in the $[H^+]$.

Acids:

An acid is a compound that, on dissolving in water, releases H^+. An alkali is the converse; on dissolving in water, alkalis combine with H^+, thus decreasing their concentration (or raising the $[OH^-]$). Pure water is defined as neutral; the $[H^+]$ and the $[OH^-]$ are equal at 100 nmol/l.

$$HA \text{ (acid)} \longleftrightarrow H^+ + A^-$$
$$\text{(conjugate base)}$$
$$NaOH \text{ (alkali)} \longleftrightarrow Na^+ + OH^-$$
$$H_2O \longleftrightarrow H^+ + OH^-$$

Questions to Consider

How is CO_2 transported from tissues to the lungs?

What does loss of CO_2 do to the $[H^+]$?

Are there signals that help O_2 bind to or be released from *hemoglobin?*

How would you measure the $[H^+]$?

What is required for H^+ to bind to a *buffer*?

Can the lungs excrete H^+ or *acids*?

How much ATP is produced per unit of CO_2 when CO_2 is produced from carbohydrates or fatty acids?

BACKGROUND

THE CENTRAL IMPORTANCE OF H⁺

> *Hydrogen ions can impair bodily functions by changing the ionic charge on proteins.*

Plasma always has a very low [H^+] (0.000040 mmol/l, or 40 nmol/l). This concentration must not rise or fall appreciably because H^+ bind avidly to proteins (including all enzymes), changing their net charge and causing alterations in their shapes and functions. For this reason, the control of [H^+] in the body is essential. Table 2·2 shows that minor variations from this value (halving or doubling) are clinically very significant or life-threatening.

Table 2·2
Range of [H^+] in blood

Condition	[H^+] nmol/liter	pH	Importance
Acidosis	>100	<7.0	Life-threatening
	50–80	7.1–7.3	Clinically significant
Normal	40 ± 2	7.4 ± 0.02	Normal
Alkalosis	25–30	7.4–7.6	Clinically significant
	<20	>7.7	Life-threatening

Conditions causing *acidosis* and *alkalosis* are discussed later in Chapter 2 and also in Chapters 3, 4, and 6.

pH:
The negative logarithm of the [H^+].

Acidosis:
Conditions (respiratory, metabolic) causing the [H^+] in plasma to increase.

Alkalosis:
Conditions (respiratory, metabolic) causing the [H^+] in plasma to decrease.

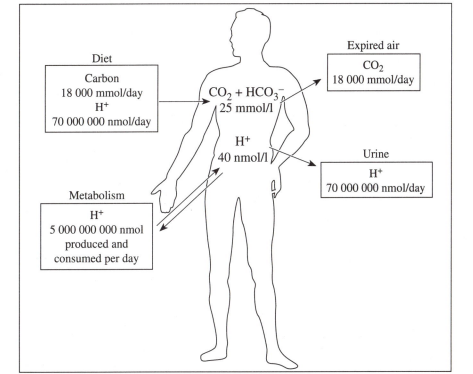

Figure 2·1
Daily turnover of H^+ and CO_2. The [H^+] in the body is tiny (40 ± 2 nmol/l), and only minor fluctuations are acceptable. In contrast, great quantities of H^+ (5 000 000 000 nmol/day) flow through the pool of H^+. Thus, buffering of H^+ must be very effective. The CO_2 system also has a very high daily turnover (10 mmol/min, or 14–20 mol/day) despite a tiny concentration in body fluids (25 mmol/l).

Figure 2·1 indicates that each day enormous quantities of H^+ (approximately 5 000 000 000 nmol) are formed and consumed compared with the total amount present in the body at any one time. To maintain a normal [H^+] in the face of such an enormous turnover, the body must rely on the mechanisms for quick temporary removal of H^+ (buffers) as well as long-term solutions to this problem (elimination of H^+).

BUFFERS

> *Buffers minimize changes in the [H+] caused by addition of acids or alkalis.*

Buffers respond to changes in the level of H^+ in the blood, binding H^+ when their concentration increases and releasing them from the bound form when the concentration decreases. Equations 1 and 2 describe the buffering action of an anion (A^-) and a neutral compound (B°), which act as H^+ acceptors (conjugate bases); HA and HB^+ act as H^+ donors (acids).

$$H^+ + A^- \longleftrightarrow HA \qquad\qquad (1)$$
$$H^+ + B^\circ \longleftrightarrow HB^+ \qquad\qquad (2)$$

The [H^+] at which a buffer is effective depends on the equilibrium constant for the reaction (equation 3).

$$K = \frac{[H^+][B^\circ]}{[HB^+]} \qquad\qquad (3)$$

Figure 2·2 indicates that buffers are most effective in binding or releasing a large quantity of H^+ only when the ratio of bound H^+ to free conjugate base is close to 1.0 (i.e., when the solution has a [H^+] that is approximately equal to the value of K). Obviously, the concentration of the buffer also affects its efficiency in stabilizing the [H^+].

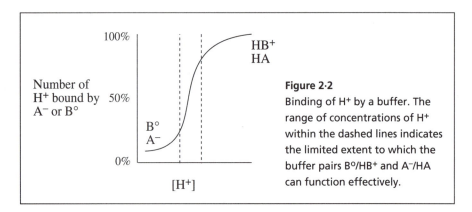

Number of
H^+ bound by
A^- or B°

Figure 2·2
Binding of H^+ by a buffer. The range of concentrations of H^+ within the dashed lines indicates the limited extent to which the buffer pairs B°/HB^+ and A^-/HA can function effectively.

THE BICARBONATE BUFFER SYSTEM (BBS)

> *Because the lungs can control the Pco_2 of plasma, the BBS is essential for adjusting the [H+] in the extracellular fluid (ECF) and the quantity of H+ bound to proteins in the intracellular fluid (ICF).*

Components of the BBS

The BBS, the most important buffer system in the body, consists primarily of HCO_3^-, which can bind H^+, and of carbonic acid (H_2CO_3), which can release H^+.

In the BBS, the concentrations of H^+, HCO_3^-, and CO_2 are linked together through the reactions shown in equation 4.

$$H_2O + CO_2 \longleftrightarrow H_2CO_3 \longleftrightarrow H^+ + HCO_3^- \qquad (4)$$

Carbonic anhydrase rapidly catalyzes the interconversion between H_2CO_3 and CO_2 plus H_2O. Carbonic acid, which always has a very small concentration, spontaneously and rapidly equilibrates with H^+ and HCO_3^-. Always at chemical equilibrium, the concentrations of the chemicals in equation 4 must obey the dictates of equation 5, where K_{eq} is the equilibrium constant.

Carbonic anhydrase:
The enzyme that catalyzes the interconversion of CO_2 and water to carbonic acid.

$$K_{eq} = \frac{[H_2O][CO_2]}{[H^+][HCO_3^-]} \qquad (5)$$

Equation 5 may also be written as the *Henderson equation* (equation 6), which we feel is more useful for the health professional. For a description of units, see Table 2·3.

$$[H^+] = 23.9 \times \frac{Pco_2}{[HCO_3^-]} \qquad (6)$$

Henderson equation:
A description of the BBS that doesn't contain logarithms. When the $[H^+]$ and the dissociation constant (the equilibrium constant for the reaction in which a compound breaks down into components—usually two ions) are expressed as logarithms, the description of the BBS is called the Henderson-Hasselbalch equation.

In this equation, $[H^+]$ is expressed in nmol/l, Pco_2 is in mm Hg, and $[HCO_3^-]$ is in mmol/l. Normal values are shown in Table 2·3. The number 23.9 is a composite constant that takes into account values for the K_{eq} (equation 5), $[H_2O]$ (which can be presumed to be constant), the solubility of CO_2 in water, the influence of carbonates (which at physiological pH are relatively minor components of the BBS that arise from the following reaction:
$$HCO_3^- \longleftrightarrow H^+ + CO_3^{2-}),$$
and the units used in equation 5.

Table 2·3
Normal concentrations of components of the BBS in plasma and in cells

		Plasma	Cells
$[HCO_3^-]$	mmol/l	24 ± 2	12
Pco_2*	mm Hg	40 ± 2	42
$[H^+]$	nmol/l	40 ± 2	80
pH	units	7.4 ± 0.02	7.1

* If the Pco_2 is expressed in kilopascals, K_{eq} is 180.4, and Pco_2 values are 5.3 and 6.1, respectively.
Note that the $[HCO_3^-]$ is approximately 10^6-fold greater than the $[H^+]$.

Effectiveness of the BBS

The BBS has three unique aspects. Buffers act most efficiently when the K_{eq} is close to the $[H^+]$. In the BBS, however, the K_{eq} is an order of magnitude higher than the $[H^+]$ (10^{-3} M for the dissociation of carbonic acid $\times 10^{-3}$ M for the reaction catalyzed by carbonic anhydrase). The BBS works in spite of this unfavorable K_{eq} because the $[HCO_3^-]$ is close to six orders of magnitude higher than the $[H^+]$ and therefore has a large capacity to buffer H^+.

A second feature that permits the BBS to be effective is that the lungs control the $[CO_2]$. When H^+ are added to the body, they react with HCO_3^- to form CO_2 and water. The elevated $[H^+]$ and the increased level of CO_2 stimulate the respiratory center to excrete CO_2, thereby causing a fall in the ratio of Pco_2 to $[HCO_3^-]$ and thus a decrease in the $[H^+]$ (equation 6). The ability of the lungs to change the Pco_2 (the acid form of the buffer pair in the BBS) is unique to the BBS; in conventional buffers, such as proteins described below, the sum of the concentrations of the hydrogen ion acceptors and donors must remain constant.

The third unique feature of the BBS is that HCO_3^- are removed when H^+ are buffered. Accordingly, for this buffer system to return to normal, another organ must add new HCO_3^- to the body. The kidneys perform this task largely through the excretion of NH_4^+ (see pages 62–64). The following modification of equation 6 provides an overview of the BBS.

$$[H^+] = 23.9 \times \frac{Pco_2 \quad \text{(controlled by the lungs)}}{[HCO_3^-] \;\text{(controlled by the kidneys)}}$$

The BBS and Body Compartments

In the extracellular fluid (ECF), the BBS is the only important buffer. For H^+ to be buffered, they must bind to HCO_3^-, causing the $[HCO_3^-]$ to fall. Because not all of the H^+ bind to HCO_3^- during buffering, some increase in the $[H^+]$ is observed. Hyperventilation occurs in response to the raised $[H^+]$, and the steady elimination of CO_2 drives equation 4 to the left. Since the $[H^+]$ is six orders of magnitude lower than that of HCO_3^-, and they react in a 1:1 stoichiometry, the fall in Pco_2 will cause a large proportional fall in the $[H^+]$ and a very small fall in the $[HCO_3^-]$.

In the intracellular fluid (ICF), the BBS is one of two major buffers (protein is the other). Its response to a load of H^+ is similar to that of the BBS in the ECF. A rise in the $[H^+]$ causes the $[HCO_3^-]$ to fall and more CO_2 to be formed; hyperventilation then causes the Pco_2 to fall, leading to a fall in the $[H^+]$. The difference in the ICF is that, when the $[H^+]$ begins to fall, H^+ bound to proteins are released according to the equilibrium described in equation 7. These H^+ combine with additional HCO_3^-, leaving the proteins free to bind more H^+ if the $[H^+]$ rises at a later time. The net result is a smaller fall in the $[H^+]$, a larger fall in the $[HCO_3^-]$, and a commensurate rise in the concentration of proteins without bound H^+ (B°) in the ICF after hyperventilation (Figure 2·3).

$$H^+ + Protein (B^\circ) \longleftrightarrow H \bullet Protein (HB^+) \tag{7}$$

Figure 2·3

Hyperventilation and the bicarbonate buffer system (BBS) in the ECF and ICF. In the figure, the BBS in the ECF and ICF are contained in the dashed rectangles. Since the ECF has virtually no proteins to act as buffers, hyperventilation causes a major fall in the $[H^+]$ in this compartment. In the ICF, hyperventilation causes H^+ to be released from proteins, resulting in a greatly lowered $[HCO_3^-]$ and an increase in the number of groups on proteins (B°) that no longer have bound H^+.

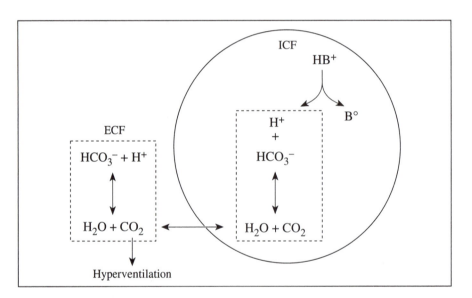

BUFFERING OF H⁺ DUE TO METABOLIC REACTIONS

Phosphate as P_i, but not CrP, binds H^+ in cells.

Phosphate is an important buffer only if it is in its free or inorganic form (P_i) because its pK is too low for effective buffering when bound in organic compounds such as ATP or creatine phosphate (CrP). Skeletal muscle has large quantities of CrP, which breaks down to P_i during a sprint (Chapter 3, page 81). This sequence can then be viewed as the creation of a buffer as a result cf metabolic events.

pK:

The negative logarithm of the dissociation constant. For the BBS, it includes the equilibrium constant of the carbonic anhydrase reaction.

Question

(Discussion on page 253)

2·1 The lab sends you the following results for a blood sample: $[H^+] = 60$ nmol/l, pH = 7.22, $Pco_2 = 50$ mm Hg, and $[HCO_3^-] = 32$ mmol/l. What do you conclude?

PROTEIN BUFFERS

Proteins and the BBS are the main buffers inside cells.

The ICF has high concentrations of groups on proteins (specifically, the imidazole groups on the amino acid histidine) that buffer H^+. Because these proteins have values for K (equation 3, page 44) near a $[H^+]$ that is close to 80 nmol/l (the normal value for the ICF), they are effective physiological buffers in the ICF. In the ECF, however, proteins contribute little to buffering because of their low concentrations in that compartment (Table 2·4).

When the $[H^+]$ rises in the ICF, proteins buffer these H^+ by driving equation 7 to the right; these protein buffers can only limit the rise in the $[H^+]$ and thus are passive buffers. In contrast, the BBS, described above, can actually lower the $[H^+]$ because of independent regulation of the concentration of carbonic acid by actions of the lungs; the BBS is thus a dynamic buffer system.

Table 2·4
Buffer capacities in the ECF and ICF

		ECF	ICF
$[H^+]$	nmol/l	40	80
Buffers			
- Bicarbonate	mmol	375	300
- Protein	mmol	<10	400

The values are representative for muscle in a 70 kg person.

CONTROL OF THE $[H^+]$ IN BLOOD THROUGH RESPIRATORY RATE

The body transports CO_2 from organs to lungs through the BBS.

By regulating the Pco_2, the lungs exert close control over the $[H^+]$ in the ECF and the net charge on proteins in the ICF.

The Pco_2 in blood, a major determinant of the $[H^+]$ in blood (equation 6, page 45), results from the balance between the rate at which metabolism forms CO_2 and the rate at which the lungs excrete it; both rates are enormous (close to 10 mmol/min, as shown in Figure 2·1).

Formation of CO_2

The rate at which CO_2 is formed in mitochondria depends on the rate of turnover of ATP.

In the ATP generation system in mitochondria, CO_2 is produced directly during the oxidation of fuels to regenerate ATP. The rate of formation of CO_2 normally depends on the demand for regeneration of ATP, which is primarily affected by the degree of muscular activity. During vigorous exercise, CO_2 is also produced indirectly when the H^+ from lactic acid are buffered by HCO_3^- in the ECF and ICF. This anaerobic formation of CO_2 can be very rapid but lasts for less than a minute (see Chapter 3, pages 82 and 84).

Excretion of CO_2

> *Excretion of CO_2 occurs through the lungs; the $[CO_2]$ in alveolar air is close to 2 mmol/l (Pco_2 is 40 mm Hg).*

The excretion of CO_2 normally depends on the respiratory rate and depth, which determine the rate at which the air in the lungs is exchanged with outside air (Pco_2 less than 1 mm Hg). More specifically, the rate of excretion of CO_2 depends on the Pco_2 in each liter of alveolar air. At the usual Pco_2 of 40 mm Hg, the $[CO_2]$ in alveolar air is close to 2 mmol/l.

If the rate of formation of CO_2 by the body is kept constant, a decrease in the rate of respiration (*hypoventilation*) will temporarily slow the rate of excretion of CO_2 from the blood until the Pco_2 in blood increases to the new steady state; as a result, the $[H^+]$ in blood will rise (pH will fall). Thus, hypoventilation leads to respiratory acidosis. (A calculation with a 50% reduction in alveolar ventilation is provided in the margin.) Conversely, hyperventilation leads to respiratory alkalosis.

This ability of the lungs to control the $[H^+]$ in blood does not mean that the lungs can excrete H^+; by exerting control over Pco_2, the lungs can only affect the ratio of Pco_2 to $[HCO_3^-]$, and hence the $[H^+]$ (equation 6).

TRANSPORT OF CO_2 IN THE BLOOD

> *Transport of CO_2 involves conversion to H^+ and HCO_3^-.*

A person burning 2100 kcal/day produces approximately 14 mol of CO_2 per day, or 10 mmol/min. This value depends primarily on the quantity of ATP that must be regenerated to perform biological work.

Quantitative Considerations for Transporting CO_2 Through Blood

- Venous blood must carry 2 mmol/l more CO_2 than arterial blood (10 mmol of CO_2 in 5 liters of blood pumped per minute).

- Most of this extra CO_2 is carried in the form of HCO_3^-; the H^+ formed are buffered by histidines on hemoglobin (Hgb).

- During vigorous exercise, almost all the O_2 in arterial blood (about 8 mmol/l) will be extracted and converted to CO_2 (about 8 mmol/l) to be transported to the lungs as CO_2 (refer to the equations above). This process requires a very high Pco_2 in venous blood (about 90 mm Hg).

- Since the *cardiac output* can rise as high as 20–25 liters/min in vigorous exercise, the quantities of O_2 and CO_2 exchanged in lungs and the peripheral circulation are 160–200 mmol/min (8 mmol/l × 20–25 liters, Table 2·5).

Hypoventilation:
A lower rate or depth of breathing that causes the Pco_2 of blood to be higher than needed.

Sample calculation:

CO_2 production = CO_2 excretion
  = 10 mmol/min

Normally, alveolar ventilation is 5 liters/min, so each liter of alveolar air contains 2 mmol of CO_2 (or a Pco_2 of close to 40 mm Hg). If alveolar ventilation lowers to 2.5 liters/min, then each liter of alveolar air will contain 4 mmol of CO_2, which is equivalent to a Pco_2 of close to 80 mm Hg (double 40). Hence, arterial Pco_2 will be 80 mm Hg.

Cardiac output:
The volume of blood pumped by the heart per unit time.

Note:
More than 170 mmol of CO_2 can be produced in vigorous exercise if there is anaerobic glycolysis and H^+ react with HCO_3^-.

Table 2·5
Calculations for delivery of O_2

		Rest	Exercise
[Hgb] in blood	mmol/ (g/l)	2.2 (140)	2.2 (140)
O_2/Hgb		4	4
$[O_2]$ in blood	mmol/l	8.8	8.8
Cardiac output	l/min	5	20
O_2 delivered to periphery	mmol/min	44	176
O_2 consumed	mmol/min	12	170
O_2 consumed/delivered	%	27	97
CO_2 produced	mmol/min	10	170+

- During vigorous exercise, the brain seems to know that it must increase alveolar ventilation to maintain the Pco_2 in arterial blood at about 40 mm Hg. The signals to the brain are derived from muscle work and include H^+ and K^+, *neural* influences, and hormones such as adrenaline (Figure 2·4). Probably the most important of these signals in aerobic exercise is the $[K^+]$ in plasma.

- At a Pco_2 of 40 mm Hg, alveolar air contains close to 2 mmol of CO_2 per liter. To excrete the normal 10 mmol of CO_2 per minute, 5 liters of alveolar air are exchanged per minute; this exchange must rise to close to 100 liters/min for exhalation of the 200 mmol of CO_2 produced during vigorous exercise.

Neural:
Pertaining to the nervous system.

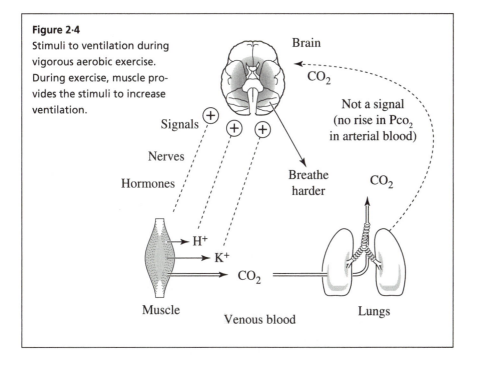

Figure 2·4
Stimuli to ventilation during vigorous aerobic exercise. During exercise, muscle provides the stimuli to increase ventilation.

THE BBS AND TRANSPORT OF O_2

The CO_2 produced in peripheral organs helps hemoglobin in blood to release O_2; the low level of CO_2 in the lungs helps the blood to absorb O_2.

There is a "competition" between H^+ and O_2 for binding to a histidine group in hemoglobin (Hgb). When the $[H^+]$ in blood increases, Hgb tends to release O_2 because of its high affinity for H^+. Since an increase in the $[CO_2]$ leads to an increase in the $[H^+]$, a tissue producing CO_2 (utilizing O_2) will encourage blood to give up O_2. Further, when the supply of O_2 is not sufficient to meet the demands of organs for O_2, ATP is regenerated anaerobically, yielding lactic acid. The H^+ released will also bind to Hgb and thereby promote release of additional O_2. When blood returns to the lungs, removal of CO_2 from blood (in the lungs) lowers the $[H^+]$, thereby encouraging the transfer of O_2 into blood (in the lungs).

Discussion of Case 2·1
Ursula Prepares for a Sprint
(Case presented on page 41)

Ursula's trainer was correct in saying that her arterial blood was already almost completely saturated with O_2 at the normal Po_2 in alveolar air. In more detail, Figure 2·5 shows that hemoglobin is more than 95% saturated with O_2 at a Po_2 greater than 80 mm Hg; the Po_2 of humidified air is close to 150 mm Hg, while that of arterial blood and alveolar air is close to 100 mm Hg. Since the *aqueous* phase of blood holds only an extremely small quantity of O_2, the net increase in the $[O_2]$ in arterial blood as a result of hyperventilation (raising the Po_2 of arterial blood to 115 mm Hg, for example) is not appreciable.

Aqueous:
The water phase of a solution.

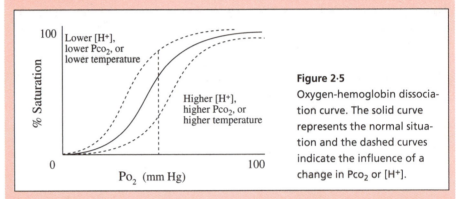

Figure 2·5
Oxygen-hemoglobin dissociation curve. The solid curve represents the normal situation and the dashed curves indicate the influence of a change in Pco_2 or $[H^+]$.

Although the quantity of O_2 in arterial blood is not increased by hyperventilation, blowing off CO_2 and lowering its concentration can help Ursula regenerate more ATP in her muscles during anaerobic metabolism. The net effect of hyperventilation before the race is to remove H^+ bound to intracellular proteins by forcing them to combine with intracellular HCO_3^- (Figure 2·3 on page 46, and Figure 2·6). The driving force for this sequence of events is the fall in $[CO_2]$ which shifts the BBS equation (equation 4, page 45) to the left. Since the $[H^+]$ is 10^6-fold lower than the $[HCO_3^-]$ in the ECF, and H^+ and HCO_3^- react with a 1:1 stoichiometry, H^+ must be pulled off buffers when the $[CO_2]$ is lowered.

Figure 2·6
Summary of events resulting from hyperventilation before the race. Hyperventilation before the race removes H^+ bound to intracellular proteins and forces them to combine with HCO_3^-. During the race, lactic acid is produced to regenerate ATP, but, because of hyperventilation before the race, Ursula will have more "sites on her intracellular proteins" to bind these extra H^+. She can therefore regenerate more ATP before muscle function deteriorates.

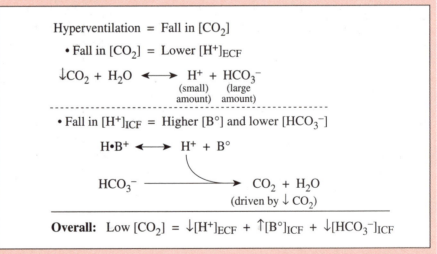

As soon as hyperventilation leads to a fall in the $[H^+]$ in cells, H^+ are released from their bound form on intracellular proteins (designated as HB^+ in Figure 2·6), leaving many of these proteins with fewer bound H^+ (designated as $B°$ in Figure 2·6). Intracellular proteins will then have a greater capacity to bind new H^+.

During the 10-second sprint, Ursula's muscles regenerate much of their ATP from anaerobic glycolysis (see Event 3·1, page 88), producing 0.67 mmol of H^+ per mmol of ATP. When the $[H^+]$ rises in her muscle cells, H^+ bind to intracellular proteins, causing an increase in their net positive charge. This action impedes the function of proteins, adversely affecting the performance of muscle. With hyperventilation before the race, more of Ursula's intracellular proteins are in the B^o form, so they have extra capacity to bind H^+ before the quantity of HB^+ becomes so great that it inhibits muscle function.

Buffering by the BBS during the sprint rather than before it would not be nearly as effective because there is insufficient time to remove the large amounts of H_2CO_3 formed in skeletal muscle from the combination of H^+ with HCO_3^- in these cells. Thus, only hyperventilation before the sprint helps in buffering a larger load of lactic acid.

A note of caution is in order. Excess hyperventilation before the race can cause the $[H^+]$ to fall to such a low level that intracellular proteins will bear too little net positive charge, and bodily functions may become compromised. You can discover for yourselves that forced hyperventilation causes lightheadedness. Further, the low P_{CO_2} and $[H^+]$ caused by excessive forced hyperventilation depresses the natural rate of respiration. If the supply of O_2 becomes inadequate for cells with a higher rate of regeneration of ATP, damage to these cells might occur.

Discussion of Case 2·2
Claude Has Chronic Lung Disease
(Case presented on page 41)

> *The $[H^+]$ is almost 10^6-fold lower than the $[HCO_3^-]$. When the CO_2 rises, H^+ and HCO_3^- are produced in a one-to-one ratio.*

When the concentration of CO_2 rises, the BBS (equation 4) is driven to the right, forming equal amounts of H^+ and HCO_3^-. Since the $[H^+]$ is about 10^6-fold lower than the $[HCO_3^-]$, the rise in the $[H^+]$ should be proportionately much larger than that of the $[HCO_3^-]$. Hence, the $[H^+]$ will rise markedly unless some process leads to an appreciable rise in the $[HCO_3^-]$. During the slow development of COPD, the rise in P_{CO_2} was accompanied by a smaller rise in $[H^+]$ because of the higher $[HCO_3^-]$. It is evident that some other process must have acted to raise the $[HCO_3^-]$. Extra HCO_3^- are generated by the kidneys as part of the process of excreting NH_4^+ (see page 62). Because the kidneys take several days to increase the rate of excretion of NH_4^+, an acute rise in P_{CO_2} will raise the $[H^+]$ appreciably, since the P_{CO_2}—but not the $[HCO_3^-]$—will rise.

It is also of interest to consider whether Claude's lungs removed CO_2 at a different rate when the P_{CO_2} rose during his acute lung infection. Assuming that Claude had the same metabolic rate before and after the acute infection and that there was no change in the fuels he was oxidizing (see Case 2·3), he must have been excreting the same amount of CO_2. The increase in P_{CO_2} is due to the decrease in the rate of exhalation of alveolar air, which caused an increase in the $[CO_2]$ of alveolar air (he had a higher $[CO_2]$ in exhaled air but exhaled less air per minute).

Discussion of Case 2·3
Nutrition in Very Serious Lung Disease
(Case presented on page 41)

To discover which fuel benefits Claude more—carbohydrates or fatty acids—relate the quantity of ATP regenerated to the rate of production of CO_2. The production of ATP per molecule of CO_2 released can be calculated from knowledge of the metabolic pathways; the numbers obtained from such calculations are shown in Table 2·6. Oxidation of fatty acids yields 25% more ATP per unit of CO_2 and hence is preferable for a patient, such as Claude, who has difficulty excreting CO_2.

Table 2·6
Stoichiometry of oxidation of glucose and fatty acids

Reaction	ATP/O_2	ATP/CO_2
Glucose (C_6) + $6\,O_2 \rightarrow 6\,CO_2$ + 36 ATP	6.0	6.0
Fatty acid (C_{16}) + $23\,O_2 \rightarrow 16\,CO_2 + 16\,H_2O$ + 129 ATP	5.6	8.0

Consider what Claude's preference for fuels would be if his problem were the delivery of oxygen to tissues. Table 2·6 shows that the oxidation of glucose yields slightly more ATP per O_2 consumed than does oxidation of fatty acids; therefore, if the delivery of oxygen is marginal, glucose may appear to be the preferred fuel. However, there are other considerations. When excess carbohydrates are consumed, utilization of energy becomes faster because the body uses ATP to store glucose as glycogen or as triacylglycerols, causing a parallel increase in consumption of O_2 and production of CO_2. Refer to question 2·2 below for a possible treatment.

Questions

(Discussions on pages 253–54)

2·2 How would you decrease the quantity of CO_2 excreted by a patient suffering from chronic obstructive pulmonary disease (COPD)?

2·3 Water and CO_2 are both produced in energy metabolism. What relative proportions are produced?

How might a clinician use this information in calculating water balance?

2·4 What can be done to help the lungs excrete CO_2 if the rate of alveolar ventilation is low and fixed and you want to minimize the degree of acidemia?

PART B
FORMATION AND CONSUMPTION OF H⁺ IN NORMAL ENERGY METABOLISM AND IN METABOLIC ACIDOSIS

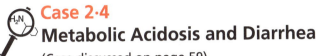

Case 2·4
Metabolic Acidosis and Diarrhea
(Case discussed on page 59)

Peggy, a relief worker returning from a typhoon disaster area, was brought from the airport to the emergency room. She was confused and looked very ill. While abroad, she had had severe diarrhea, for which she had received an antibiotic and a gut motility depressant. On the long journey she had consumed a large quantity of fruit juice.

Her blood chemistry (Table 2·7) showed marked acidosis and a large increase in the *anion gap in plasma*. The only other physical or chemical abnormalities in a routine examination were abdominal distention and scanty bowel sounds. What caused the acidosis? Could it become lethal?

Anion gap in plasma:
A calculation used to determine if acids have accumulated (see pages 56–57 for more discussion).

Table 2·7
Peggy's blood chemistry and urine chemistry

		Normal values	Peggy's values	
		Plasma	Plasma	Urine*
pH		7.40	7.22	5.0
P_{CO_2}	mm Hg	40	27	No data
Na^+	mmol/l	140	138	110
K^+	mmol/l	4.0	3.8	10
Cl^-	mmol/l	103	103	10
HCO_3^-	mmol/l	25	11	0
Glucose	mmol/l (mg/dl)	5 (90)	6.0 (108)	No data
Creatinine	μmol/l (mg/dl)	90 (1.0)	103 (1.2)	No data
Albumin	g/l	40	28	No data
Osmolality**	mosmol/kg	290	289	906
L-lactic	mmol/l	1.0	1.0	0
Ketoacids	mmol/l	<0.05	Negative	Negative

* Peggy produced 0.2 liters of urine in two hours.
** See pages 57–58.

Case 2·5
Jim, After a Wild Party
(Case discussed on page 60)

Jim, normally healthy, attended a lively party. The next morning he was disoriented and complained of not seeing properly. His blood chemistry (Table 2·8) showed significant acidosis and an increase in the anion gap in plasma. How would you diagnose and treat Jim's condition?

Table 2·8
Jim's blood chemistry on admission

		Normal	**Jim's values**
pH		7.40	7.30
Pco_2	mm Hg	40	30
HCO_3^-	mmol/l	25	15
Anion gap	*mEq/l*	12	22
Na^+	mmol/l	140	140
K^+	mmol/l	4.0	4.0
Cl^-	mmol/l	103	103
Osmolality	mosmol/kg H_2O	240	350
Glucose	mmol/l	5.0	5.0
Urea	mmol/l	4.0	5.0
Creatinine	µmol/l	90	100

mEq/l:
Milliequivalents per liter, the term used to describe ions; milliequivalents equal millimoles when the ion carries one positive or negative charge, but one millimole contains *n* milliequivalents if the valence of the ion is *n* net positive or negative charges.

Questions to Consider

What energy fuels can add acid or base to the blood?

How fast can H^+ be produced?

How many H^+ can the body buffer?

What energy fuels have no effect on the amount of acid or base in the blood?

What would you add to the diet of a patient who had a limited ability to excrete acid?

What rapid tests might be used to indicate how fast H^+ were being produced?

BACKGROUND

METABOLIC FORMATION AND CONSUMPTION OF H⁺

Complete Combustion of Dietary Fuels

> *Hydrogen ions are produced or consumed in metabolic processes in which the net ionic charge of the products is different from that of the substrates.*

No effect on H⁺. Most of energy metabolism involves conversion of fuels that have no net ionic charges (glucose, glycogen, triacylglycerols, neutral amino acids) to end products that are not charged (CO_2, H_2O, urea). These pathways therefore neither produce nor consume H⁺.

$$\text{Glucose or triacylglycerols} + O_2 \longrightarrow CO_2 + H_2O$$
$$\text{Neutral amino acids} + O_2 \longrightarrow CO_2 + H_2O + \text{Urea}$$

Producing H⁺. Seven of the 20 common amino acids are consumed for energy in pathways that consume or release H⁺. Any pathway in which the products have a lower net positive charge (or higher net negative charge) than the substrates will produce H⁺, and vice versa. Pathways that produce H⁺ include those that convert the uncharged sulfur-containing amino acids (methionine and cysteine) to CO_2, H_2O, urea, and sulfuric acid, and those converting positively charged amino acids (lysine, arginine and some histidine) to the uncharged products CO_2, H_2O, and urea.

$$\text{Methionine or Cysteine} + O_2 \longrightarrow CO_2 + H_2O + \text{Urea} + H_2SO_4$$
$$\text{Amino acid}^+ + O_2 \longrightarrow CO_2 + H_2O + \text{Urea} + H^+$$

Consuming H⁺. Pathways that consume H⁺ include those that convert negatively charged amino acids (glutamate or aspartate) or organic anions (e.g., citrate or malate) to the common neutral end products.

$$\text{Amino acid}^- + H^+ + O_2 \longrightarrow CO_2 + H_2O + \text{Urea}$$
$$\text{Organic anion}^- + H^+ + O_2 \longrightarrow CO_2 + H_2O$$

Net effect of the diet on H⁺. Since the normal diet contains more energy fuels that produce H⁺ than consume them, the body typically must excrete approximately 70 mmol of H⁺ per day in an adult in order to maintain normal acid-base balance. Only the kidneys can excrete free H⁺ or NH_4^+; renal malfunctions therefore cause acidosis.

Acid-Base Effects of Circulating Fuels

Neutral fuels, such as glucose and triacylglycerols, that are burned to the neutral products CO_2 and H_2O can be metabolized to intermediate forms (lactic acid, pyruvic acid, ketoacids, and fatty acids). In normal metabolism, more than 5 mol of lactic acid, fatty

$$\text{Neutral substrates} \xrightarrow{V_1} \text{Acids} \xrightarrow{V_2} \text{Neutral products}$$

acids and ketoacids are produced per day without imposing a net acid load because they are rapidly consumed, often in another organ. In some metabolic conditions, such as ketoacidosis due to normal fasting or lack of insulin, or lactic acidosis resulting from hypoxia or inhibited glucogenesis, these acids can accumulate in the blood (Figure 2·7). Since the concentrations of lactic acid or ketoacids can rise to 10 mmol/l or more, compared with the 40 nmol/l of H⁺ normally present in the extracellular fluid, the body removes more than 99.99% of these loads by buffering the H⁺. It also

Figure 2·7

Net accumulation or consumption of acids. V = velocity of the reaction. If $V_1 = V_2$, there is no accumulation of acid. If $V_1 > V_2$, acids accumulate (e.g., anaerobic muscle in severe exercise). If $V_1 < V_2$, acids are consumed (e.g., recovery from lactic acidosis after severe exercise).

excretes H^+ as NH_4^+ in the urine (see pages 62–63) and metabolizes anions to neutral end products.

Other Causes of Acid-Base Imbalances

Acid-base imbalances can also result from vomiting, diarrhea, gastrointestinal disorders, or metabolism of abnormal fuels.

Vomiting and diarrhea. Normal digestion first releases strong into the stomach (pH is close to 1) and then $NaHCO_3$ into the intestines, making the contents slightly alkaline (pH = 7–8). Vomiting causes loss of stomach acid and thus a lower $[H^+]$; diarrhea causes loss of $NaHCO_3$ and thus an increase in $[H^+]$. Chronic vomiting or diarrhea can cause serious acid-base problems.

Gastrointestinal disorders. The normal digestive process—with largely neutral foodstuffs, feces, and absorbed fuels—does not create a large load of acid. However, changes in the normal bacterial content of the intestines (from new flora or antibiotics), or in the composition of the diet (which may alter bacterial flora), or in the rate at which the food moves through the intestines (from an obstruction, blind loops of the bowel, narcotic actions on intestinal muscles, or infections) can all cause serious changes in the normal actions of bacteria in the intestines. As a result, unusual organic acids may enter the body. The most common is D-lactic acid (the physiological isomer is L-lactic acid).

Metabolism of Abnormal Fuels. Some abnormal fuels are oxidized to acids that often cannot be metabolized further; hence, H^+ accumulate. Examples of such fuels are methanol (forming formic acid), ethylene glycol (forming glycolic, glyoxylic and oxalic acids), and toluene (forming hippuric acid).

Questions

(Discussions on pages 254–55)

2·5 A normal person cannot buffer more than 1000 mmol of H^+. How can an elite athlete whose muscles, red blood cells, and brain produce 400 mmol of lactic acid per hour survive for more than four hours?

2·6 In hemodialysis, the solution used may contain an appreciable quantity of sodium acetate. Why?

DIAGNOSIS OF ACIDOSIS RESULTING FROM ACCUMULATION OF METABOLIC ACIDS

The Anion Gap

> *The anion gap provides a convenient means of determining if unknown anions are present in blood or urine. In plasma, it is calculated from measured values of the major ions ($Na^+ - Cl^- - HCO_3^- = 12 \pm 2$ mEq/l).*

In every solution, the total number of positive charges (cations) must equal the total number of negative changes (anions). If measurements indicate an imbalance, either the measurements were wrong, or not all the ionized materials were measured.

Although physiological fluids may contain many anions and cations, only a few are present in significant amounts, and the concentrations of others usually change little. By measuring the concentrations of only a few cations or anions and making some assumptions about the ions normally found in the fluid, it is possible to obtain a rough measure of whether large amounts of unsuspected anions or cations are present.

Serum normally contains the anions and cations shown in Table 2·9. Of these, the concentrations of potassium (K^+), calcium (Ca^{2+}), and magnesium (Mg^{2+}) are usually constant (minor variations are life-threatening). The concentrations of proteins can also be assumed to be constant (though some clinical situations lead to significant changes, e.g., Case 2·9). Hence, subtraction of the measured values for [Cl^-] and [HCO_3^-] from that for [Na^+] will normally yield a value of 12 ± 2 mEq/l, which is the expected value for the anion gap in normal individuals.

Table 2·9

Normal concentrations of cations and anions in serum

Cations in serum (mEq/l)		Anions in serum (mEq/l)	
Na^+	140	Cl^-	103
K^+	4	HCO_3^-	25
Ca^{2+}	5	Proteins	19
Mg^{2+}	2	Phosphate, sulfate, lactate, and others	4
Total	151	Total	151

The values in this table are approximate because some, such as Na^+, represent values per liter of plasma (really 152 mEq/l of water) and others, such as HCO_3^-, are per liter of water.

Values for the measured anion gap that are significantly larger than normal indicate the presence of one or more unknown anions, with the difference between elevated and normal values approximating the total concentration of the unknown anions in plasma; Table 2·10 shows typical values for the anion gap in normal individuals and in patients with lactic acidosis.

Similar calculations can be performed for the urine, in which the normal major cations are Na^+, K^+, and NH_4^+ and the major anions are Cl^- and HCO_3^-. The formula for the anion gap in urine ($Na^+ + K^+ - Cl^-$) is used to detect unusual anions, as illustrated in Cases 2·4 and 2·5, and to estimate the [NH_4^+] (in urine, NH_4^+ are not measured routinely); obviously, the NH_4^+ salts of organic acids are not detected using this clinical shortcut.

Table 2·10

Typical values for the anion gap in plasma

		Normal	Lactic acidosis
Na^+	mmol/l	140	140
Cl^-	mmol/l	103	103
HCO_3^-	mmol/l	25	10
Anion gap	mEq/l	12	27

The Osmolal Gap

The osmolal gap is a convenient diagnostic method to detect unknown, uncharged compounds in plasma by using values in plasma (measured osmolality minus 2 [Na+] + [glucose] + [urea], all in mmol/l).

The osmolal gap is a useful means of detecting the presence of unknown, uncharged molecules in the plasma or urine. The osmotic pressure of a solution is determined by the concentration of dissolved particles in the water; a protein molecule, a molecule of glucose, and a sodium ion make roughly equal contributions to the osmotic pressure. Since the numbers of anions and cations must be equal in any solution, and the concentrations of K^+, Ca^{2+} and Mg^{2+} are approximately constant but the $[Na^+]$ varies, the value of $2 \times [Na^+]$ will account for the osmotic pressure of normal anions plus cations. Glucose and urea are the two major non-ionized molecules in plasma that are likely to change in concentration. Hence, the calculated osmolality is $2 \times [Na^+] +$ [glucose] + [urea] (all measurements in mmol/l); the difference between the measured and the calculated osmolality is the osmolal gap. A high osmolal gap indicates the presence of an unmeasured compound—probably an alcohol (Table 2·11).

Table 2·11

Typical values for calculation of the osmolal gap in plasma and urine

		Normal	Methanol intoxication
Glucose	mmol/l	5	5
Na^+	mmol/l	140	140
Urea	mmol/l	5	5
Measured osmolality	mosmol/l	290	350
Osmolal gap	mosmol/l	0	60

Discussion of Case 2·4
Metabolic Acidosis and Diarrhea
(Case presented on page 53)

Table 2·7 (page 53) shows that neither L-lactic acid nor ketoacids caused Peggy's acidosis. The medications taken, the history of diarrhea, the distended bowel, and the recent diet all suggest excess production of acids from GI flora. These bacteria always produce low levels of organic acids, mainly D-lactic acid, which can normally be handled by the liver and kidneys. Disturbance of the bacterial flora (from antibiotics), accumulation of bacteria (caused by bowel stasis from the motility depressant), and intake of sugar (fruit juice on the plane provided substrates for the bacteria) can markedly affect the metabolism of the GI flora. All the above can lead to excessive production of unusual organic acids, which can account for the acidosis with large anion gaps in plasma and urine.

Peggy's diarrhea could also cause acidosis, by loss of $NaHCO_3$. If this loss is faster than the rate of excretion of NH_4^+ by the kidneys (the maximal rate is approximately 200 mmol/day), H^+ will accumulate. The loss of HCO_3^- in diarrhea fluid is usually not excessive because the $[HCO_3^-]$ in this fluid is only 40–50 mmol/l. Hence, over 4–5 liters of diarrhea are needed to overwhelm the ability of the kidneys to generate new HCO_3^- (excrete NH_4^+). Obviously, compromised renal function can also contribute to the degree of acidosis.

SEVERITY OF ACIDOSIS

> *To determine if the acidosis is potentially life-threatening, count the anions in blood and urine and compare the value to 1000 mmol.*
>
> *Consider also the decline in $[HCO_3^-]$ in the ECF.*
>
> *Measure the number of liters of diarrhea fluid if losses are very large.*

Acidosis becomes lethal when H^+ are added to blood faster than they are consumed or excreted and when the buffering capacity of the body (1000 mmol) is exceeded. The $[HCO_3^-]$ in Peggy's plasma had fallen to half of the normal level, suggesting that close to half of the overall buffering capacity had been used (500 mmol).

To determine whether this case requires urgent attention, the rate of production of new H^+ (as organic acids) must be assessed. Since organic anions are always formed with H^+, the quantity of H^+ added to the body can be measured by counting the number of organic anion molecules that appear in blood or urine over a certain period. To measure the accumulation of new organic anions in the body, multiply the changes in the anion gap in plasma by a number that is close to half of the body weight (the volume of distribution of organic anions). The rate of excretion of anions in the urine without H^+ or NH_4^+ (measured by the anion gap multiplied by the urine volume) indicates an additional load of H^+. These calculations (110 mmol/l × 0.2 liters) show that the rate of accumulation of H^+ in the body was very low (22 mmol over 2 hours), indicating that the acidosis was under control. Treatment therefore need focus only on the GI disturbance.

Question

(Discussion on page 255)

2·7 How can therapy with insulin lead to a decrease in the concentration of D-lactate in Peggy (Case 2·4)?

Discussion of Case 2·5
Jim, After a Wild Party
(Case presented on page 54)

Jim's anion gap and osmolal gap in plasma are both very elevated (Table 2·8, page 54), indicating an unknown anion as well as an unknown uncharged compound. The case history and symptoms (general good health but visual impairment from excess consumption of alcohol) should lead you to suspect methanol intoxication. Both methanol and ethanol are metabolized using the same alcohol dehydrogenase. Whereas ethanol is metabolized to CO_2 and H_2O, metabolism of methanol produces formaldehyde, which is highly toxic; further metabolism of formaldehyde yields formic acid, a reasonably strong acid that cannot be metabolized further. Hence, the goals of treat-

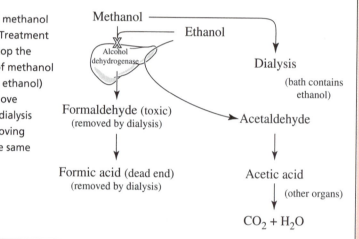

Figure 2·8

Treatment of methanol intoxication. Treatment should first stop the metabolism of methanol (by providing ethanol) and then remove methanol by dialysis (without removing ethanol at the same time).

ment should be to prevent methanol from being converted to its toxic metabolites, followed by removal of methanol (Figure 2·8).

Hemodialysis can help the kidneys remove methanol and its metabolites, but the process is rather slow. Because oxidation of methanol must be stopped immediately, the patient should first be treated with compounds that inhibit the metabolism of methanol. Alcohol dehydrogenase uses ethanol in preference to methanol; hence, high levels of ethanol will inhibit metabolism of methanol. Ethanol can be administered by mouth or intravenously and should also be added to the hemodialysis fluid. A further option for therapy is the drug methylpyrazole, an inhibitor of alcohol dehydrogenase.

Questions

(Discussions on pages 255–56)

2·8 Why does ethanol create less of an acid load than methanol?

2·9 If both glucose and methanol are present at the same concentration in the blood (50 mmol/l), why will glucose but not methanol cause an osmotic diuresis?

PART C
EXCRETION OF ACID AND NITROGENOUS WASTES: THE ROLE OF THE KIDNEYS

Case 2·6
Shirley Has Serious Kidney Disease
(Case discussed on page 65)

Shirley has poor kidney function. She does not want to undergo hemodialysis or a kidney transplant at this time. She feels reasonably well, and her disease is progressing slowly. She does not have *hypertension*, *hyperkalemia*, or acidosis. How could she modify her diet to preserve lean body mass and avoid the accumulation of toxic materials?

Hypertension:
High blood pressure.

Hyperkalemia:
High levels of potassium in plasma.

Case 2·7
Where Is Horace's Muscle Going?
(Case discussed on page 66)

Horace, extremely obese and desperate to lose weight, has started a complete starvation regimen (he consumes no fuels, but drinks water and takes vitamins and minerals). Now, in the fourth week of his fast, his weight loss is 300 g/day (see Case 1·9, pages 29–31). His energy consumption is 1500 kcal/day and he excretes 5 g of nitrogen per day. During the fast, his blood sugar has increased from 3.3 mmol/l (60 mg/dl) in the first week to 5 mmol/l (90 mg/dl) in the fourth week. Why has Horace catabolized lean body mass—to supply energy or for some other purpose?

Veterinary Case 2·1
The Hibernating Bear
(Case discussed on page 67)

Bears can hibernate for very long periods without urinating. What does this ability indicate about their rate of production of urea?

Veterinary Case 2·2
The Alligator and Excretion of Water
(Case discussed on page 67)

An alligator in fresh water tries to swallow a frog. It is successful on the fourth try, after swallowing a lot of fresh water, which lacks the particles necessary for excretion (an alligator's kidneys are too primitive to excrete urine that is hypo-osmolal to plasma). How might the alligator obtain "extra particles" to excrete in its urine?

Questions to Consider

Why don't the kidneys simply filter the blood and then reabsorb the molecules that should be kept? Think about the $[H^+]$ in this regard.

How do the kidneys add new HCO_3^- to the body?

In ketoacidosis, the urinary filtrate contains more ketoacid anions than can be reabsorbed. What is the accompanying cation in the filtrate, and what is it in the urine? How is the urinary cation related to maintaining acid-base balance?

BACKGROUND

ROLE OF THE KIDNEYS IN ACID-BASE BALANCE

> *Only the kidneys can continually excrete acid or base.*
> *The kidneys control excretion of acid or base by adjusting the rates of excretion of NH_4^+ or HCO_3^-.*

The formation of acid in the body consumes HCO_3^-, yielding CO_2 and H_2O, which are excreted. Some acids that are produced provide conjugate base for excretion of H^+ (e.g., phosphate) or metabolism of H^+ (e.g., lactate anion); formation of these acids does not usually lead to chronic acidosis. In contrast, overproduction of other acids (e.g., sulfuric) whose conjugate base cannot help in the removal of H^+ by excretion or metabolism can lead to chronic acidosis.

The kidneys remove H^+ and thereby generate new HCO_3^- in the blood by metabolizing glutamine (a neutral amino acid) to NH_4^+ and HCO_3^-. Other products of this reaction are CO_2, glucose, and H_2O—all neutral compounds. The NH_4^+ are excreted into the urine and the HCO_3^- are returned to the blood (Figure 2·9).

Figure 2·9

Renal pathway for the addition of new HCO_3^- to the body. (a) The metabolism of glutamine produces NH_4^+ and HCO_3^-. The NH_4^+ are excreted in the urine (they are reabsorbed in exchange for filtered Na^+) and HCO_3^- are delivered to the body via the renal vein. (b) The H^+ that titrate HPO_4^{2-} are derived from CO_2 and are produced with HCO_3^- (equation 4, page 45).

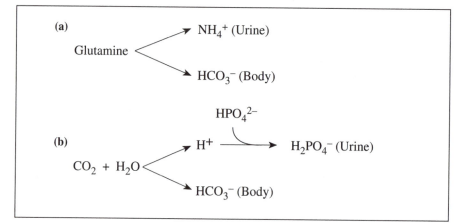

EVENTS IN THE KIDNEYS

> *The kidneys adjust the internal environment by saving just the right amount of essential materials and excreting any surplus of these compounds, plus all wastes.*

Filtrate:

The fluid entering the renal tubule (nephron) from the glomerulus; it contains a solution of all the small molecules in the plasma but not proteins as large as albumin.

Glomerulus:

The structure at the head of the renal tubule (nephron) in which the blood capillary filters its contents (25% of the volume entering the capillary) into the urinary filtrate.

The kidneys excrete almost all metabolic wastes from the body except CO_2 (excreted by the lungs) and some material secreted into the intestines and not reabsorbed (e.g., bilirubin, a bile pigment). The kidneys help in the control of blood pressure and blood volume by excreting H_2O, H^+, and salts. They secrete hormones such as erythropoietin (which controls synthesis of red blood cells) and the active forms of Vitamin D (required for calcium homeostasis).

The kidneys' excretory functions begin with the formation of a *filtrate* in the *glomerulus*. All soluble compounds with a molecular weight that is less than 68 000 daltons are filtered. Valuable compounds are reabsorbed from the filtrate as the urine passes through the renal tubule. The sites of reabsorption of the large volume of filtrate (approximately 150 l/day, 50 times more than the normal plasma volume) are shown in Table 2·12. The general functions of the nephron are illustrated in Figure 2·10. In essence, virtually all the valuable nutrients (glucose, amino acids, organic anions, phosphate, carnitine, etc.), 60–75% of the water, Na^+, and Cl^-, and 85% of the HCO_3^-

are reabsorbed at the earliest (proximal) nephron sites. The next section of the nephron (loop of Henle) reabsorbs close to two-thirds of the NaCl delivered here without reabsorbing much water; another function of the loop of Henle is to help excrete NH_4^+. The last parts of the nephron (distal convoluted tubule and the collecting duct) make all the fine-tuning adjustments for excretion, reabsorbing most of the remaining water, Na^+, and Cl^-, playing a major role in excretion of K^+, and helping to excrete more NH_4^+.

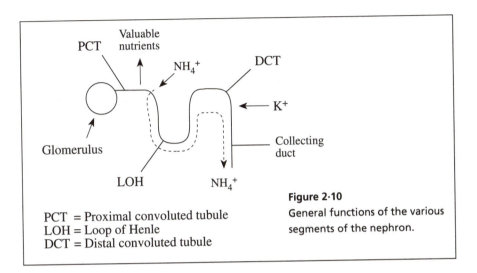

Figure 2·10
General functions of the various segments of the nephron.

PCT = Proximal convoluted tubule
LOH = Loop of Henle
DCT = Distal convoluted tubule

Table 2·12
Sites of reabsorption of the filtrate

	Liters/day	Major solutes
Blood flow through kidney	1500	Has same contents of small molecules and ions as plasma
Filtrate formed in glomerulus	150–200	Has same contents of small molecules and ions as plasma
Fluid leaving proximal tubule	50	Contains NaCl that is virtually isotonic; most nutrients are reabsorbed
Dilute urine leaving loop of Henle	40	Has low concentrations of Na^+ and Cl^-
Isotonic urine leaving renal cortex	5	Contains Na^+, Cl^-, NH_4^+, and urea, plus K^+ that will be excreted
Final urine leaving collecting ducts	0.5–2.5	Contains wastes, NH_4^+, and extra salts

EXCRETION OF NITROGEN AS UREA

The vast majority (98%) of nitrogen from the metabolic breakdown of dietary or body proteins is normally excreted as urea in the urine. Urea is a non-toxic, non-charged compound. The production of urea is a good measure of the rate of oxidation of proteins.

Almost all urea is made in the liver. Nitrogen waste is carried to the liver as NH_4^+ and amino acids (mainly alanine and glutamine, which have carbon skeletons that can be converted into glucose). Synthesis of urea in the liver rapidly removes NH_4^+, which are toxic; this pathway has a 10-fold excess capacity. Synthesis of urea from amino acids is closely linked to synthesis of glucose from amino acids, since these pathways share a common key intermediate (see Chapters 8 and 12, pages 202–3 and 242–44).

In the renal medulla (which contains the loop of Henle and collecting ducts), a high extracellular concentration of urea helps the kidneys reabsorb water. The renal medulla has a discrete and low blood supply; urea becomes trapped between its cells, leading to a very high concentration of urea (and salt). When the body must retain water, antidiuretic hormone causes the collecting duct membrane to become permeable to water. Water is then "sucked out" of the lumen by the more concentrated urea and salt solution in the cortical and medullary interstitium and then flows back to the blood.

Urea, which crosses cell membranes rapidly, appears in all body secretions. For example, 20–25% of the urea produced enters the intestines and is broken down to NH_4^+, which re-enter the blood and are then converted back to urea in the liver; refer to Veterinary Case 2·1, page 67. Usually 50% of the urea that is filtered is reabsorbed in the nephron.

WAYS OF ASSESSING KIDNEY FUNCTION

Glomerular Filtration Rate

Creatinine:
A small nitrogen-containing organic compound formed at a constant rate from the breakdown of creatine phosphate (mainly from skeletal muscle). Its rate of formation can be used to determine the amount of filtrate formed in the kidneys.

Creatinine, an organic compound containing nitrogen, is formed at a fairly constant rate from creatine phosphate (the ATP reservoir in muscle) and contributes 1% of urinary nitrogen. Freely filtered in the glomerulus, creatinine is not reabsorbed; it is also secreted to a minor extent in the proximal tubule. The amount of creatinine excreted in urine per unit time divided by its concentration in plasma allows an estimate of the *glomerular filtration rate* (GFR). Estimating the GFR using creatinine clearance is particularly useful in long-term follow-up of cases (e.g., in following the functions of a transplanted kidney). The concentration of urea in plasma also provides a rough index of the GFR.

Glomerular filtration rate (GFR):
The rate of formation of filtrate in the glomeruli.

Renal Blood Flow

Certain substances are not reabsorbed from the glomerular filtrate but are extracted from plasma and are secreted into the urine by cells lining the proximal tubule. This active secretion ensures that compounds such as para-aminohippurate (PAH) are maintained at low concentrations in plasma. The rate of excretion of PAH in urine divided by its concentration in plasma provides a measure of the rate at which plasma flows through the kidneys (for an example of a case with overproduction of a hippuric acid analog, see Case 2·8, Acidosis from Sniffing Glue, on page 68).

Question

(Discussion on page 256)

2·10   How might clinicians use the rate of excretion of urea to indicate whether a patient is losing too much muscle mass (i.e., is in a very catabolic state)?

Discussion of Case 2·6
Shirley Has Serious Kidney Disease
(Case presented on page 61)

Renal insufficiency causes inadequate excretion of Na^+, K^+, and H^+. Accumulation of these substances can be life-threatening.
1. Excess Na^+ may lead to hypertension.
2. Hyperkalemia may cause abnormalities of cardiac function and rhythm (arrhythmias).
3. Acidosis is dangerous by itself and may also contribute to life-threatening hyperkalemia (H^+ enter cells and K^+ exit).

Consider these problems when planning a course of treatment (Table 2·13). The primary aims of treatment are to minimize the production of metabolic acids and restrict the intake of salts while providing an optimal diet.

Minimizing the production of metabolic acids entails watching the diet, controlling metabolic acidosis, and ensuring that gastrointestinal upsets do not cause an acid load. The dietary content of fuels that yield H^+ when they are metabolized (e.g., sulfur-containing and positively charged amino acids; see page 55) should be kept as low as possible; they cannot be omitted from the diet because some are essential amino acids. Control of metabolic acidosis involves keeping lactic acid and ketoacid levels at their normal low levels by maintaining caloric balance (unrestricted intake of carbohydrates and triacylglycerols), preventing hypoxia, and ensuring that any diabetes mellitus is well controlled (diabetes mellitus is a major cause of kidney failure). GI disturbances must be carefully monitored.

Minimizing the intake of salts involves avoiding seafood, meat and broths, and K^+-rich vegetables as much as possible (intracellular fluid contains a lot of K^+). An adequate diet must be preserved while keeping all these restrictions in mind.

Renal insufficiency:
Inability of the kidneys to excrete wastes properly.

Table 2·13
Aims of treatment

Aim	Method
1. Minimize the load of H^+.	Restrict proteins with sulfur-containing amino acids.
2. Minimize the load of K^+.	Limit foods rich in K^+.
3. Lower the intake of Na^+.	Consume no added table salt.

Questions

(Discussions on pages 256–57)

2·11 Would administration of $NaHCO_3$ be a good way to treat metabolic acidosis in a patient with renal failure?

2·12 How long would it take an acid load to accumulate to lethal levels in patients who suffer from total anoxia, diabetic ketoacidosis, or complete renal failure?

How important are the kidneys in controlling the acidosis from anoxia or diabetes? Assume that a) death occurs after accumulation of 1000 mmol of H^+; b) the body normally consumes 12 mmol of O_2 per minute; c) ketoacids are produced at 1 mmol/min; d) the renal "excretion of H^+" is 70 mmol/day.

Discussion of Case 2·7
Where Is Horace's Muscle Going?
(Case presented on page 61)

Early in fasting, some breakdown of proteins (50–180 g/day) from lean body mass is required to supply the brain with glucose (via gluconeogenesis; see pages 29–31 and Chapter 8). As the ketoacidosis of fasting develops, the brain can obtain energy from ketoacids, and the oxidation of proteins drops to low rates (almost 30 g/day). Two hypotheses might explain why Horace is using lean body mass (Figure 2·11). He may be consuming proteins to make glucose for his brain, or he may be using proteins to make glutamine so that he can excrete NH_4^+. If breakdown of proteins is needed to make glucose, then the signal should be a low concentration of glucose in blood. Horace is not hypoglycemic and the concentration of glucose in his blood has been increasing progressively over the past three weeks. Perhaps, later in fasting, proteins are broken down so that excretion of NH_4^+ can help eliminate part of the acid load.

Figure 2·11

Breakdown of proteins during chronic fasting. The major reasons to catabolize proteins are (1) to provide glucose for the brain, yielding urea as a waste product or (2) to provide glutamine to yield NH_4^+ for the urine (to match the excretion of ketoacid anions) and HCO_3^- for the body (to titrate the load of H^+ with excretion of ketoacid anions; note the H^+ and HCO_3^- in the ovals in the diagram).

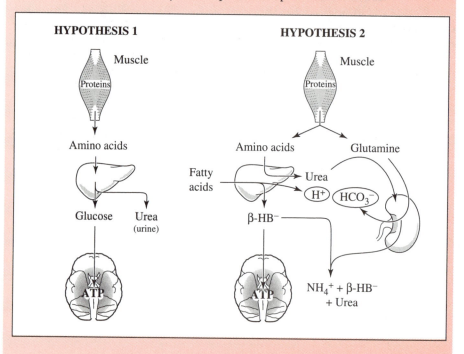

Ketoacidosis is a form of acidosis. In fasting, ketoacid anions appear in the urine; therefore, NH_4^+ must also be excreted to achieve acid-base and electrolyte balance (no other cations can be spared). Because excretion of NH_4^+ requires nitrogen, the body must break down proteins. During consumption of proteins, the concentration of glucose in blood may rise because 60% of the carbon in proteins can be converted to glucose. In acidosis, a larger proportion of nitrogen is excreted as NH_4^+ in urine; this excretion rises to as much as 50% of the total excretion of nitrogen during chronic fasting, compared with less than 5% in a non-acidotic person.

The second hypothesis can be tested by giving an oral load of alkali ($NaHCO_3$) equivalent to the rate of excretion of NH_4^+. This treatment will remove excess H^+, thereby reducing the need to excrete nitrogen; it will also provide Na^+ to excrete with ketoacid anions. If the NH_4^+ hypothesis is true, total urinary nitrogen should decrease, both as NH_4^+ and as urea. This decrease was observed in a direct test; the excretion of nitrogen was halved by feeding oral $NaHCO_3$. Thus, increased breakdown of proteins in the ketotic phase of fasting seems to be due to the need to excrete NH_4^+; the synthesis of glucose can be viewed as a by-product in the ketotic phase of fasting (as opposed to the protein phase of fasting, when it is the essential product; see Chapter 1, pages 29–31).

Discussion of Veterinary Case 2·1
The Hibernating Bear
(Case presented on page 61)

For optimal efficiency, the bear should use endogenous triacylglycerols in adipose tissue to regenerate virtually all of its ATP during hibernation. Nevertheless, some proteins will be catabolized and urea will be formed. The bear must eliminate this urea—not via the urine—and the bear cannot metabolize urea directly. In the hibernating bear, urea enters the lumen of the gastrointestinal tract along with the usual secretions. Because of the suppressed motility in the GI tract, bacteria normally restricted to the colon move upstream and colonize the duodenum. These bacteria secrete the enzyme urease, which splits urea into NH_4^+ and HCO_3^-. The bacteria then convert the NH_4^+ to essential and nonessential amino acids and then to bacterial protein. As these bacteria are propelled slowly down the bear's intestinal tract, they are digested by secretory enzymes, and the amino acids are absorbed and "resynthesized" into new proteins in the bear. Hence, the small rate at which the bear catabolizes proteins is matched by the synthesis of new proteins via the bacteria in its GI tract.

This story has several spin-offs. First, urease is secreted by bacteria that invades the stomach of some people, leading to the formation of NH_4^+, local toxicity, and a propensity for gastritis and duodenal ulcers. Second, urea can be used as a nitrogen supplement to foods for ruminants because microorganisms in their stomachs release urease and convert urea into a biologically useful form. Third, when examining the stoichiometry of ureagenesis, half the nitrogen destined for urea must first be converted to NH_4^+ (Chapter 12, page 244). Much of this conversion occurs when 20–25% of the urea formed is secreted into the GI tract and, if urease is present in the luminal fluid, is hydrolyzed to NH_4^+.

Discussion of Veterinary Case 2·2
The Alligator and Excretion of Water
(Case presented on page 61)

The alligator must excrete 300 particles with every liter of water because its primitive kidneys cannot separate particles from water (excrete a dilute urine) as human kidneys can do. Thus, the alligator needs a source of spare particles when fresh water is ingested and must be excreted in its urine. Lipids will not do since they are sparingly soluble in water. Carbohydrates will not do since glucose is too valuable and excretion of lactate imposes a threat of excess H^+. Thus, proteins are the material to use, but, again, excretion of amino acids is too costly. The solution is to use nitrogen wastes. Instead of excreting urea (one particle), the alligator excretes its precursors, 2 NH_4^+ and 2 HCO_3^- (four particles) in the urine. When the alligator has a need to excrete fresh water, its kidneys reabsorb Na^+ in exchange for NH_4^+ and reabsorb Cl^- in exchange for HCO_3^- from the filtrate. Another special feature of the alligator's kidneys is that their luminal membranes are relatively impermeable to NH_3, a necessity when the urine contains high concentrations of NH_4^+ and HCO_3^-.

PART D
CASES FOR REVIEW

Case 2·8
Acidosis from Sniffing Glue
(Case discussed on pages 71–72)

Connie, a drug addict, was very low in cash. To maintain her habit for a few days, she sniffed glue. Friends brought her to the hospital because she was unusually confused and had an unsteady gait. Laboratory results from a blood sample show a degree of metabolic acidosis (high [H$^+$] and low [HCO$_3^-$]) but normal anion and osmolal gaps. The level of NH$_4^+$ in her urine was normal for chronic acidosis, but the level of urea was 30% of normal (Table 2·14). How could she have metabolic acidosis with a normal anion gap in plasma?

Table 2·14

Connie's blood chemistry and urine chemistry on admission

Plasma		Normal	Connie
Na$^+$	mmol/l	140	140
K$^+$	mmol/l	4	2.5
Cl$^-$	mmol/l	103	113
HCO$_3^-$	mmol/l	25	15
pH		7.40	7.30
Pco$_2$	mm Hg	40	30
Osmolality	mosmol/kg H$_2$O	285	285
Glucose	mmol/l (mg/dl)	5 (90)	5 (90)
Urea	mmol/l (mg/dl)	4 (11)	2 (5.6)

Urine		Normal	Connie
NH$_4^+$	mmol/l	200*	200
Na$^+$ + K$^+$ – Cl$^-$	mmol/day	40	40**
Osmolal gap	mosmol/day	100	400
Hippurate	mmol/day	<10	140
Urea	mmol/day	500	150

* During chronic acidosis
** Urine should contain more Cl$^-$ than Na$^+$ + K$^+$ in chronic acidosis.

Case 2·9
Ketoacidosis with a Very Low Pco$_2$ in a Cachectic Patient

(Case discussed on page 73)

Robert, 32 years old, came to the emergency room alert but in distress. His severe emaciation—23 kg (50 lb) body weight—was consistent with his long hospital record of debilitating juvenile rheumatoid arthritis. He stated that he had neither eaten nor taken medication nor drugs for three days but that he had had a double vodka four hours ago. His heart and respiratory rates were high and his breath reeked of acetone. His blood and urine chemistry (Table 2·15) showed severe acidosis, a large anion gap (consistent with the very low [HCO$_3^-$] in his plasma), mild hypoglycemia, severe ketosis, and the presence of some ethanol.

Robert was treated with saline plus HCO_3^-, K^+, glucose, insulin, and thiamine—all administered intravenously. Within 13 hours, his serum creatinine fell from 51 μmol/l (about half the normal level of a person of normal weight) to 18 μmol/l, and his anion gap in plasma had fallen to 8 mEq/l, (zero unidentified anions), though a moderate acidosis remained.

Can the double vodka explain the acidosis? For purposes of calculation, assume that his total body water is 17 liters and that an anion gap of 8 mEq/l would be normal for him.

Table 2·15
Robert's blood chemistry and urine chemistry on admission

Plasma		Normal	Robert
Na^+	mmol/l	140	147
K^+	mmol/l	4	5
Cl^-	mmol/l	103	115
HCO_3^-	mmol/l	25	2
pH		7.40	7.05
Pco_2	mm Hg	40	8
Osmolality	mosmol/kg H_2O	285	305
Glucose	mmol/l (mg/dl)	5 (90)	3.4 (61)
Urea	mmol/l (mg/dl)	4 (11)	1.1 (3.1)
Creatinine	mmol/l (mg/dl)	18 (0.2)*	51 (0.6)
Albumin	g/l	40	21
Urine			
NH_4^+	mmol/day	200**	Calculate from osmolal gap
$Na^+ + K^+ - Cl^-$	mmol/day	40	–***
Urea	mmol/l (mg/dl)	500 (1400)	20 (56)
pH		5–6	5.0
Extra tests (plasma)			
Lactate	mmol/l	1.0	0.5
Ketoacids		0	Strongly positive
Ethanol	mmol/l (mg/dl)	0	7 (32)

  * This value is normal for Robert.

  ** This value is normal for a person with chronic metabolic acidosis.

 *** Ketonuria prevents prediction of NH_4^+ from this value.

Case 2·10
Shirley Has Severe Acidosis Between Hemodialysis Treatments

(Case discussed on page 74)

Shirley (Case 2·6) needs hemodialysis every other day because her kidneys have now failed. She missed her dialysis yesterday because she was stranded in a rural area. She ate a large steak for dinner, developed some diarrhea, and arrived at the emergency room with rather severe metabolic acidosis (pH = 7.10, $[H^+]$ = 80 nmol/l, HCO_3^- = 5 mmol/l). The $[K^+]$ in her plasma is very high (6.8 mmol/l) and there will be a considerable time before she can be dialyzed. How much alkali ($NaHCO_3$) should be administered, and what are the side effects of this treatment?

Case 2·11
Diabetic Ketoacidosis After a Broken Wrist
(Case discussed on page 74)

For the past seven years, Malcolm, aged 12 years, has had diabetes mellitus, for which he requires daily injections of insulin. He has had four admissions for diabetic ketoacidosis in the past two years. Yesterday, while roller skating, he fell and broke his wrist. For the pain, he took aspirin with codeine. He did not eat because of possible surgery. Knowing that insulin could cause a hypoglycemic reaction, the intern did not give him his regular injections of insulin. Since he was very thirsty, breathing heavily, and passing a very large quantity of urine, the intern drew a blood sample expecting to see values typical of diabetic ketoacidosis. She was somewhat surprised that the lab reports failed to show an elevated value for the anion gap in his plasma, yet metabolic acidosis and hyperglycemia were present (pH = 7.2; $[H^+]$ = 62 nmol/l; $[HCO_3^-]$= 10 mmol/l; anion gap = 12 mEq/l; glucose = 50 mmol/l, or 900 mg/dl). How can you explain these findings if the cause of the metabolic acidosis was ketoacidosis? What extra tests would you do to confirm your suspicions?

Case 2·12
Diarrhea After Consuming Milk Products
(Case discussed on page 75)

Mrs. Albright usually avoided milk. She hosted a birthday party for her three-year-old son and ate the same food as the others—milk, cake, and ice cream. Several hours later, she developed severe, crampy abdominal pain and explosive diarrhea. She feared food poisoning, but everybody else who attended the party felt fine. Twenty-four hours later, she had recovered completely. She did not take medication at any time. What do you think was responsible for her symptoms? What advice would you give?

Veterinary Case 2·3
Neptune and the Seal
(Case discussed on pages 75–76)

Neptune, reincarnated in today's biotechnological world, is increasingly frustrated by his inability, when in human form, to keep up with seals. Although both humans and seals are air-breathing mammals with substantially similar internal organs, seals can swim underwater 20–25 times longer than humans and also much faster. What metabolic changes could Neptune make to match a seal's diving abilities?

Discussion of Case 2·8
Acidosis from Sniffing Glue
(Case presented on page 68)

Connie's metabolic acidosis, indicated by the low pH and $[HCO_3^-]$ (see Table 2·14, page 68), means that excess H^+ remain in her blood. As shown in Figure 2·12, Connie's liver reacted to the presence of toluene by converting it to hippurate anions plus H^+. Concurrently, her kidneys attempted to maintain acid-base balance by breaking down glutamine to NH_4^+ and HCO_3^-. For each ammonium ion that is excreted in the urine, one bicarbonate ion remains, neutralizing a hydrogen ion. Unfortunately for Connie, this process occurred at a slower rate than the rate at which her liver converted toluene to hippuric acid. Thus, her acidosis persisted.

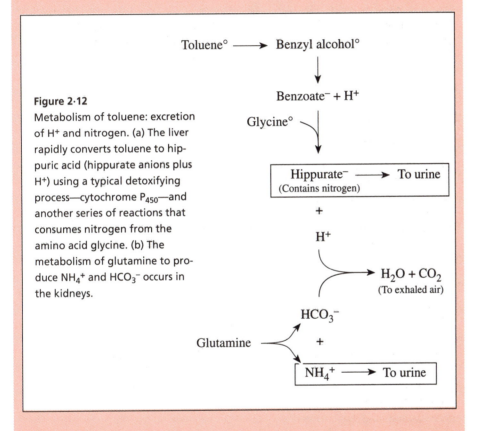

Figure 2·12
Metabolism of toluene: excretion of H^+ and nitrogen. (a) The liver rapidly converts toluene to hippuric acid (hippurate anions plus H^+) using a typical detoxifying process—cytochrome P_{450}—and another series of reactions that consumes nitrogen from the amino acid glycine. (b) The metabolism of glutamine to produce NH_4^+ and HCO_3^- occurs in the kidneys.

Why did Connie have a normal anion gap in plasma with metabolic acidosis? Refer to Table 2·14. She had a large osmolal gap in urine, reflecting the excretion of 150 mmol/l of NH_4^+ salts. Since the $[Na^+]$ and $[K^+]$ in her urine greatly exceeded the $[Cl^-]$, she had a large quantity of unmeasured anions in her urine. Very few of these anions remained in her plasma, so they were probably secreted by the kidneys. She would have had an elevated anion gap in plasma if the anions were not filtered (as in renal failure) or if they were reabsorbed effectively (as in acidosis resulting from excess lactic acid or ketoacids). Instead, she excreted hippurate anions rapidly, dragging out Na^+ and K^+ into the urine (in addition to NH_4^+).

ADDITIONAL QUESTIONS FOR CASE 2·8

Why did she excrete NH$_4^+$ so rapidly?

Her kidneys merely performed their acid-base function—excreting the expected quantity of NH$_4^+$. Hence, this case involves no special disease of the kidneys.

Why were the levels of urea low in her urine and plasma?

Drug addicts typically limit their consumption of proteins, but Connie's level of urea was low, even for the fasted state. However, her urine contained abundant amounts of nitrogen, as NH$_4^+$ and hippurate. Excretion of equal quantities of hippurate and NH$_4^+$ met most of her body's needs for excreting nitrogen (equivalent in terms of acid-base and nitrogen to equimolar excretion of urea per day).

Question

(Discussion on pages 257–58)

2·13   Some patients cannot synthesize urea because of an inborn defect in the urea cycle. How might they be helped to excrete waste nitrogen?

Discussion of Case 2·9
Ketoacidosis with a Very Low Pco₂ in a Cachectic Patient

(Case presented on page 68)

The double vodka should contain 36 g (770 mmol) of ethanol, of which 120 mmol (7 mmol/l × 17 liters of body water = 120 mmol) remain in his blood. Hence, approximately 650 mmol had been metabolized (assuming that all had been absorbed); this amount could have yielded a maximum of 325 mmol of ketoacids (two acetyl groups per ketoacid) if none were oxidized or excreted. Ketoacids are normally distributed in two-thirds of total body water (11 liters in Robert's case), yielding a possible concentration of 29 mmol/l.

The brain and kidneys can normally consume approximately 0.5 and 0.25 mmol of ketoacids per minute, respectively. Over the four hours since the double vodka, this consumption could have accounted for 180 mmol of ketoacids, potentially leaving about 15 mmol of ketoacids per liter from ethanol (180 mmol/11 liters). Because the anion gap (30 mEq/l) indicates 22 mEq/l of organic anions (the correction for albumin is 8 rather than 12 because Robert is hypoalbuminemic) and the L-lactic acid levels are low, the calculated ketosis from ethanol is less than the level of ketoacids in blood. However, the marked drop in his serum creatinine after treatment suggests that Robert's kidney function was impaired when he arrived at the hospital. Although the ethanol might be solely responsible for the ketoacidosis, his prior fasting and renal insufficiency could also have added to the acidosis. If his double vodka was generous, though, ethanol could easily account for his ketoacidosis.

ADDITIONAL QUESTION FOR CASE 2·9

How might the cachexia contribute to the severity of his acidosis?

Robert's appearance indicates very low muscle mass, as does his very low serum creatinine level (creatinine comes from breakdown of creatine phosphate in muscle). Muscle normally forms a major part of the body's buffering capacity for H^+, but because Robert has so little muscle, the BBS in his ECF must take a larger than normal share of the load of H^+. Because of his very emaciated state, Robert's ability to buffer H^+ is well below the 1000 mmol capacity for a normal adult male.

The very low Pco₂ (one-fifth of normal) indicates a very low rate of formation of CO_2; Robert's hyperventilation alone could not have achieved this level, particularly in view of his weakened state. Low formation of CO_2 is consistent with his very low muscle mass, since muscle is a major source of CO_2. Also, the metabolism of fatty acids or ethanol to ketoacids produces ATP in the liver without formation of CO_2 (equation below). Oxidation of fatty acids yields less CO_2/ATP than does oxidation of carbohydrates, as shown in Table 2·6, page 52.

$$\text{Palmitate}^- + H^+ + 27\ (ADP + P_i) \longrightarrow 4\ \text{Ketoacids} + 27\ \text{ATP}$$

Other factors that could decrease the rate of formation of CO_2 are a lower metabolic rate resulting from hypothermia or from depression of the central nervous system (CNS). Robert is not hypothermic and his condition is not consistent with a low metabolic rate from depression of the CNS (he is alert and has only a moderate level of ethanol in his blood).

Discussion of Case 2·10
Shirley Has Severe Acidosis Between Hemodialysis Treatments
(Case presented on page 69)

Shirley's metabolic acidosis is unduly severe. It could have several causes, including loss of HCO_3^- from the gastrointestinal tract or production of acids from ingested proteins. In addition, the kidneys synthesize few if any new HCO_3^- during renal failure because they excrete few NH_4^+. Given the likelihood of more diarrhea, the acidosis might worsen. Not only might the acidosis impair cardiac function, but the associated hyperkalemia could threaten Shirley's life by causing a cardiac arrhythmia. The hyperkalemia probably arose because K^+ in the ICF exchanged with H^+ from the ECF in order to buffer some of the H^+, and the kidneys could not excrete the K^+.

Until Shirley can be put onto hemodialysis, the goals of treatment are to reduce the acidosis and the accompanying hyperkalemia. Administering $NaHCO_3$ to reduce the $[H^+]$ could cause K^+ to re-enter cells, thus reducing the degree of hyperkalemia. Consider the extent to which the $[HCO_3^-]$ has fallen when planning the amount of $NaHCO_3$ to administer. To raise the $[HCO_3^-]$ by 10 mmol/l, for example, assume that H^+ are distributed through almost two-thirds of body water or half of body weight (30–40 liters in a 70 kg person). Thus, you might have to give 300–400 mmol of sodium salts, enough Na^+ to expand the ECF volume by more than two liters. The side effects (against which precautions must be taken) include severe hypertension and congestive heart failure. Obviously, without a means to remove this load of Na^+ (dialysis therapy), this treatment cannot be given without jeopardizing Shirley's life.

Discussion of Case 2·11
Diabetic Ketoacidosis After a Broken Wrist
(Case presented on page 70)

There are several explanations for metabolic acidosis despite a normal anion gap in plasma. One of the most obvious—a direct loss of $NaHCO_3$—is unlikely in this case because Malcolm had normal bowel sounds, no diarrhea, and a low urine pH, which rules out a loss of $NaHCO_3$ in the urine at this time.

A second explanation is that the calculation of a normal anion gap is invalid because of unrecognized low levels of albumin in plasma or the presence of an abnormal protein with a net positive charge. Additional laboratory tests ruled out these possibilities.

Figure 2·13

Metabolic acidosis following the production of acids. After ketoacids are produced in the liver, the resulting H^+ are titrated primarily by buffers (e.g., the BBS in the ECF). The anion β-hydroxybutyrate (β-HB$^-$) can either be reabsorbed (causing an increased anion gap in plasma) or excreted without NH_4^+ or H^+. In the latter case, metabolic acidosis occurs, but the urine—not the body—contains the β-HB$^-$.

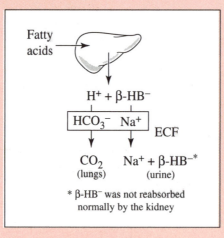

Fatty acids

$H^+ + β\text{-}HB^-$

HCO_3^- Na^+

ECF

CO_2 (lungs) $Na^+ + β\text{-}HB^{-*}$ (urine)

* β-HB$^-$ was not reabsorbed normally by the kidney

A third explanation might be a large indirect loss of $NaHCO_3$. If many ketoacid anions were excreted without H^+ or NH_4^+, the concentration of $Na^+ + K^+$ in the urine would exceed that of Cl^-. Later analysis confirmed the presence of high levels of ketoacid anions in the urine. Thus, Malcolm had an overproduction of ketoacids along with an abnormally high rate of urinary excretion of ketoacid anions without H^+ or NH_4^+ (Figure 2·13). High rates of excretion of these anions could be due to a higher filtered load (not so for Malcolm because the level of ketoacid anions in plasma is low, as reflected by the anion gap in plasma) or less reabsorption (perhaps aspirin and codeine have inhibited the reabsorption of ketoacid anions).

Discussion of Case 2·12
Diarrhea After Consuming Milk Products
(Case presented on page 70)

One possible diagnosis is a nervous stomach resulting from anxiety about the party; a second is food poisoning. The former explanation seems a bit facile, and the latter is unlikely since no one else had the same problems.

Everyone drank a lot of milk at the party. Many adults lose their ability to digest milk sugar (lactose) because of low levels of the digestive enzyme lactase in the small intestine; normal bacteria in the large intestine then metabolize this lactose to a variety of organic acids, which irritate the intestinal walls, causing cramps and diarrhea. The acids do not cause acidosis since the body can oxidize them to neutral end products at a sufficiently rapid rate. The symptoms abate when all of the lactose has been consumed. Hence, this case involves overproduction of acids with local irritation of the gastrointestinal tract but no overall problem of excess H^+.

You should advise Mrs. Albright to avoid milk products and have other members of her family tested for lactase deficiency.

Discussion of Veterinary Case 2·3
Neptune and the Seal
(Case presented on page 70)

The ability to swim underwater (without a continuing supply of O_2) ultimately depends on how long the essential organs of the body can meet their demand for regenerating ATP. To match the seal's metabolic capacity for swimming underwater, Neptune should consider the following alterations:

1. Regenerate ATP Without O_2

   Neptune should improve his ability to regenerate ATP using anaerobic glycolysis. Two features are important—a large supply of glycogen in muscle and an increased ability of muscle to buffer H^+. Neptune should be advised to shut down blood flow to muscle (to save O_2 for use by other essential organs), to have a large quantity of glycogen and creatine phosphate in muscle, and to have a large buffering capacity (on proteins or histidine-containing dipeptides) to buffer the expected load of H^+. Note that the BBS does not help much with the buffering because CO_2 cannot be exhaled.

2. Save Energy by Eliminating Unnecessary Work

   Advise Neptune to lower his metabolic rate. He should not perform work that is not essential in the dive. Functions that can be closed down are kidney work (i.e., shut off glomerular filtration so there will be no need to reabsorb filtered Na^+), GI work (i.e., stop peristalsis and absorption), and metabolic work involving interconversion of fuels (e.g., glucogenesis). Cardiac work should be cut to a minimum by slowing the heart rate, lowering blood pressure, and delivering blood only to essential organs. Brain work should also be minimized (have the brain pump as few ions as possible).

   In quantitative terms, refer to Figure 1·5 (page 11), which shows the consumption of O_2 in organs. To have a 10-fold increase in ability to survive underwater, Neptune would have to diminish his consumption of O_2 to 10% of normal by excluding muscle and reducing the work performed by other organs.

3. Have a High Storage Capacity for O_2

   Since some bodily functions require O_2, survival can be increased by augmenting the reserve of O_2 in the blood and lungs. At rest, Neptune normally extracts about one-fourth of the O_2 in his blood each minute (the entire volume of blood circulates in one minute). His supply of O_2 will therefore allow him to survive for four minutes. When he lowers his metabolic rate, the length of this time will increase.

Neptune can increase his store of O_2 by producing a greater supply of red blood cells (to match that of the seal). The seal has a huge reserve of oxygenated red blood cells in its spleen. As the seal dives, its spleen contracts and releases enough red blood cells to raise the concentration of hemoglobin to 240 g/l and double the volume of red blood cells. Hence, the seal can almost double the length of time that it can survive underwater.

The volume of O_2 in Neptune's lungs is close to his volume of blood and contains a similar amount of O_2 per liter. Hence, by extracting this extra O_2, he could double the amount of O_2 available—a strategy that will work provided that the load of CO_2 is not lethal.

PART E
SUMMARY OF MAIN POINTS

- Although the body contains a tiny amount of H^+, the daily turnover is great. Since minor changes in the $[H^+]$ can be life-threatening, very efficient buffer systems are required. The mechanism for excreting CO_2 provides an effective way to adjust the $[H^+]$; it involves the bicarbonate buffer system (BBS).

- The net production of H^+ occurs when neutral compounds are converted to anions or when cations are converted to neutral compounds. Conversely, H^+ are consumed when anionic fuels are converted to neutral end products. In daily metabolism, the net load of H^+ is produced from the oxidation of sulfur-containing and cationic amino acids. These H^+ are removed when NH_4^+ are excreted in the urine.

- The body can buffer close to 1000 mmol of H^+. The BBS buffers just over half of this load, with two-thirds of the buffering by the BBS occurring in the ECF and one-third in the ICF. Buffering by the BBS occurs earlier than buffering by intracellular proteins because hyperventilation helps the body to adjust to a load of acid.

- The body produces H^+ at different rates. The rate can be extremely fast when lactic acid is produced during hypoxia and not nearly as fast during ketoacidosis (when there is a relative lack of the actions of insulin).

- To adjust acid-base balance, the kidneys excrete extra HCO_3^- or synthesize new HCO_3^- by excreting NH_4^+.

- The anion gap is a measurement that can reveal the presence of unusual ions in the blood and urine. The osmolal gap can identify uncharged molecules in the blood and urine.

CHAPTER 3
ENERGY METABOLISM IN EXERCISE

PART A
METABOLIC DEMANDS OF MUSCLE DURING EXERCISE

Muscle can vary its demands for energy over a very wide range—from resting, through normal daily activity, to sustained exercise and the explosive energy demands of a 100 m dash by a trained athlete. These different demands for energy pose a variety of requirements not only on muscle but also on the rest of the body.

PART B
METABOLIC DISORDERS IN MUSCLE

PART C
SUMMARY OF MAIN POINTS

PART A
METABOLIC DEMANDS OF MUSCLE DURING EXERCISE

Event 3·1
The 100 m Sprint
(Race discussed on page 88)

In sprints, the athlete goes as fast as possible, with no thought for endurance. The demand for regeneration of ATP is maximal, faster than can be met by oxidative metabolism. For the most part, ATP is regenerated anaerobically, causing H^+ to accumulate. Muscle cells must buffer these H^+ before excessive fatigue sets in.

Event 3·2
The 1500 m Race
(Race discussed on page 89)

In middle distance races, the athlete must consider endurance. The body regenerates ATP primarily by oxidizing glycogen and glucose, with CO_2 as the carbon product. The body must eliminate a large load of CO_2 but has no major problem with H^+ (except during the finishing kick, which resembles a sprint).

Event 3·3
The Marathon
(Race discussed on page 90)

Very long distance races exhaust the body's reserves of carbohydrate, so muscle cells oxidize fatty acids for much of their energy, thereby limiting the rate of regeneration of ATP. This race has no acid-base problem other than during the finishing kick.

Questions to Consider

In each race, which fuels are used to regenerate ATP?

What metabolic factors limit the athlete's performance in each race?

Why does the average speed of a race decline as the distance increases?

What nutritional (or other) preparation before the race or intake of food during the race could help the athlete achieve a personal best?

How do organs other than muscle help in the regeneration of ATP in muscle?

How could muscle's high demand for fuel affect the availability of fuel for other organs?

Will the availability of fuels and oxygen or the removal of wastes affect the performance of the athlete?

What types of muscle fiber are best suited to (a) strenuous, short-term activities and (b) activities requiring endurance?

What steps might an ambitious and somewhat amoral coach use to help his or her athletes?

BACKGROUND

TYPES OF MUSCLE

> *Some types of skeletal muscle are best suited to violent bursts of activity; others are adapted to endurance.*

There are three major types of muscle, each with a unique biochemistry and set of functions (Table 3·1). We shall consider only skeletal muscle and describe its energy metabolism in the sprint, middle distance race, and marathon. The critical factors limiting performance are the rates of regeneration of ATP (which depend on the fuels available, the supply of oxygen, and the speed of the metabolic pathways involved) and the rates of removal of wastes or other by-products. Figure 3·1 (see page 82) provides an overview.

Table 3·1

Types of muscle tissue

Type	Major function	Metabolic features
Heart	Pumps blood to organs	Aerobic metabolism: burns fatty acids or lactate; burns glucose less commonly
Smooth muscle	Controls diameters of hollow organs; propels contents	None to speak of
Skeletal muscle		
- Fast twitch (white fibers)	Enables activities (sprinting, jerking, lifting, etc.) that require a sudden burst of energy	High glycolysis, large load of H^+, and high buffer capacity; fastest rate of synthesis of ATP
- Slow twitch (red fibers)	Is used in activities that require endurance (e.g., longer races)	Aerobic, longer duration, not quite as fast in making ATP

REGENERATION OF ATP IN SKELETAL MUSCLE

Creatine Phosphate

> *Creatine phosphate is the fastest source of ATP and also removes H^+.*

Creatine phosphate (CrP) regenerates ATP in muscle; because the reaction is at chemical equilibrium (equation 1), it yields ATP faster than any other pathway. Creatine phosphokinase (CPK), which catalyzes this reaction, is the only route for synthesis or breakdown of CrP. At rest, the high ratio of ATP to ADP yields a high ratio of CrP to Cr, and [CrP] can be as high as 25 mmol/kg, sufficient to meet the body's need for ATP during the first three to four seconds of sprinting. At the start of exercise, the greatly increased rate of consumption of ATP causes the ratio of ATP to ADP to fall, pulling equation 1 to the right. The ATP formed from CrP is immediately used (equation 2), and the net reaction is hydrolysis of CrP (equation 3). The *inorganic phosphate* (P_i) formed helps absorb H^+ produced in anaerobic glycolysis (discussed below).

Creatine phosphate:

A compound found mainly in skeletal muscle that is used to regenerate ATP quickly when levels of ADP rise.

Inorganic phosphate (P_i):

A compound that can exist in a monovalent ($H_2PO_4^-$) or divalent (HPO_4^{2-}) form, depending on the $[H^+]$.

$$CrP^{2-} + ADP^{3-} + H^+ \longleftrightarrow Cr + ATP^4 \qquad (1)$$

$$\underline{ATP \longrightarrow ADP + P_i + \tfrac{1}{2}H^+ + \text{biological work}} \qquad (2)$$

$$CrP^{2-} + \tfrac{1}{2}H^+ \longrightarrow Cr + P_i + \text{biological work} \qquad (3)$$

The $\frac{1}{2}$ H$^+$ in equations 2 and 3 arises because inorganic phosphate exits as $H_2PO_4^-$ and HPO_4^{2-} in equal proportions at rest (when the pK for P_i is near the pH of the intracellular fluid). When the intracellular [H$^+$] rises (during a sprint), a total of 1 H$^+$ can be buffered because almost all of the HPO_4^{2-} will be converted to $H_2PO_4^-$ at the higher [H$^+$].

Figure 3·1

Regeneration of ATP in muscle.

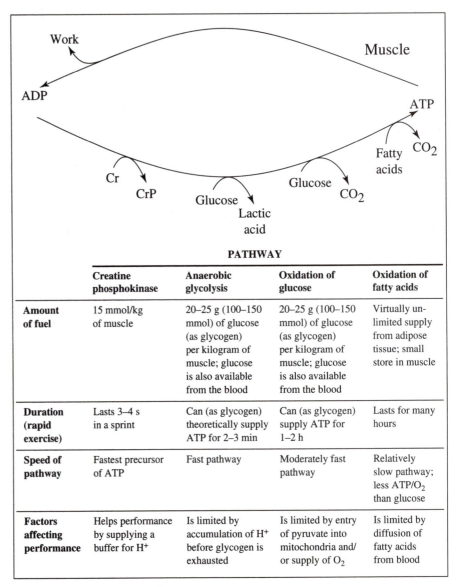

PATHWAY			
	Creatine phosphokinase	**Anaerobic glycolysis**	**Oxidation of glucose**	**Oxidation of fatty acids**
Amount of fuel	15 mmol/kg of muscle	20–25 g (100–150 mmol) of glucose (as glycogen) per kilogram of muscle; glucose is also available from the blood	20–25 g (100–150 mmol) of glucose (as glycogen) per kilogram of muscle; glucose is also available from the blood	Virtually un-limited supply from adipose tissue; small store in muscle
Duration (rapid exercise)	Lasts 3–4 s in a sprint	Can (as glycogen) theoretically supply ATP for 2–3 min	Can (as glycogen) supply ATP for 1–2 h	Lasts for many hours
Speed of pathway	Fastest precursor of ATP	Fast pathway	Moderately fast pathway	Relatively slow pathway; less ATP/O_2 than glucose
Factors affecting performance	Helps performance by supplying a buffer for H$^+$	Is limited by accumulation of H$^+$ before glycogen is exhausted	Is limited by entry of pyruvate into mitochondria and/ or supply of O_2	Is limited by diffusion of fatty acids from blood

Anaerobic Glycolysis

> *Anaerobic glycolysis produces ATP rapidly but imposes a large load of H$^+$.*

Anaerobic glycolysis from stores of glycogen is the next fastest source of ATP in muscle. Glucose (mainly from stores of glycogen but also from the circulation) can be metabolized very rapidly to lactic acid with the production of ATP (three molecules of ATP per glucose from glycogen and two per glucose from blood; see Chapter 8, page 194). This anaerobic metabolism is necessary because the demands for ATP in the explosive energy consumption of the sprint are far greater than can be supplied by oxidative metabolism. Muscle can generate much more ATP from glycogen than it can from CrP, but at a slower rate. At maximal rates of use of glycogen in the sprint, this store could last about three minutes, but the accumulation of H$^+$ prevents muscle from functioning this long anaerobically.

The overall reaction in anaerobic glycolysis is shown in equation 4 (see Chapter 8 for more information). Equation 5 summarizes the net reaction for anaerobic glycolysis, (disregarding the turnover of ATP because hydrolysis of ATP was required to initiate glycolysis). Each molecule of ATP generated through anaerobic glycolysis requires formation of 1 mmol of H^+. If the glucose comes from glycogen, the proportion is 0.67 mmol of H^+ per ATP.

$$\text{Glucose} + 2\,\text{ADP}^{3-} + 2\,\text{P}_i \longrightarrow 2\,\text{Lactate}^- + 2\,\text{ATP}^{4-} \tag{4}$$

$$\text{Glucose} \longrightarrow 2\,\text{Lactate}^- + 2\,\text{H}^+ \tag{5}$$

Oxidation of Glucose and Glycogen

> *Oxidation of glucose and glycogen to CO_2 is a less rapid source of ATP but does not produce a load of H^+.*

Complete oxidation of glucose yields 36–40 mmol of ATP per molecule of glucose (depending on the source of glucose and the details of the pathway). Despite the much larger yield of ATP per glucose consumed, the overall maximal rate of synthesis of ATP by oxidation of glucose to CO_2 is slower than in anaerobic glycolysis. The rate of aerobic metabolism may be limited by the supply of O_2 to muscle or by the rate of entry of pyruvate into the mitochondria; endurance may also be limited by the rate of removal of CO_2 from muscle.

The overall reaction for aerobic glycolysis is shown in equation 6; because ATP is used immediately, the net reaction (equation 7) is simple conversion of glucose to neutral end products.

$$\text{Glucose} + 6\,\text{O}_2 + 36\text{–}40\,(\text{ADP} + \text{P}_i) \longrightarrow 6\,\text{CO}_2 + 6\,\text{H}_2\text{O} + 36\text{–}40\,\text{ATP} \tag{6}$$

$$\text{Glucose} + 6\,\text{O}_2 \longrightarrow 6\,\text{CO}_2 + 6\,\text{H}_2\text{O} \tag{7}$$

Trained athletes running in middle distance races use ATP more slowly than in the sprint and therefore have longer lasting supplies of glycogen, which can provide oxidative energy for 70–100 minutes of racing.

Oxidation of Fatty Acids

> *Oxidation of fatty acids is an even slower way to regenerate ATP.*

Oxidation of fatty acids consumes the most abundant fuel in the body. Fatty acids derived from triacylglycerols in adipose tissue are transported in plasma bound to the protein albumin. Because fatty acids must cross many barriers before being oxidized in muscle cells (Figure 3·2), production of ATP is comparatively slow and may limit performance. No such barriers exist for the fatty acids already present in muscle cells, but the quantity of triacylglycerols stored in muscle is small compared with the large store in adipose tissue.

To provide ATP needed in a race, muscle must extract about 50% of the fatty acids in each liter of plasma delivered to muscle. The net quantity of fatty acids in 1 liter of blood is 0.36 mmol (1 liter of blood contains 0.6 liters of plasma, and the concentration of fatty acids in plasma is close to 0.6 mmol/l). Since 1 liter of blood contains 8–9 mmol of O_2, it has enough O_2 to oxidize almost 0.3 mmol of fatty acids. The stoichiometry for the oxidation of fatty acids (palmitate) follows:

$$\text{Fatty acid (C}_{16}) + 23\,\text{O}_2 \longrightarrow 16\,\text{CO}_2 + 16\,\text{H}_2\text{O} + 129\,\text{ATP}$$

Figure 3·2

Transport of fatty acids from adipose tissue to muscle. Fatty acids are very sparingly soluble in water. Hence, their transport through aqueous environments requires binding to the protein albumin. (TG = triacylglycerols.)

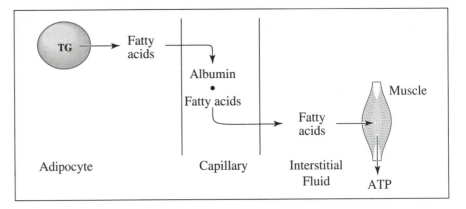

Note that when compared with glucose, oxidation of fatty acids produces approximately 5% less ATP per O_2 consumed but approximately 25% more ATP per CO_2 produced (Table 2·4, page 47).

SUPPLY OF O_2 AND REMOVAL OF WASTES

Two aspects of skeletal muscle metabolism that limit performance are the management of H^+ (e.g., in a sprint) and the delivery of O_2 plus the removal of CO_2 (e.g., during a 1500 m race).

The Management of H^+

1. Production of H^+

> *Skeletal muscle can produce H^+ very rapidly.*

In anaerobic glycolysis with turnover of ATP, H^+ are produced when lactate anions accumulate (equation 5, page 83). Quantitatively, two H^+ will accumulate every time three ATP are regenerated during the conversion of glycogen to lactate; rates of accumulation of close to 10 mmol of lactate per second may occur during a sprint. This rate of production of H^+ exceeds the body's capability of removing them.

2. Removal of H^+

> *Removal of H^+ by metabolism is a slow process.*

The body can remove H^+ by metabolism (conversion of an anion to a neutral end product) or by buffering. During a sprint, lactate is produced faster than it is removed. However, the hydrolysis of CrP, which provides the ATP needed for the first three to four seconds of the race, produces divalent inorganic phosphate ions (HPO_4^{2-}), which absorb some of the H^+ formed in anaerobic glycolysis.

The H^+ produced in the rest of the sprint can be buffered by histidines in proteins or on the *dipeptides* carnosine or anserine in muscle. The absorption of H^+ by proteins changes the charge on proteins, possibly inhibiting their function. In quantitative terms, these buffers can account for the binding of close to 3 mmol of H^+ per 0.1 unit fall in pH per kilogram of skeletal muscle.

Dipeptides:

Molecules that contain two amino acids joined together by a peptide bond (the same bond as in proteins). Carnosine and anserine are dipeptides that appear mainly in muscle; because they contain histidine, they act as effective buffers.

Respiration, the O_2/CO_2 story

1. Delivery of O_2

> The rate of delivery of O_2 can limit the capacity for prolonged severe exercise.

Table 3·2 indicates that muscle can increase its consumption of O_2 by as much as 50-fold from rest to exhaustive exercise and that the heart responds to this increased demand by delivering more blood to peripheral organs. With aerobic training, skeletal muscle can increase its capacity to extract O_2 from blood, thereby enhancing performance. The following adaptations help:

(i) More capillaries are situated around skeletal muscle fiber bundles (shortening the distance for diffusion of fuels and wastes).

(ii) More effective delivery of O_2 from hemoglobin (Hgb) to mitochondria results from:
   - more effective delivery of blood to muscles;
   - easier unloading of O_2 from Hgb (a higher $[H^+]$ from a high Pco_2 and the production of some lactic acid decreases the affinity of Hgb for O_2);
   - more *myoglobin* in skeletal muscle cells to bind O_2 and therefore help its movement from the cell membrane to mitochondria;
   - a lower Po_2 in mitochondria as a result of higher NADH and ADP levels and a higher concentration of enzymes in the TCA cycle.

Table 3·2
Cardiac output and peripheral consumption of O_2 at rest and in strenuous sustained exercise in an elite athlete

	Rest	1500 m race (approximate figures)
Cardiac stroke volume (ml/beat)	100	140
Cardiac rate (beats/min)	50	150
Cardiac output (l/min)	5	21
O_2 delivered (mmol/min)	45	180
O_2 used by body (mmol/min)	12	160
O_2 used by muscle (mmol/min)	3	150
Respiration volume (l/breath)	0.5	1.0
Respiration rate (breaths/min)	14	100
Alveolar ventilation (l/min)	5*	80*
CO_2 formed by body (mmol/min)	10	160
CO_2 formed by muscle (mmol/min)	2.5	150

* Alveolar ventilation is less than the volume of air breathed because of the dead space of ventilation.

Myoglobin:
A protein similar to hemoglobin that is found in skeletal muscle (the darker the muscle the more myoglobin). It helps transfer O_2 from blood to mitochondria.

Cardiac stroke volume:
The volume of blood pumped per heart beat. The normal volume is close to 70 ml/beat in an untrained person; training can raise the resting stroke volume to 100 ml/beat, and exercise can result in a value of 130 ml/beat.

Cardiac rate:
The number of beats per unit time. The resting rate without training is in the range of 60–80 beats/min. Training can reduce the resting rate to 45–55 beats/min. In severe exercise, it can rise to 150–180 beats/min.

2. Removal of CO_2

Carbonic anhydrase in red blood cells converts CO_2 and water to H^+ and HCO_3^- (equation 8). The H^+ formed from this reaction bind to hemoglobin (Hgb) and displace O_2 (equation 9). Hemoglobin also has the capacity to bind four O_2 (which can form four CO_2). Some CO_2 is transported on one of the four N-terminal amino groups ($-NH_2$ groups) as carbamino compounds (equation 10).

$$CO_2 + H_2O \longleftrightarrow H_2CO_3 \longleftrightarrow H^+ + HCO_3^- \qquad (8)$$

$$Hgb(O_2)_4 + 4\,H^+ \longrightarrow Hgb(H^+)_4 + 4\,O_2 \qquad (9)$$

$$Hgb\text{-}NH_2 + CO_2 \longleftrightarrow HgbNHCOO^- + H^+ \qquad (10)$$

The reaction in equation 8 can also move to the left. The body generates CO_2 when H^+ from anaerobic glycolysis react with HCO_3^- in muscle and blood. This formation of CO_2 can be very rapid indeed and is a sign that the $[H^+]$ is rising fast and that performance is starting to decline.

Most (85%) of the CO_2 in the body is carried as HCO_3^- (see Chapter 2, page 48). Carrying all this CO_2 requires a very high Pco_2 in venous blood. For CO_2 to diffuse from muscle mitochondria to blood, the Pco_2 in muscle must be even higher.

In strenuous aerobic exercise, the body can consume 160 mmol of O_2 per minute, thereby producing 160 mmol of CO_2 per minute. To maintain the Pco_2 in arterial blood at close to 40 mm Hg, the $[CO_2]$ in alveolar air must be kept at 2 mmol/l. The rate of alveolar ventilation will therefore have to be 80 liters/min.

COORDINATION OF HEART AND LUNGS DURING EXERCISE

> *During vigorous exercise, the need to eliminate CO_2 drives respiration.*

When a person breathes at rest, the capacity of the blood to carry oxygen happens to match the concentration of oxygen in the air (air and blood each contain 8–9 mmol of O_2 per liter). Also, *cardiac output* and *alveolar ventilation* (delivery of air to air sacs in the lungs) are each 5 liters/min. Strenuous exercise alters this balance. Alveolar air contains almost three times more O_2 than CO_2 in millimoles per liter (6 vs 2 mmol/l). When the body uses O_2 rapidly (consuming almost all of the O_2 available in the blood), an equally rapid production of CO_2 results (oxidation of carbohydrates produces 1 mol of CO_2 for every mole of O_2, since the *respiratory quotient* for oxidation of carbohydrates is 1.0). Unless the lungs compensate, the $[CO_2]$ in alveolar air will rise to a dangerous level and cause a compromise in the function of muscle. In a system that is perfectly efficient, an increase in alveolar ventilation to four times the cardiac output will prevent the buildup of CO_2 in arterial blood and thereby in organs other than skeletal and cardiac muscle.

BODY FLUIDS AND EXERCISE

> *The body must keep as much water in the extracellular fluid (ECF) as possible for optimal performance during vigorous exercise.*

During vigorous exercise, a large volume of blood must be delivered to skeletal muscle (to deliver O_2), to the skin (to dissipate heat), and to the heart (a large stroke volume is needed for the high cardiac output). This large volume is achieved by adding and retaining extra fluid in the circulation; "sports anemia" is a result. The kidneys of athletes retain more sodium and water, thereby increasing the volume of the ECF and creating a larger volume of plasma for the same quantity of hemoglobin. After strenuous activity, the capacitance vessels (veins) dilate, creating a safe storage place for the larger volume of blood that the body will require during the next bout of exercise.

Water also shifts between body compartments during vigorous exercise. When large chemical particles in muscle cells (e.g., glycogen) break down to many smaller molecules (e.g., lactic acid), the osmolality in muscle rises almost 10%. This increase could cause water (about 2 liters) to enter these cells by osmosis. If all the water gained in muscle shifted from the ECF, the volume of blood would decline considerably and poorer performance would result. However, the body stores glycogen with water (3 g of water per gram of glycogen). If this water were stored without ions and released when the content of glycogen in muscle decreases, there would be little need for a shift of water from the ECF (removal of 450 g of glycogen could release 1.35 liters of water; refer to Question 8·9, page 209).

Cardiac output:

The volume of blood pumped by the heart per unit time; the normal rate is 5 liters/min, which can rise fourfold in severe exercise. Cardiac output is the product of the cardiac stroke volume and the cardiac rate.

Alveolar ventilation:

The rate at which atmospheric air is exchanged with air in the alveoli (the small sacs in the lungs in which the major exchanges of O_2 and CO_2 occur).

Respiratory quotient (RQ):

The molar ratio of CO_2 produced to oxygen consumed when a fuel is oxidized. Carbohydrates have an RQ of 1; triacylglycerols have an RQ of 0.7. The RQ can be used in whole-body studies to determine the major type of fuel being burned.

PROCEDURES USED TO HELP ATHLETIC PERFORMANCE

Blood Doping

Some coaches strive to increase the ability of their runners to carry O_2 in their blood by infusing red blood cells collected from the athletes several weeks earlier or by giving injections of erythropoietin, which increases the production of red blood cells. Although blood doping will probably have little effect on a sprinter, it may increase the performance of other athletes by a few percentage points. However, the balance between gain or loss of performance is delicate. Though each liter of blood will have an added capacity to carry O_2, as the viscosity of the blood rises (because of the higher *hematocrit*), the cardiac output declines. Also, a higher Pco_2 in venous blood and muscle may limit performance. The real benefit may be psychological rather than physical.

Hematocrit:
The percentage of blood volume occupied by red blood cells.

Glycogen Stripping (Vigorous Exercise Several Days Before) and Supercompensation

Consumption of meals rich in carbohydrates after depletion of glycogen in muscle can considerably increase the amount of glycogen stored in muscle. This buildup should benefit marathon runners because a low level of glycogen in muscle will probably limit their performance (see Event 3·3, page 90).

Soda Doping

After consuming large amounts of sodium bicarbonate ($NaHCO_3$) to prepare for a load of acid (lactic acid), some athletes seem to show improved performance over short distances but probably not for the reason expected. An increase in cardiac output resulting from the load of Na^+ may have a greater effect than the influence of HCO_3^- per se (not enough HCO_3^- reach muscle cells to make a significant difference). The risks include a paradoxical loss of extracellular fluid volume caused by a shift of water into the intestines before the absorption of $NaHCO_3$ can occur. Also, the body will convert the extra HCO_3^- to CO_2, which the lungs must remove. A benefit may arise from neutralization of HCl in the stomach; burping of acidic stomach contents during a race is not fun!

Discussion of Event 3·1
The 100 m Sprint
(Race presented on page 80)

An elite sprinter uses approximately 60 mmol of ATP per kilogram of muscle during the 10 seconds of the race (Table 3·3). This ATP comes from four sources:

1. During a sprint, the concentration of ATP in muscle falls by approximately 20% (normal content is 5 mmol/kg of muscle); this amount cannot power even one second of the race.

2. Creatine phosphate in muscle (which provides about 25 mmol of ATP per kilogram) largely disappears in the first three to four seconds of the race but can supply the ATP needed for this period and can buffer some of the H^+ produced in anaerobic glycolysis.

3. Anaerobic glycolysis from glycogen in muscle provides most of the ATP needed (approximately 40 mmol/kg of muscle, largely in the later two-thirds of the sprint).

4. Although some of the ATP comes from the aerobic ATP generation system, the amount is probably small since there is insufficient time to increase the supply of O_2 rapidly enough to meet the body's demands for O_2 during the sprint (instead, the body relies on the small amount of O_2 present in the muscles and on hemoglobin in muscle capillaries).

The major source for regenerating ATP in muscle is glycogen. Each kilogram of muscle has 20–25 g of glycogen (equivalent to 110–140 mmol of glucose), which has the capacity to yield up to 330–420 mmol of ATP from anaerobic metabolism, well in excess of the amount needed for the duration of the race. Sprinters will not be able to deplete this source of ATP because anaerobic glycolysis from glycogen produces 0.67 mmol of H^+ per ATP formed; this accumulation of H^+ causes fatigue in muscle by inhibiting glycolysis and thus the rate of regeneration of ATP. A high $[H^+]$ also interferes with interactions between actin and myosin (the contractile elements of muscle) and causes the sarcoplasmic reticulum to bind Ca^{2+} more readily, thereby inhibiting muscle contraction (release of Ca^{2+} signals contraction).

Buffers within muscle minimize the accumulation of H^+. In addition to CrP (which consumes H^+ as it breaks down), proteins and special dipeptides in muscle (carnosine and anserine) can buffer about 10 mmol of H^+ per kilogram of muscle with a decline in intracellular pH from 7.1 to 6.8 (see Chapter 2, page 47). Also, as discussed in Case 2·1 (pages 50–51), hyperventilation before a race helps produce intracellular "parking spots" for H^+.

After the race, the large amount of lactic acid produced must be removed, either by conversion back to glycogen in muscle or the liver (Chapter 8) or by oxidation to CO_2 to yield ATP (Chapter 10).

Table 3·3
Data from muscle during a sprint

Constituent	Before exercise	After sprint	Net ATP
ATP (mmol/kg)	5	4	1
CrP (mmol/kg)	25	7	18
Glycogen (mmol of glucose/kg)	56	42	42*

* Whereas CrP yields 1 mmol of ATP per CrP, glycogen yields 3 mmol of ATP per unit of glucose in glycogen.

Discussion of Event 3·2
The 1500 m Race
(Race presented on page 80)

The 1500 m race has three phases.

1. In the first phase, which resembles the start of the sprint, the runner produces lactic acid at a rapid rate but only for a short time.

2. In phase two, the longest part of the race, the runner regenerates ATP aerobically from glycogen. Turnover of ATP slows greatly and much of the lactic acid that built up in the first phase is oxidized. Runners consume much less ATP per second during the major part of the 1500 m race (1 mmol/s/kg of muscle) than during a sprint (up to 5 mmol/s/kg of muscle).

3. Phase three of the race is the finishing kick, in which the athlete's only concern is high speed until the end of the race. During this phase, anaerobic glycolysis can be used, provided that H^+ do not accumulate enough to limit performance.

Maximal performance for the 3.5 minutes of phase two depends on the supply of oxygen to the muscles needing it most and on the delicate balances required to ensure that the demands for regenerating ATP can just be met by—but not exceed— the availability of oxygen. Too little demand for ATP means not enough speed, and too high a demand for ATP causes a requirement for rapid anaerobic glycolysis, involving accumulation of H^+ and decreased performance of muscle (see Event 3·1 above).

Efficient lungs are fully able to keep arterial blood saturated with O_2. The increased cardiac output combined with delivery of most of the blood to the muscles needing it (without rendering other organs hypoxic or anoxic) helps in the transport of O_2 to areas with the greatest need. Blood flow through muscle in prolonged exercise can reach 15–20 liters/min, which is equivalent to the delivery of the full blood volume every 15–20 seconds.

Removal of CO_2 from muscle requires that the P_{CO_2} of muscle and venous blood be very high to carry back to the lungs the 8–9 mmol of CO_2 per liter produced from the 8–9 mmol of O_2 per liter carried to the muscles (respiratory quotient = 1.0). This high P_{CO_2} helps the blood discharge O_2 at the muscles (see Chapter 2, page 49).

The quantities of glycogen used can be assessed from the rate of consumption of O_2—at least for phase two of the race. Glycogen in muscle should be sufficient for up to 80–100 minutes of running at the speed achieved in phase two of the race (Table 3·4).

Table 3·4
Calculation of the amount of ATP supplied by glycogen in muscle

O_2 consumed by runner	180 mmol/min
O_2 consumed by muscle	9 mmol/min/kg of muscle
ATP produced by muscle	60 mmol/min/kg of muscle
Glucose used by muscle (38–40 mmol of ATP per glucose)	1.5 mmol/min/kg of muscle
Glycogen content of muscle	110–140 mmol of glucose per kilogram of muscle
Duration of muscle glycogen as fuel	80–100 min

Discussion of Event 3·3
The Marathon
(Race presented on page 80)

The marathon requires muscle to use fuels from the fat system for energy since there is not enough glycogen to permit regeneration of ATP for the whole race of 125–135 minutes. Carbohydrates provide most of the energy used by the athlete during the first third of the race; glycogen in muscle is the main source, but glucose from the blood also contributes some energy. The factors affecting performance during this part of the race are very similar to those in Event 3·2.

Since stores of glycogen become depleted, triacylglycerols must provide increasing amounts of energy as the race progresses (Table 3·5). Muscle has small reserves of triacylglycerol, which can provide fatty acids to oxidize for the regeneration of ATP, but these stores are rapidly depleted. Fatty acids released from triacylglycerols in adipose tissue therefore become an important source of energy. Since delivery of fatty acids to mitochondria in muscle slows the rate at which muscle regenerates ATP (Figure 3·2), the maximal rate of muscle activity decreases as fatty acids contribute more to the production of energy.

The remaining aspects of the marathon race are similar to those for the 1500 m race. Table 3·5 shows that significant hypoglycemia may develop late in the race. A sufficient quantity of glucose remains in the blood for much of the race because elevated levels of adrenaline and low levels of insulin stimulate breakdown of triacylglycerols in adipose tissue (see Chapter 11, page 231). Complications from hypoglycemia may arise if runners enter the final sprint with a depleted supply of glycogen in muscle; they will use glucose from the blood, resulting in an insufficient quantity of glucose for both muscle and the brain (refer to Case 1·3, page 14).

Table 3·5
Fuels used in a marathon

Time (hours)	O_2 Consumed (mmol/h)	Glycogen (moles used)	Fatty acids (moles used)	[Glucose] in blood (mmol/l)
0–1	10	50	12	5.6
1–2	10	26	37	4.6
2–3	10	20	40	3.8
Total	30	96	89	–

Questions

(Discussions on pages 258–60)

3·1 After finishing a race on two identical occasions, the concentration of lactate in an athlete's blood was 10 mmol/l. On the first occasion, the athlete lay down and rested immediately after the race; the concentration of lactate remained high for 30 minutes. On the second occasion, after the athlete "cooled-down" by jogging around the track for 30 minutes, the concentration of lactate was close to normal. How would you explain this difference in the concentration of lactate?

3·2 We stated that about 60 mmol of ATP per kilogram of muscle was used during the 10-second sprint. How could the consumption of ATP have been measured?

3·3 A runner must decide when to initiate the final sprint to the finish line. What are the trade-offs?

3·4 Each kilogram of muscle contains 20–25 g of glycogen, or 110–140 mmol of glucose. The molecular weight of glucose is 180. Explain the discrepancy.

3·5 During vigorous exercise, a signal causes an increase in the rate of respiration without causing the P_{CO_2} and $[H^+]$ of arterial blood to shift out of the normal range. What might this signal be?

3·6 How does arterial blood differ in composition from venous blood during exercise?

3·7 How does the activity of PDH in resting muscle compare with that in vigorously exercising muscle?

3·8 What is the metabolic reason for a marathon runner "hitting the wall"?

PART B
METABOLIC DISORDERS IN MUSCLE

Case 3·1
Bob Suffers from Muscle Pain and Fatigue
(Case discussed on page 94)

Bob, aged 17 years, complains of extreme fatigue and pain in his muscles only when he does strenuous work for a protracted period of time. Otherwise, he feels quite well. These symptoms, easily reproduced when he performs work of long duration, do not result from fasting or a quick sprint. He has never suffered from hypoglycemia. What is wrong?

Case 3·2
Eleanor Can Hike Long Distances but Cannot Sprint
(Case discussed on page 94)

Eleanor, aged 12 years, suffers from fatigue and painful cramps in her muscles when she tries to sprint or perform strenuous exercise, but she is able to keep up with her friends pretty well on long hikes and in usual play, and she is normal in other respects. What is wrong?

Veterinary Case 3·1
The Racehorse
(Case discussed on page 95)

Horses have enormous endurance. They can trot steadily for long periods and gallop for several minutes, especially if trained. Table 3·6 shows data obtained from blood samples taken from a horse at rest and after exercising for three minutes on a treadmill.

Relevant information:

- The ECF volume in this horse is 200 liters—20% of body weight (1000 kg).

- The rise in hematocrit and blood volume is due to the addition of red blood cells.

- The rate of breathing is 100 liters/s (equivalent to 75 liters/s alveolar ventilation).

- Glycogen in muscle provides the fuel for muscle activity.

From the data provided, calculate how long, on average, it would take a single red blood cell to complete a full circuit from left ventricle to left ventricle during the exercise. You are essentially calculating the cardiac output and relating it to the new blood volume.

92

Table 3·6

Blood chemistry of a horse at rest and during maximal exercise

Blood		Rest		Exercise	
		Arterial	Venous	Arterial	Venous
pH		7.40	7.36	6.90	6.80
Pco_2	mm Hg	40	50	55	110
HCO_3^-	mmol/l	25	27	10	10.4
Po_2	mm Hg	95	65	65	10
Hematocrit	%	40	40	58	58
Glucose	mmol/l	6.2	6.2	7.5	6.6
Lactate	mmol/l	1	1	32	33
Volume	liters	70	70	100	100

Questions to Consider

When Bob (Case 3·1) and Eleanor (Case 3·2) suffer from fatigue and pain during exercise, what fuels are likely to be in demand and why might they not be available?

What advice can you give Bob and Eleanor about minimizing the effects of their disorders?

Discussion of Case 3·1
Bob Suffers from Muscle Pain and Fatigue

(Case presented on page 92)

Bob handles sprints well and has never been hypoglycemic. Hence, it appears that he can break down glycogen in his liver and muscle and that he is able to make glucose in his liver. His problem, therefore, does not seem to be related to the carbohydrate system, the pyruvate dehydrogenase (PDH) system, or the ATP generation system (see Chapter 1, page 26). The fat system remains to be evaluated.

Because he makes glucose normally, his liver's ability to oxidize fatty acids seems normal (see Case 5·2, pages 128–29). The fact that he suffers when he performs strenuous work for a long time suggests that the ability of his muscle to oxidize fatty acids for energy is compromised. A muscle biopsy shows a low level of carnitine-palmitoyl transferase (see Chapter 11, page 232), the enzyme that allows fatty acids to enter the mitochondria for oxidation. This enzyme deficiency cannot be corrected, and Bob's condition will worsen progressively.

Bob should be advised to avoid strenuous exercise and eat a diet high in carbohydrates, with small, frequent meals. Excess exercise, which demands oxidation of fatty acids for ATP in muscle, may cause breakdown of muscle cells, releasing muscle contents (e.g., myoglobin) into the blood. Myoglobin in blood and urine can cause kidney failure.

Discussion of Case 3·2
Eleanor Can Hike Long Distances but Cannot Sprint

(Case presented on page 92)

Eleanor's case history indicates that she cannot get fuel for anaerobic muscle function and hence that her problem stems from rapid breakdown of glycogen in muscle. This condition results from an inherited defect in activating muscle phosphorylase, an enzyme required for the breakdown of glycogen (see Chapter 8, page 208); this defect was recognized by McArdle, who showed that strenuous exercise did not cause the normal increase in lactic acid in the blood.

Question

(Discussion on page 260)

3·9 Will the pH in Eleanor's muscle differ from that of normal individuals during a sprint? If so, why?

Discussion of Veterinary Case 3·1
The Racehorse
(Case presented on page 92)

The steps to calculate are the production of CO_2, the delivery of O_2, and the transit time for a red blood cell.

PRODUCTION OF CO_2

- The arterial P_{CO_2} remains essentially constant during the later stages of exercise (55 mm Hg).

- The total CO_2 produced can be calculated by multiplying the alveolar ventilation rate (75 liters/s) and the $[CO_2]$ in alveolar air (3 mmol/l, assumed from equivalence to the P_{CO_2} in arterial blood; see the sample calculation on page 48).

Thus, 225 mmol of CO_2 are produced per second; this CO_2 comes from two sources.

1. H^+ (formed with lactic acid) titrate HCO_3^- in the ECF and ICF.

   The ECF (200 liters) loses 15 mmol of HCO_3^- per liter, thus forming 3000 mmol of CO_2 in three minutes of exercise.

   The very high $[H^+]$ likely in the ICF of muscle (400 liters) will probably titrate almost all the HCO_3^- there (10 mmol/l), thus forming 4000 mmol of CO_2 in three minutes of exercise.

   In total, 7000 mmol of CO_2 (35 mmol/s) are formed via titration of HCO_3^-.

2. Aerobic metabolism forms CO_2.

   The total of 225 mmol produced minus 35 mmol from titration of HCO_3^- results in 190 mmol/s.

DELIVERY OF O_2

- Since glycogen is the fuel used (RQ = 1.0), the production of CO_2 from metabolism should equal the use of O_2. Thus, 190 mmol of O_2 must be delivered per second.

- With a hematocrit of 58%, each liter of blood can carry 12 mmol of O_2 ($58 \times 8.25 \div 40$; see Chapter 2, page 48).

- To deliver 190 mmol of O_2 per second, the horse's heart must pump 16 liters of blood per second.

TRANSIT TIME FOR A RED BLOOD CELL

- With a blood volume of 100 liters, delivery of 16 liters of blood per second requires each red blood cell to circulate in 6 seconds, on average.

OTHER DATA

- The horse breathes 6000 liters of air per minute, and its heart pumps close to 1000 liters of blood per minute.

  These enormous rates combined with the high blood viscosity (58% hematocrit) result in very high pulmonary arterial blood pressure, which can cause bleeding into the lungs. Administration of a pulmonary vasodilator (e.g., furosemide, or Lasix) before a race can minimize such bleeding.

Questions

(Discussions on pages 260-61)

3·10   If the horse in Veterinary Case 3·1 were to gallop for 24 hours, how much CO_2 would it produce?

3·11   If the horse obtained extra red blood cells from its spleen (the original blood doper) what is the minimum weight of its spleen at rest, and what signal do you think caused its spleen to contract?

3·12   In what ways might the production of lactic acid be advantageous to the horse during the race? Are there also disadvantages?

3·13   What adaptations occurred in the horse's muscle to permit it to extract so much O_2 during the race?

PART C
SUMMARY OF MAIN POINTS

- ATP can be regenerated from four major pathways, each with a different impact on the body.

  1. Creatine phosphate generates ATP at the fastest rate and yields a means of removing H^+.

  2. Anaerobic glycolysis regenerates ATP quickly but at a high cost—production of H^+.

  3. Aerobic glycolysis, which produces ATP at a modestly fast rate, requires delivery of O_2 but does not pose a problem with H^+.

  4. Oxidation of fatty acids proceeds at the slowest rate and is limited, for the most part, by the delivery of fatty acids.

- During a sprint, H^+ are produced rapidly. They can be buffered in muscle but bind to proteins in the process, diminishing the functions of the proteins.

- The lungs and heart operate in a coordinated, integrated fashion for optimal delivery of O_2 and removal of CO_2.

- Energy metabolism in skeletal muscle changes according to the length and intensity of exertion (Table 3·7).

  1. In the sprint, ATP is regenerated very quickly from CrP and anaerobic glycolysis. So many H^+ are produced in the latter pathway that their accumulation can limit performance.

  2. In the 1500 m race, glycogen is the major fuel, and ATP is regenerated by aerobic glycolysis in red muscle fibers (which are designed for endurance). Performance is limited primarily by the delivery of O_2 (accumulation of H^+ does not occur until the finishing kick).

  3. In the marathon, glycogen fails to meet the full needs for ATP, so fuels from the fat system must be used. As in the 1500 m race, H^+ do not cause problems. The rate at which fatty acids enter muscle cells is likely to affect performance.

Table 3·7
Metabolic considerations in three races

Event	Speed	Fuel used	Metabolic considerations	Limit
100 m sprint	10 m/s for 10 s	CrP	Lack of O_2	Amount of CrP; load of H^+
		Glycogen in muscle	Need for buffering	Buffering capacity
1500 m race	6 m/s for 240 s	Glycogen in muscle	Rate of delivery of O_2	Rate of aerobic regeneration of ATP from carbohydrates and possibly the removal of CO_2
		Glucose from the circulation	Rate of removal of CO_2	
Marathon	5 m/s for 130 min	Glycogen in muscle and the liver (earlier) and fatty acids (later)	Need for fatty acids to be delivered from adipose tissue	Rate of aerobic regeneration of ATP from fatty acids

SECTION TWO

Clinical Applications

CHAPTER 4
DIABETES MELLITUS

PART A
INSULIN AND METABOLISM OF GLUCOSE IN DIABETES MELLITUS

Diabetes mellitus, both insulin-dependent (IDDM) and noninsulin-dependent (NIDDM), results from a relative deficiency of insulin, the body's major coordinating signal for energy metabolism. Insufficient insulin causes the body to act as if it were in an uncontrolled state of starvation despite the intake of food; fuels are released from stores in excess of the body's needs, resulting in hyperglycemia, wasting of lean body mass, and, ultimately, ketoacidosis.

PART B
FACTORS AFFECTING THE SEVERITY OF HYPERGLYCEMIA AND KETOACIDOSIS

Hyperglycemia occurs only when the levels of insulin are so low that the metabolic pathways that remove glucose from the blood are inactivated. Provision of glucose from the diet or from catabolism of body proteins exacerbates the degree of hyperglycemia, as does decreased excretion of glucose in the urine.

Ketoacidosis is seen only when a lack of insulin and high levels of fatty acids in blood permit a rapid rate of ketogenesis. Ketoacidosis becomes very severe when oxidation of ketoacids by the brain or kidneys is diminished or when excretion of ketoacids in urine is reduced.

PART C
COMPLICATIONS OF HYPERGLYCEMIA

Persistent hyperglycemia damages tissues by producing osmotic effects and nonenzymatic *glycation* of proteins.

Glycation:
Covalent reaction of a sugar, such as glucose or fructose, with another compound.

PART D
CASE FOR REVIEW

PART E
SUMMARY OF MAIN POINTS

PART A
INSULIN AND METABOLISM OF GLUCOSE IN DIABETES MELLITUS

Case 4·1
Insulin Was Withheld from a Type I Diabetic
(Case discussed on page 105)

In the 1930s, Fred, a 19-year-old who had had insulin-dependent diabetes for 12 years, agreed to be taken off insulin so that the development of hyperglycemia and ketoacidosis could be followed. Changes in water, electrolyte, and metabolic balances were recorded before, during, and after withdrawal of insulin. Before the experiment, Fred had maintained good control of his weight and his glycemia by carefully monitoring both his diet and his daily doses of insulin. He maintained the same diet throughout the 12 days of the experiment, which included a four-day control period (during which insulin treatments were maintained), four days without insulin, and a four-day recovery period when insulin was readministered. What do you predict happened to the levels of glucose, ketoacids, sodium, potassium, urea, and ammonium in Fred's blood and urine during the course of the experiment?

Cases 1·5, 1·6, 1·7, 1·8, and 1·9 also examine the control by insulin of energy metabolism.

Questions to Consider

How important is the *antilipolytic action* of insulin in causing hyperglycemia?

How much muscle must be broken down to cause the excretion of 1 liter of urine through *osmotic diuresis*?

What dietary restrictions would be advisable for a patient with noninsulin-dependent diabetes mellitus?

What are the likely metabolic effects of withholding insulin from an insulin-dependent diabetic?

Why do some patients with a mild degree of hyperglycemia subsequently develop a very marked elevation in their concentrations of glucose?

Why does ketoacidosis become more severe when the patient is comatose?

How does hyperglycemia affect the distribution of water in the body? Can hyperglycemia cause direct damage?

Antilipolytic action:
An action that inhibits lipolysis, the breakdown of triacylglycerols in adipose tissue to release fatty acids and glycerol.

Osmotic diuresis:
Osmosis occurs across semipermeable membranes through which water, but few dissolved materials, can pass quickly. If solutions containing different numbers of particles (ions or any dissolved material) are placed on opposite sides of a semipermeable membrane, water will flow through the membrane from the weaker to the stronger solution, unless prevented by a pressure applied in the opposite direction.

Osmotic diuresis occurs when urine contains a large quantity of particles, such as glucose, that cause excessive excretion of water and electrolytes.

BACKGROUND

TYPES OF DIABETES MELLITUS

> *Diabetes mellitus is caused by a relative lack of insulin or resistance to its actions.*

Diabetes mellitus is a syndrome resulting from a relative lack of the metabolic actions of insulin (or resistance to these actions from elevated concentrations of counter-insulin hormones, such as glucagon). Diabetes mellitus has two major forms (Table 4·1). The crucial difference between these two types seems to be the degree to which the actions of insulin are lacking or the intensity of response to counter-insulin hormones.

- Type I, or insulin-dependent diabetes mellitus (IDDM), the more aggressive syndrome, occurs in younger patients and can cause diabetic ketoacidosis (DKA) if insulin is not administered or if the levels of counter-insulin hormones rise (e.g., due to infection, stress, associated illness, drugs).

- Type II, or noninsulin-dependent diabetes mellitus (NIDDM), tends to occur in obese, more elderly patients. The disease has an insidious onset, and these patients suffer primarily from the more long-range complications associated with *atherosclerosis;* it is very rare to find the acute metabolic condition of DKA in Type II diabetes mellitus.

The metabolic hallmark of both types of diabetes mellitus is hyperglycemia; a significant degree of ketoacidosis is a major factor in Type I diabetes mellitus but is rare in the Type II subgroup.

Atherosclerosis:

A degenerative disease in which lipids accumulate in the walls of blood vessels, narrowing their diameters and thereby compromising blood flow to the organs they serve.

Table 4·1

Main types of diabetes mellitus

Disease classification	Distinguishing features
Type I: insulin-dependent diabetes mellitus (IDDM) (lack of insulin)	Under 40 years at onset, usually a teenager or younger; Usually thin; Onset is acute or subacute; Prone to ketoacidosis; Requires insulin to maintain life
Type II: noninsulin-dependent diabetes mellitus (NIDDM) (resistance to the actions of insulin)	Over 40 at diagnosis; Generally obese; Gradual onset; Rarely ketotic; Positive family history; May not need insulin unless very strict control is desired
Pancreatic destruction (e.g., pancreatitis)	Symptoms of the underlying disease
β cell suppression (low insulin, but not really diabetes mellitus)	α-adrenergics (e.g., from a low ECF volume); Phaeochromocytoma (a tumor that produces adrenaline); Drugs (e.g., Dilantin, diuretics); Certain endocrinopathies

Question

(Discussion on pages 261–62)

4·1 Why is ketoacidosis a rare event in patients with Type II diabetes mellitus?

EFFECTS OF INSUFFICIENT INSULIN ON ENERGY METABOLISM

Lack of insulin signals the body to release fuels from energy stores despite dietary intake.

Insulin signals the body to store energy fuels and not to release them. Lack of insulin signals the body not to synthesize stores of energy fuels despite dietary intake. The levels of insulin in blood are normally closely controlled by—and also control—the concentration of glucose in blood (Figure 4·1).

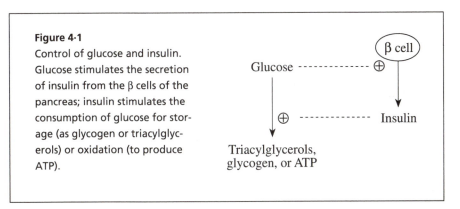

Figure 4·1

Control of glucose and insulin. Glucose stimulates the secretion of insulin from the β cells of the pancreas; insulin stimulates the consumption of glucose for storage (as glycogen or triacylglycerols) or oxidation (to produce ATP).

The effects of counter-insulin hormones (e.g., ACTH, adrenaline, and glucagon) offset the effects of insulin; we shall therefore use the term *low net concentration of insulin* to indicate a hormonal status in which the actions of insulin are reduced, either from low levels of insulin or from high levels of counter-insulin hormones. When the net concentration of insulin is low, excess glucose, fatty acids, and amino acids are released from stores, resulting in hyperglycemia (see Figures 1·9 and 1·10, page 17, and pages 29–31).

Low net concentration of insulin: The hormonal status that indicates a lack of insulin or resistance to its metabolic actions by high levels of counter-insulin hormones.

Hyperglycemia

The most important effect of a lack of insulin is inhibition of the metabolic mechanisms to remove glucose.
Any formation or ingestion of glucose is too much when removal of glucose has slowed considerably.

A low net concentration of insulin causes fatty acids to be released at an increased rate from triacylglycerol stores in adipose tissue (see Chapter 11, page 231), thus providing an excess of fat-derived fuels (fatty acids and ketoacids) for regeneration of ATP. These fat-derived fuels inhibit the rate of oxidation of glucose in the regeneration of ATP in muscle and the brain (Figure 4·2), causing glucose to accumulate in the blood. A low net concentration of insulin directly inhibits storage of the excess glucose as either glycogen in the liver or as triacylglycerols in adipose tissue (see Chapter 8, page 207 and Chapter 11, page 229).

Not only does a low net concentration of insulin prevent glucose from being oxidized or stored but it also increases the rate at which glucose is formed by promoting breakdown of glycogen in the liver and of proteins in muscle (Figure 4·3); the liver releases glucose directly into the blood until glycogen reserves are exhausted, and muscle provides the substrate (amino acids) for gluconeogenesis. The increased rate of breakdown of triacylglycerols also releases glycerol, which is converted to glucose by the liver.

Figure 4·2

Control of the oxidation of glucose. A relative insufficiency of insulin inhibits regeneration of ATP from glucose.

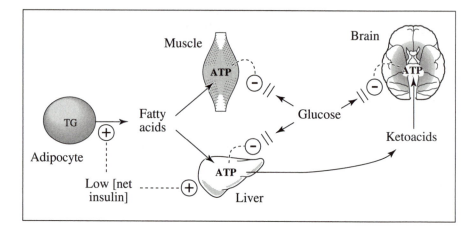

These effects of a low net concentration of insulin—inhibiting consumption of glucose and increasing production of glucose—set the stage for hyperglycemia. As long as the amount of insulin remains too low to control the concentration of glucose in the blood, hyperglycemia will persist. Because the total amount of glucose in the body is quite small (Table 4·2), minor changes in the rates of addition to this pool or removal from this pool have rapid effects.

Figure 4·3

Control of the production of glucose. A relative insufficiency of insulin stimulates the production of glucose.

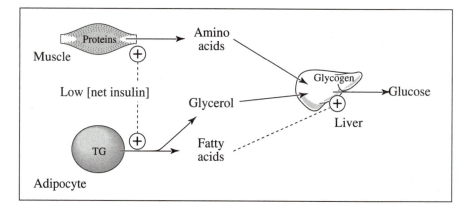

Ketoacidosis

Because it permits hormone-sensitive lipase in adipose tissue to be active, a low net concentration of insulin results in increased levels of fatty acids in blood. This hormonal setting also affects hepatocytes, stimulating the rapid formation of ketoacids in the liver (see Chapter 11).

Table 4·2

Daily supply of glucose, consumption of glucose, and pool size of glucose in a normal adult and in a diabetic in poor control

Glucose	Normal mmol/day*	Diabetic in poor control mmol/day*
Supply		
Diet (carbohydrates)	1500	1500
Breakdown of proteins	350	300
From glycerol	5	10
Consumption		
Oxidation to CO_2	1800	< 200
Conversion to triacylglycerols	50	0
Excretion in urine	0	0–2000
Pool size		
Free in body	80 (5 mmol/l)	800 (50 mmol/l)
In liver glycogen	550	Very low
In muscle glycogen	2500	Low or normal

* Values are approximate.

Questions

(Discussions on pages 262–63)

4·2 Under what circumstances might there be excessive breakdown of proteins in the body?

4·3 The urine of a hyperglycemic patient contains 300 mmol of glucose per liter. How much muscle must this patient be breaking down to permit the excretion of one liter of urine if all the glucose is derived from proteins in muscle?

4·4 Hyperglycemic patients usually excrete a predictable amount of glucose in their urine; why might the amount excreted be less than expected?

Discussion of Case 4·1
Insulin Was Withheld from a Type I Diabetic

(Case presented on page 101)

Fred's withdrawal of insulin caused rapid release of fatty acids from triacylglycerols in adipose tissue and rapid inhibition of pathways consuming glucose—both direct effects of a low net concentration of insulin.

Table 4·3 shows Fred's blood chemistry and values for excretion of fuels, ions, and nitrogenous wastes in urine. After withdrawal of insulin, he developed hyperglycemia, a high level of ketoacids, hyperkalemia, and a high level of urea; hyponatremia was also present.

Table 4·3

Values in Fred's blood and urine during the course of the experiment

	Control period Blood	Control period Urine*	Withdrawal of insulin Blood	Withdrawal of insulin Urine**	Recovery period Blood	Recovery period Urine*
			All values are in mmol/l.			
Glucose	6	0	15	300	7	0
Ketoacid anions	<0.05	0	8	110	0.1	0
Na^+	140	150	137	48	140	138
K^+	4.0	60	5.2	21	3.6	10
Urea	5.0	400	8.1	200	4.1	280
NH_4^+	<0.1	30	<0.1	80	<0.1	48

* The 24-hour urine volume was 1 liter.
** The 24-hour urine volume was 2 liters on the third day of withdrawal of insulin.

PART B
FACTORS AFFECTING THE SEVERITY OF HYPERGLYCEMIA AND KETOACIDOSIS

Case 4·2
Aunt Agnes Has Diabetes Mellitus in Poor Control
(Case discussed on page 112)

Agnes, 70 years old and a noninsulin-dependent diabetic for over 20 years, has never needed insulin but has had to watch her diet. Last week, her doctor noted hypertension and prescribed a thiazide diuretic to lower her blood pressure. When her urine output increased, she became very thirsty so she drank a lot of apple juice. Her weight fell by several pounds, and she complained of dizziness when she stood. No acetone was detected on her breath.

This week, routine laboratory tests found severe hyperglycemia, hyperosmolality, and hyperkalemia but only mild ketoacidosis (Table 4·4) and essentially normal excretion of urea. How do you explain these observations? What treatment should she receive?

Case 4·3
Jimmy Did Not Take His Insulin Yesterday
(Case discussed on pages 112–13)

Jimmy, 17 years old, has diabetes mellitus but did not take insulin yesterday because he was feeling too sick to eat. Ketoacidosis developed (Table 4·4). Why did ketoacids accumulate? What could cause the degree of ketoacidosis to become much more severe?

Case 4·4
Why Did the Concentration of Glucose in Lillian's Blood Fall?
(Case discussed on pages 113–14)

Lillian, 23 years old, had diabetic ketoacidosis—very high concentrations of glucose, fatty acids, and ketoacids in her blood (Table 4·4) and a very low extracellular fluid (ECF) volume. She received an infusion of insulin and 6 liters of isotonic saline, which expanded her ECF volume from 10 to 15 liters. She then excreted 1 liter of urine over four hours. The levels of fatty acids and ketoacids in her blood were still elevated four hours later, but the concentration of glucose had fallen to 24 mmol/l. What primarily caused the fall in the concentration of glucose in Lillian's blood?

Table 4·4
Laboratory values on admission in cases 4·2, 4·3, and 4·4

Values in plasma		Cases		
		4·2 Agnes	4·3 Jimmy	4·4 Lillian
Glucose	mmol/l (mg/dl)	50 (900)	25 (450)	50 (900)
Urea	mmol/l (mg/dl)	30 (84)	20 (56)	30 (84)
Ketoacids	mmol/l	3	20	8
HCO_3^-	mmol/l	25	5	12
Anion gap	mEq/l	14	32	25
pH		7.40	7.00	7.25

BACKGROUND

CONTROL OF THE CONCENTRATION OF GLUCOSE IN BLOOD

Changes in the concentration of glucose in blood result from removal or addition of glucose. Glucose can be removed from blood to regenerate ATP, it can be converted to stores of glycogen or triacylglycerols, and it can be excreted in the urine. Glucose can be added to blood from the diet or from *glucogenesis* from amino acids, lactate (derived from glycogen, not glucose), or glycerol.

When the net concentration of insulin is normal, the concentration of glucose in blood remains between 3 and 8 mmol/l (Table 4·5), even while a heavy, carbohydrate-rich meal is being absorbed (see Cases 1·8 and 1·11). Excess dietary sources of glucose or its precursors will not cause the concentration of glucose in blood to exceed 8–10 mmol/l in a healthy person. Chronic hyperglycemia and *glycosuria* arise only in conditions involving a low net concentration of insulin; acute hyperglycemia may arise from intestinal disorders such as dumping syndrome (Case 1·11, pages 32 and 34–35) or from rapid intravenous infusion of glucose.

Glucogenesis:
Synthesis of glucose from precursors.

Glycosuria:
The presence of glucose in the urine.

Table 4·5
Normal and abnormal concentrations of glucose in blood and urine

	Blood		Urine	
	mmol/l	mg/dl	mmol/l	g/l
Healthy person				
Before eating meal	3–4	55–70	0	0
Maximum while eating	7–8	130–150	0	0
2–3 hours after absorption	3–5	55–90	0	0
Diabetic patient				
Untreated Type I diabetes mellitus	18–50*	180–900	300	55
Untreated Type II diabetes mellitus	10–15*	180–270	0–300	0–55

* Representative values.

With a low net concentration of insulin, the increased supply of fat-derived fuels (fatty acids and ketoacids) can provide almost all of the body's requirements for regeneration of ATP; the brain still oxidizes glucose, but for only about 20% of its needs for energy (see Chapter 1, page 30). The body's capacity to store glucose carbon as glycogen or triacylglycerols is largely lost, yet its ability to make glucose is enhanced so that glucogenic substrates (amino acids, glycerol, lactate) are converted to glucose. Thus, any provision of glucose or its precursors—from the diet or from breakdown of stored proteins—in excess of the small amount used for the brain will lead to hyperglycemia.

Figure 4·4 (see page 108) shows that the degree of hyperglycemia depends on the rates at which glucose is added from the diet and synthesized from body proteins. The former can be measured, and synthesis from dietary and body proteins can be calculated from the nitrogen balance (see Question 4·6). When the level of ketoacids in plasma is greater than 4 mmol/l, all but approximately 25 g of the glucose that is ingested or synthesized must appear in the urine if the concentration of glucose in blood is to achieve a steady state (only the brain uses glucose in this condition, and oxidation of ketoacids provides most of the energy required by the brain). Limitation of dietary glucose (or its precursors) and active urinary excretion of glucose are therefore essential for maintaining some degree of control over hyperglycemia caused by a low net concentration of insulin.

Figure 4·4

Illustrative values for daily disposition of glucose among tissues. When there is a low net concentration of insulin, the body's ability to remove glucose from the blood decreases, and glucose accumulates. Some glucose is oxidized in the brain; the rest either accumulates in the body (largely in the ECF) or is excreted in the urine.

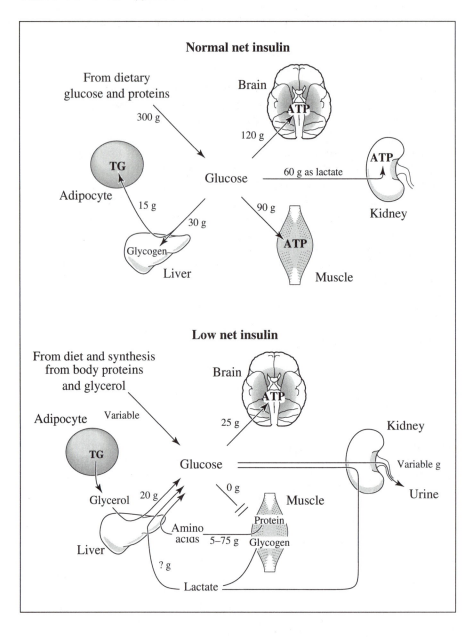

Questions

(Discussions on page 263)

4·5 Assume that a compound X inhibits its own formation and stimulates its own use (see page 16 and also Chapter 7, pages 171–74) and that the capacities for its formation (V_1) and use (V_2) are also subject to external controls.

What influences on V_1 and/or V_2 will cause the [X] to rise or fall? What conditions would lead to an uncontrolled increase in the [X]? What conditions are required for the [X] to change to new steady states?

4·6 Nitrogen makes up one-sixth of the weight of protein, and 100 g of protein can yield 60 g of glucose. How many millimoles of glucose can be formed per millimole of urea or ammonium excreted in urine?

CONTROL OF THE CONCENTRATION OF KETOACIDS IN PLASMA

Rate of Formation of Ketoacids

> *Ketoacids are not formed rapidly, even at the maximal rate. Thus, for ketoacids to reach high levels, their normal rate of removal must be compromised.*

Ketoacids are formed during the partial oxidation of fatty acids in mitochondria of liver cells. Since oxidation of fatty acids requires generation of ATP (equation 1), the liver's rate of consumption of ATP places an upper limit on the rate of ketogenesis. Ketogenesis is influenced both by the amount of hepatic work and by the rate of regeneration of ATP from other fuels. Hence, ketogenesis, even at its maximum, is a relatively slow pathway (close to 1 mmol of ketoacids are formed per minute).

Since ketoacids are formed from neutral substrates (triacylglycerols), the $[H^+]$ in blood rises, requiring buffering by the bicarbonate and non-bicarbonate buffer systems (see Chapter 2, pages 44–48).

$$\text{Fatty acid } (C_{16}) + 30\text{--}40 \text{ ADP} + P_i \longrightarrow 4 \text{ Ketoacids} + 30\text{--}40 \text{ ATP} \quad (1)$$

Rate of Removal of Ketoacids

> *Ketoacids are mainly removed via oxidation in the brain.*

When the concentration of ketoacids in blood has built up to approximately 5 mmol/l, ketoacids can be oxidized to regenerate ATP in two major organs—the brain and the kidneys. Normally, the brain removes about 750 mmol of ketoacids per day (half the normal rate of synthesis), and the kidneys oxidize approximately 250 mmol/day (Figure 4·5). The kidneys can also excrete ketoacids—mainly as their NH_4^+ salts—to eliminate H^+ (approximately 150 mmol/day).

In addition, acetoacetate, one of the ketoacids, decomposes spontaneously to acetone, which accounts for the fruity odor on the breath of a person with florid diabetic ketoacidosis. Not only does this decomposition remove ketoacid anions (about 150 mmol/day), it also removes an equivalent amount of H^+ (equation 2).

$$\text{Acetoacetate} + H^+ \longrightarrow \text{Acetone} + CO_2 \quad (2)$$

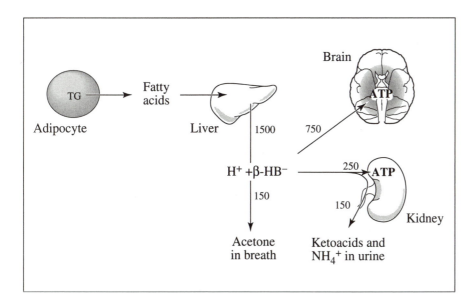

Figure 4·5
Daily ketoacid balance with a low net concentration of insulin. The brain and the kidneys remove most of the ketoacids (1150 mmol/day) that the liver forms in response to a low net concentration of insulin. Acetone in breath removes another 150 mmol/day. Other organs, such as muscle and the intestines (not shown in the figure), consume the remaining 200 mmol of ketoacids per day.

Question

(Discussion on page 264)

4·7   What major pathways besides ketogenesis lead to the regeneration or utilization of ATP in the liver? How might changes in the rates of these pathways affect the rate of ketogenesis?

Factors Determining the Degree of Ketoacidosis

Ketoacidosis poses an acid-base threat when ketoacids are formed at a rate that is more rapid than their oxidation. Ketogenesis achieves its maximal potential rate when the concentration of fatty acids in blood exceeds 800 µmol/l (as seen in conditions involving a low net concentration of insulin, such as fasting or diabetes mellitus). Under such conditions, the rate of ketogenesis is affected by the liver's need to regenerate ATP. Liver damage or alternative sources of ATP (e.g., from gluconeogenesis from amino acids) can limit ketogenesis and hence ketoacidosis.

Because the brain and the kidneys are the major routes for consumption of ketoacids, decreased brain function (from a coma, anesthesia, or a stroke) or kidney function (from contraction of ECF volume, kidney damage, or a nephrectomy) will decrease oxidation of ketoacids and hence exacerbate ketoacidosis.

INFLUENCE OF KIDNEY FUNCTION ON THE DEGREE OF HYPERGLYCEMIA AND KETOACIDOSIS

In making urine, the kidneys first prepare a filtrate of blood and then reabsorb the fuels and ions that are valuable for the body (Chapter 2, pages 62–63). The kidneys are unable to reabsorb all the glucose or ketoacids in the filtrate when the levels of these fuels in blood rise above 10 and 2–3 mmol/l, respectively; these concentrations are called the *renal thresholds* for these fuels. Tables 4·6 and 4·7 show that these fuels appear in urine when their concentrations in blood rise above threshold levels.

Renal threshold:

When the concentration of a compound in plasma reaches this level, the compound is excreted in the urine.

Excretion of these fuels diminishes the degree of hyperglycemia and ketoacidosis (if the anions are excreted with H^+ or NH_4^+) but at the cost of an increased urine volume (diuresis), which carries with it an increased excretion of inorganic ions (Na^+, K^+, Cl^-). This diuresis is primarily due to the osmotic effects of the high urinary concentrations of glucose and ketoacids. By reducing the extracellular fluid (ECF) volume, osmotic diuresis creates a vicious circle for the degree of hyperglycemia. A decreased ECF volume not only reduces the flow of blood to the kidneys but also lowers the glomerular filtration rate (Chapter 2, page 64) and the rate of excretion of glucose, thereby accentuating the hyperglycemia and increasing the pressure on the kidneys to excrete Na^+ and K^+. Osmotic diuresis can cause kidney failure, thus removing the body's ability to diminish hyperglycemia and ketoacidosis through excretion of excess fuels in the urine.

Table 4·6

Concentrations of ketoacids and H^+ in blood and urine

	Blood		Urine	
	[Ketoacids] mmol/l	[H⁺] nmol/l	[Ketoacids] mmol/l	[NH₄⁺] mmol/l
Healthy person				
Normal fed state	0.05	40	0	30
After 5–25 days without food	5	45	150	150
Diabetic patients				
Untreated Type I diabetic	5–20[a]	50–100+	100–300[b]	50–150[c]
Untreated Type II diabetic	1–5[a]	40–45	0–150	50–150[c]

a Representative values   b Depends on glomerular filtration rate   c Depends on underlying kidney disease

Table 4·7

Illustrative concentrations and amounts of fuels, nitrogenous wastes, and ions in the urine of people with normal and low net concentrations of insulin

Net concentration of insulin	Concentration (mmol/l)		Amount (mmol/day)	
	Normal	Low	Normal	Low
Urinary volume (liters)	1	2–5	1	2–5
Glucose	0	100–300	0	200–500
Ketoacid anions	0	50–100	0	200
Urea	500	100–200	500	400
NH_4^+	40	40–100	40	40–200
Na^+	150	50	150	100–250
K^+	50	30	50	60–150
Cl^-	150	0–50	150	100
H^+	0.001 (pH 6.0)	0.01 (pH 5.0)	0.001	0.01
HCO_3^-	0–50	0	0–50	0

DIAGNOSIS OF CAUSES OF SEVERE HYPERGLYCEMIA

Hyperglycemia occurs when a low net concentration of insulin blocks the metabolic utilization of glucose. The severity of hyperglycemia can vary widely, depending on the rate at which glucose is excreted in the urine, the amount of glucose ingested, or the rate at which glucose is formed from endogenous precursors (usually proteins).

Figure 4·6 offers a diagnostic flow chart for possible causes of severe hyperglycemia in a patient with a low net concentration of insulin. Since excretion of glucose in the urine limits the degree of hyperglycemia, the first step in diagnosis is to ascertain the rate of urine flow. Patients with a urine flow less than expected (approximately 2 ml/min in an adult—a result of osmotic diuresis induced by hyperglycemia) probably have impaired renal function; renal disease could be the cause, as could contraction of the ECF volume, involving a negative balance for Na^+ (5 mmol/kg) from prior osmotic diuresis. Observing hyperglycemia with a rapid urine flow suggests that glucose is being added to the blood. A high level of urea in urine (a strongly negative nitrogen balance) indicates that proteins have been broken down for the production of glucose. Without breakdown of proteins, excess dietary intake is the probable cause of an overproduction of glucose.

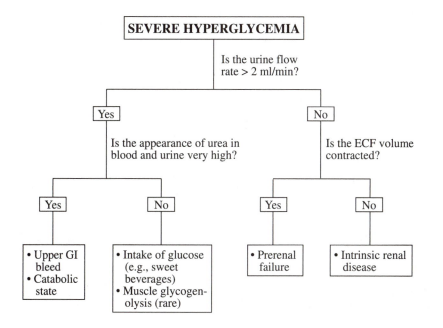

Figure 4·6

Diagnostic flow chart for causes of severe hyperglycemia accompanying a low net concentration of insulin.

Discussion of Case 4·2
Aunt Agnes Has Diabetes Mellitus in Poor Control

(Case presented on page 106)

Agnes's problem arose when she changed her medications. Her noninsulin-dependent diabetes mellitus had been controlled for many years through the diet, and though her net concentration of insulin was low, she had some insulin, which had maintained her metabolism in a stable state (with careful dietary control). Taking the thiazide diuretic lowered her insulin levels further. Her response—thirst satiated by consumption of a large quantity of glucose-containing fluid—caused her to produce urine at an increased rate. The urine contained a high concentration of glucose; the normal excretion of urea indicates that this addition was not from catabolism of proteins (Figure 4·6). Dietary ingestion of glucose from the apple juice she drank (750 mmol of glucose per liter) is a probable explanation.

Treatment for Agnes should aim to control the hyperglycemia and the hyperkalemia but avoid the possibility of making her hypokalemic. Hyperosmolality can be expected to result from the hyperglycemia, and the two should resolve together. During continued osmotic diuresis, attention must be paid to maintaining the ECF volume (administer NaCl).

The main factor lowering the concentration of glucose in her blood will be excretion of glucose in urine. A large infusion of saline is the major line of therapy. The mild ketonemia should rapidly disappear, thus making Agnes dependent on glucose for all regeneration of ATP. Hence, her blood glucose should be followed carefully, and glucose should be administered when its concentration falls toward normal levels.

Insulin will help the cells take up the extra K^+ in plasma (lack of insulin caused some K^+ to spill into the blood, and the apple juice provided an extra 25 mmol/l); levels of K^+ in plasma should be followed so that K^+ can be infused if needed.

Discussion of Case 4·3
Jimmy Did Not Take His Insulin Yesterday

(Case presented on page 106)

Why did ketoacids accumulate?

Accumulation of any metabolite results from an increased rate of production or a decreased rate of utilization.

INCREASED KETOGENESIS

Ketogenesis has two important controls mediated by a low net concentration of insulin. One control activates hormone-sensitive lipase in adipose tissue (Chapter 11), thereby increasing the rate of delivery of fatty acids to the liver. Uptake of fatty acids by the liver is directly proportional to the concentration of fatty acids in the blood (see Chapter 11). The other control mediated by a low net concentration of insulin occurs in the liver and requires an increased rate of oxidation of fatty acids. Ketogenesis has an upper limit because ATP is a necessary by-product of the process.

Assume that Jimmy's liver can produce 1600 mmol of ketoacids per day, the normal maximum for an adult. Jimmy's blood contains 20 mmol of ketoacids per liter; with a *volume of distribution* of 30 liters, he has accumulated 600 mmol. Given that the brain oxidizes close to half of the ketoacids produced and the kidneys remove 400 mmol of ketoacids per day (for a total of approximately 1150 mmol/day), net accumulation should be 450 mmol. Because Jimmy's blood contains an additional 150 mmol, his liver would have to have produced ketoacids at the maximal rate for more than a day to achieve this degree of ketoacidosis.

Volume of distribution:

The volume of body water (usually 60% of body weight) to which a substance has access. Urea, for example, is one of the few compounds that is distributed evenly in all body water. Glucose, on the other hand, is evenly distributed in the ECF and in some cells (e.g., liver), but not others (e.g., muscle); hence, its volume of distribution is close to half of body water. Other examples are Na^+ (largely in the ECF) and ketoacids (in two-thirds of body water).

DECREASED UTILIZATION OF KETOACIDS

The brain oxidizes as much as 750 mmol of ketoacids per day when levels in the blood are above 4–5 mmol/l. The brain's demand for ATP is decreased if the patient is comatose, is under anesthesia, or has taken drugs that retard cerebral metabolism (see Chapter 6, pages 156–57). The kidneys also oxidize ketoacids (250 mmol/day), though they require less ATP when the glomerular filtration rate (GFR) falls (Chapter 2, page 64). Hyperglycemia causes osmotic diuresis, which results in a loss of Na^+ in the urine. Having fewer Na^+ in the body lowers the circulating volume (and therefore blood pressure), leading to a decreased GFR. Because the kidneys then filter and reabsorb fewer Na^+, they use less ATP.

What could cause the degree of ketoacidosis to become much more severe?

A low demand for ATP in either the brain or kidneys can greatly increase the degree of ketoacidosis. This occurrence could explain why ketoacidosis becomes progressively more severe with time, when coma and contraction of extracellular fluid (ECF) volume become likely.

Question

(Discussion on page 264)

4·8 Your patient has diabetes mellitus and presents with moderately severe hyperglycemia. Two strategies have been proposed as adjuncts to insulin therapy to lower the concentration of glucose in the blood—inhibition of glucogenesis and inhibition of fatty acid oxidation. What are the dangers with these additional lines of therapy?

Discussion of Case 4·4
Why Did the Concentration of Glucose in Lillian's Blood Fall?

(Case presented on page 106)

What primarily caused the fall in the concentration of glucose in Lillian's blood?

Since Lillian still had high concentrations of fatty acids and ketoacids in her blood, they were likely to be the major fuels oxidized to regenerate ATP. Hence, oxidation of glucose was not likely to be the cause for the decreased concentration of glucose. So what was the cause? Table 4·8 shows essential data on diagnosis and four hours after treatment.

Table 4·8
Volume of distribution of glucose

	Fluid accessible to glucose (liters)	
	Before therapy	**After therapy**
ECF	10	15
Insulin-insensitive tissues	5	5*
Total volume of distribution	15	20

* Assume no change; the real change is quite small.

DILUTION OF GLUCOSE

Therapy caused her extracellular fluid (ECF) volume to increase from 10 to 15 liters. Also, free glucose was present at the same concentration as in the ECF in organs such as the liver and kidneys (another 5 liters of fluid) but not in muscle (see Figure 4·8, page 116). Before treatment, her body contained 50 mmol/l × 15 liters, or 750 mmol of glucose; diluting this amount in 20 liters caused the concentration of glucose to fall to 37.5 mmol/l.

LOSS OF GLUCOSE IN URINE

Contraction of the ECF volume had essentially stopped the formation of urine, but therapy restarted it. One liter of urine allowed excretion of 250 mmol of glucose, lowering her body's glucose content to 500 mmol/20 liters, or 25 mmol/l (Table 4·9).

REDUCED SYNTHESIS OF GLUCOSE

The amount of glucose synthesized from proteins and glycerol is very small compared with the losses occurring from urinary excretion. The amounts from protein can be calculated from the change in the total amount of urea in the body and urine over time. Table 4·9 shows that the concentration of urea in her plasma was 30 mmol/l. Since urea is distributed over the whole body (assume 40 liters), the total content was 1200 mmol. Four hours after treatment, the concentration in her plasma had fallen to 24 mmol/l, distributed over 44 liters (5 liters were infused and 1 liter was excreted), and her urine contained 250 mmol (Table 4·9). She therefore formed 106 mmol of urea. Since synthesis of glucose from proteins releases close to 1.7 mmol of urea per mmol of glucose, only 60 mmol of glucose were added from gluconeogenesis.

Treatment caused the concentration of glucose in her blood to decrease, primarily from dilution and excretion. The insulin had very little effect on the oxidation of glucose because of the high concentrations of fatty acids and ketoacid anions. Once the levels of ketoacids and fatty acids in blood decline, the concentration of glucose will decrease rapidly as a result of oxidation. To prevent this decrease, glucose should be administered about six hours after giving insulin. Case 4·6 considers the reasons for delayed oxidation of glucose.

Table 4·9
Effects of insulin plus saline on blood and urine

	On diagnosis	Four hours after treatment
Fluid volume (liters)		
ECF	10	15
Urine (excreted in four hours)	No urine	1
Concentration of glucose (mmol/l)		
Blood	50	25
Urine	No urine	250
Concentration of urea (mmol/l)		
Blood	30	24
Urine	No urine	250

PART C
COMPLICATIONS OF HYPERGLYCEMIA

 Case 4·5
Side Effects of Hyperglycemia

Over a period of three weeks, you see patients with diabetes mellitus who have developed cataracts, serious vision problems, renal failure, peripheral nerve damage, and poor peripheral circulation. All are patients who have suffered from insulin-dependent diabetes for at least 20 years. Is there a common explanation?

Discussion of Case 4·5
Side Effects of Hyperglycemia

> *Osmotic effects of hyperglycemia are loss of water and electrolytes.*
> *Metabolic complications of hyperglycemia can be due to glycation and possibly the formation of sorbitol.*

Sorbitol:
A six-carbon alcohol made from glucose.

Fructose:
A sugar (with the same chemical composition as glucose) that is readily converted to glucose by the liver. Sucrose is a disaccharide containing glucose and fructose.

Hyperglycemia has a number of important side effects that arise from osmotic effects on cells and urine and also from metabolic complications, which can result from the nonenzymatic reaction of glucose (and *fructose*) with proteins.

OSMOTIC EFFECTS OF HYPERGLYCEMIA ON CELLS

Water crosses cell membranes very rapidly to achieve osmotic equilibrium (Figure 4·7). Molecules such as urea cross the membrane rapidly, and their concentrations are approximately equal in the intracellular fluid (ICF) and the extracellular fluid (ECF). Hence, urea does not have a significant effect on the distribution of water.

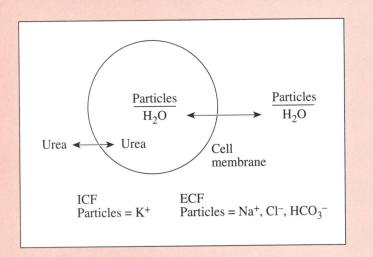

Figure 4·7
Regulation of the movement of water across cell membranes. The circle represents a cell membrane. Water crosses this membrane very rapidly to achieve osmotic equilibrium. Molecules such as urea also cross the membrane rapidly, and their concentrations are equal in the ICF and the ECF; hence, they do not alter the distribution of water. The major chemical particles restricted largely to the ECF are Na^+, Cl^-, and HCO_3^-; the particles restricted to the ICF are predominantly K^+, which are held inside cells by macromolecular anions—organic phosphates and certain proteins (minor).

Other materials have limitations. Some common inorganic materials are confined largely to either the ECF or the ICF; whereas Na^+, Cl^-, and HCO_3^- are present in much higher concentrations in the ECF, K^+ are concentrated in the ICF.

115

Glucose can cross cell membranes only on specific carriers. In some cells (e.g., liver), transport of glucose is rapid, and its concentrations inside and outside are very similar. In other tissues (e.g., muscle), transport of glucose is slow but strongly stimulated by insulin; concentrations of glucose inside muscle are always much lower than those in the ECF. Figure 4·8 shows that hyperglycemia causes muscle cells to shrink and liver cells to expand.

Figure 4·8

Effect of glucose on shifts of water in muscle and the liver. The concentration of glucose in the ECF is always much higher than that in muscle cells; hyperglycemia thus pulls water out of muscle. The concentration of glucose in liver cells is always equal to that in the ECF; hence, hyperglycemia has no effect on the distribution of water between the liver and the ECF. However, a lower concentration of Na^+ in plasma can be caused by the movement of water from muscle to the ECF. This hyponatremia causes cells of the liver to swell. The dotted lines reflect the new ICF volumes under the influence of severe hyperglycemia.

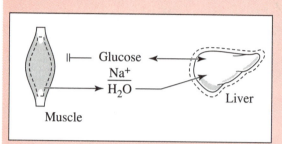

The volume of cells in the brain does not change appreciably in hyperglycemia, since two opposing forces of roughly equal magnitude are acting by the time the patient presents to the hospital. The outward force is due to the higher concentration of glucose in the ECF. The inward force is hyponatremia (lower tonicity of the ECF resulting from a low concentration of Na^+ in the ECF), which causes a shift of water from muscle cells to brain cells.

OSMOTIC EFFECTS OF HYPERGLYCEMIA ON URINE

Hyperglycemia, at levels above the renal threshold (10 mmol/ml of blood), causes glycosuria because the kidneys cannot reabsorb all the glucose that is filtered. Glucose in urine "holds" water and electrolytes, thereby preventing their normal reabsorption and causing osmotic diuresis (the loss of water, Na^+, and K^+ in the urine). Urine in uncontrolled diabetes mellitus contains, on average, 300 mmol (55 g) of glucose per liter, 50 mmol of Na^+ per liter, and a low concentration of K^+ (20–30 mmol/l). Loss of electrolytes from hyperglycemia can be a major threat to diabetics. Contraction of the ECF volume (loss of Na^+) can lead to poor circulation and, ultimately, to circulatory collapse (i.e., shock).

FORMATION OF SORBITOL

The enzyme aldose reductase, which can reduce glucose to sorbitol, has a high K_m (low affinity) for glucose and thus acts faster in the presence of hyperglycemia. Sorbitol, which is inert, is metabolized very slowly, so that it accumulates in some cells, such as those of the lens, causing them to swell (Figure 4·9). In the lens, swelling of cells prevents normal changes of shape (thereby affecting the ability of the eye to focus) and may even lead to the formation of cataracts, from which elderly diabetics commonly suffer.

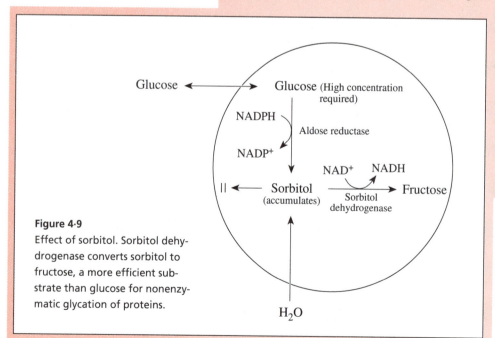

Figure 4·9

Effect of sorbitol. Sorbitol dehydrogenase converts sorbitol to fructose, a more efficient substrate than glucose for nonenzymatic glycation of proteins.

NONENZYMATIC GLYCATION

Glucose and fructose can react, nonenzymatically and ultimately irreversibly, with the exposed side chain amino group of lysine (in proteins) and can also cross-link two such lysine groups (Figure 4·10). The rate of this process, which is called glycation, increases with higher concentrations of glucose or fructose. Glycation changes the charge, structure, and function of these proteins and seems to contribute to the complications associated with diabetes mellitus. Although *macrophages* destroy the products of glycation, incomplete removal of these altered proteins may contribute to some of the degenerative changes of aging.

Macrophages:
White blood cells that eliminate bacteria, dead cells, or unwanted materials in the blood.

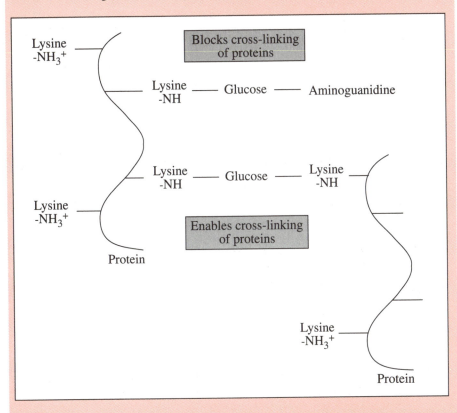

Figure 4·10
Nonenzymatic glycation. Drugs such as aminoguanidines, which resemble lysine, are being designed to prevent one lysine from cross-linking to a second lysine by providing alternate substrates. These drugs may retard the development of some of the complications of hyperglycemia.

The degree of glycation can be of diagnostic importance. Measuring the concentration of glycated albumin indicates the extent to which the body has been exposed to unusual degrees of hyperglycemia during recent weeks (albumin in serum has a half-life of one to two weeks). Likewise, the concentration of glycated hemoglobin indicates the level of exposure to hyperglycemia over the past three months (red blood cells last for approximately 120 days).

Question

(Discussion on page 265)

4·9 An analysis of two samples of blood gives the following results:

	Glycated albumin	Glycated hemoglobin
Case 1	Normal	High
Case 2	High	High

What do these results reveal?

PART D
CASE FOR REVIEW

Case 4·6
Persistent Hyperglycemia During Six Hours of Treatment

Graham, a patient in the emergency room, said that he had been drinking heavily for the past week and complained of severe pain in his upper abdomen for the past two days. He denied taking any drugs. His hospital record showed a long history of alcoholism (with resulting pancreatitis) but no diabetes mellitus. The physician found that he had a marked contraction of his extracellular fluid (ECF) volume, an extremely tender abdomen, and no bowel sounds. Neither acetone nor ethanol was detected on his breath. Laboratory data showed marked hyperglycemia and a very high concentration of urea in his plasma.

Treatment consisted of adrenaline and glucocorticoids to support his blood pressure and infusion of 6 liters of isotonic saline, which overexpanded his ECF volume. Because of the acute abdomen, surgery was performed, but it did not reveal a specific lesion. His abdominal pain was a side effect of a high concentration of triacylglycerols in plasma.

Graham's hyperglycemia persisted until at least 10 hours after the initial treatment. During six hours (from 4 to 10 hours after treatment), he passed 1.2 liters of urine, containing 300 and 480 mmol/l of glucose and urea, respectively (Table 4·10). What were his sources of glucose during this six-hour period? Was the amount of urea in his urine sufficient to account for his high rate of excretion of glucose? What if much less urea had appeared in his urine and blood?

Table 4·10

Graham's blood chemistry and urine chemistry

		On admission	At 4 hours	At 10 hours
Plasma				
Glucose	mmol/l	110	50	50
Urea	mmol/l	50	20	20
			Between 4 and 10 hours after admission	
Urine				
Glucose	mmol/l	None	300	
Urea	mmol/l	None	480	

Discussion of Case 4·6
Persistent Hyperglycemia During Six Hours of Treatment

Graham's problem was persistent hyperglycemia despite a high rate of excretion of glucose in his urine. His urine also contained high levels of urea.

What were his sources of glucose during this six-hour period?

Graham's persistent glycosuria with excretion of glucose (Table 4·10) shows that he must have been releasing glucose into his blood. He excreted 576 mmol of urea (480 mmol/l × 1.2 liters) in the six hours between 4 and 10 hours after treatment. This excretion, along with constant blood levels, suggests that the glucose was coming from proteins.

Was the amount of urea in his urine sufficient to account for his high rate of excretion of glucose?

Yes, 100 g of protein yields 60 g (330 mmol) of glucose and 16 g (1140 mmol) of nitrogen, which is equivalent to 570 mmol of urea—very close to the amount that Graham actually excreted. This breakdown of proteins could have been caused by diabetes mellitus in poor control, administration of hormones that promote catabolism (he received glucocorticoids and adrenaline, which have this effect), loss of blood into the gastrointestinal tract (a possible result of alcohol abuse), or another underlying disease.

What if much less urea had appeared in his urine and blood?

When urine contains a low level of urea, look for a source of glucose that does not contain nitrogen—either glycogen or glucose in the lumen of the GI tract (recall that Graham did not produce bowel sounds initially, but his GI peristalsis might have recovered with time). A delayed fall in the concentration of glucose may not reflect a resistance to the metabolic actions of insulin.

PART E
SUMMARY OF MAIN POINTS

- The disease process that results from a decrease in the rate of secretion of insulin is called diabetes mellitus. Serious complications can result if control of the disease is less than optimal.

- Insulin, the hormone of the fed state, promotes anabolism (storage of energy). Without enough insulin, there is less storage of fatty acids as triacylglycerols and more release of fatty acids from storage. When more fatty acids are released, they provide efficient fuels to help in the regeneration of ATP, either directly or through formation of ketoacids. These fat-derived fuels greatly decrease the oxidation of glucose and set the stage for hyperglycemia (Table 4·11).

Table 4·11

Events leading to chronic hyperglycemia after removal of the actions of insulin

Reduced use of glucose

Oxidation

> Low oxidation of glucose in muscle is due to a high rate of oxidation of fatty acids (resulting from a low net concentration of insulin) and a low transport of glucose into muscle cells (relatively minor; see Chapter 8).

> Low oxidation of glucose in the brain results from a high rate of oxidation of ketoacids (caused by high levels of fatty acids and a relative lack of the actions of insulin on hepatocytes).

Conversion to storage forms

> In the liver, low synthesis of glycogen results from an increased number of signals for its breakdown and a decrease in the number of enzymes required for its synthesis.

> Low synthesis of glycogen in muscle occurs because less glucose is transported into these cells, and signals are lacking for net synthesis.

> Low synthesis of fatty acids and triacylglycerols results from low levels of insulin.

Increased formation of glucose

Conversion of proteins to glucose

> This conversion occurs when net breakdown of endogenous proteins provides an increased number of amino acids and when signals in the liver favor gluconeogenesis.

Conversion of glycogen in muscle to glucose in the liver

> This conversion occurs when counter-insulin hormones (e.g., adrenaline) favor glycogenolysis.
> The lactate released is readily converted to glucose in the liver.

- A severe degree of hyperglycemia (> 30 mmol/l, or 540 mg/dl) implies that another factor is involved. Usually, a contracted ECF volume results in excretion of glucose at a rate lower than expected. An excessive ingestion of sweetened beverages does not cause hyperglycemia, but it may contribute to the severity in a diabetic with low levels of insulin. If hyperglycemia is seen with excessive formation of urea, look for an abnormal catabolism of proteins.

- Ketoacids accumulate with a greater lack of insulin. The maximal rate of synthesis of ketoacids is only 1 mmol/min in an adult. Ketoacids are removed primarily by oxidation in the brain. Severe ketoacidosis results from decreased brain metabolism (coma, anesthesia, stroke) or decreased renal metabolism or excretion.

- Hyperglycemia causes water to shift out of muscle but not out of the liver. It also causes large losses of Na^+, K^+, and water in the urine.

- Hyperglycemia, if chronic, can lead to changes in the charge and function of proteins (glycation).

CHAPTER 5
HYPOGLYCEMIA

PART A
DIAGNOSING THE CAUSES OF HYPOGLYCEMIA

Sympathetic overdrive is the first sign of *hypoglycemia*; dysfunction of the central nervous system occurs later. Hypoglycemia occurs if glucose is produced at a low rate (usually because of a defect in the liver) or, more commonly, if glucose is oxidized at an increased rate. For increased oxidation to occur, there must be a low level of fat-derived fuels (usually the result of actions of insulin).

Hypoglycemia:
A low concentration of glucose in the blood.

PART B
CASES FOR REVIEW

PART A
DIAGNOSING THE CAUSES OF HYPOGLYCEMIA

Case 5·1
Convulsion in a Diabetic
(Case discussed on pages 126–27)

Convulsion:
Uncontrolled, violent muscular activity that is caused by excessive firing of nerve cells in the central nervous system.

Michael, a diabetic, started to sweat and become anxious 45 minutes after giving himself what he thought was his regular injection of insulin. Fifteen minutes later, he was observed to have a *convulsion*. What treatment should Michael receive? Why did it take so long for Michael to develop symptoms?

Case 5·2
Hypoglycemia in a Healthy Tourist
(Case discussed on pages 128–29)

Jackie, a seven-year-old on holiday in Jamaica, was brought to the emergency room because she had had a convulsion in the early evening after an all-day hike with friends. Her skin was cold and clammy. Her parents said that she liked to try new foods (she had done so at lunch) and that she had never had a convulsion before. You suspect that she is suffering from hypoglycemia. A blood sample taken before treatment with glucose confirmed the diagnosis of hypoglycemia (the level of glucose was 2 mmol/l, or 36 mg/dl). When special assays were done on her blood, the level of fatty acids was high and the level of ketoacids was very low. What caused Jackie's hypoglycemia? Why did the hypoglycemia occur so long (6 hours) after her last meal?

Case 5·3
Hypoglycemia in an Alcoholic
(Case discussed on pages 129–30)

The police brought Ed, a confirmed alcoholic, to the emergency room. He had been on a binge for at least 24 hours and reeked of alcohol. Hypoglycemia was noted in blood samples drawn over several hours. How did hypoglycemia develop? What factors should be considered in his treatment? Why does hypoglycemia usually take several days to develop in this clinical setting?

Questions to Consider

Hypoglycemia may cause changes in levels of insulin in the blood, or it may be caused by such changes. How could the levels of fatty acids and ketoacids in blood help you decide between these possibilities? How would you reach your conclusions?

What organ is most directly affected by hypoglycemia? Why? What are the usual signs of hypoglycemia? What would make the symptoms more or less severe?

How would the levels of fuels in blood (e.g., fatty acids, ketoacids, lactate, and alanine) respond to chronic mild hypoglycemia caused by a) *hyperinsulinemia*; b) a low renal threshold for glucose; c) an inherited defect in glucogenesis?

How do you treat a patient who has hypoglycemia?

Hyperinsulinemia:
Excess insulin in the blood.

BACKGROUND

SIGNS AND SYMPTOMS OF HYPOGLYCEMIA

> *The earliest sign of hypoglycemia is overdrive of the sympathetic nervous system; dysfunction of the central nervous system (CNS) follows.*
>
> *Ketosis decreases or prevents symptoms of hypoglycemia.*

The effects of hypoglycemia first become evident after activation of the *sympathetic nervous system*. Sympathetic overdrive (release of adrenaline) causes anxiety, hyperventilation, and a cold sweat.

The CNS also reacts strongly to hypoglycemia because the brain always requires some glucose to regenerate the ATP needed for its function. The brain consumes energy rapidly (normally about 25% of the body's consumption of calories at rest). Brain dysfunction—the most dramatic sign of hypoglycemia—can range from confusion to *coma*, with or without convulsions. The precise signs will vary between patients. Some areas of the brain use ATP at more rapid rates and therefore require faster transport of glucose through the *blood-brain barrier*—a process that requires carriers for glucose. When the rate of this transport falls, symptoms become more evident.

Ketoacids can be an alternate fuel for the brain, supplying up to 80% of the ATP needed for functions of the CNS. Because ketosis can markedly decrease the brain's demand for glucose, and individual demands vary, the level of hypoglycemia that will cause brain dysfunction cannot be predicted precisely.

CAUSES OF HYPOGLYCEMIA

> *Hypoglycemia can be caused by an increased rate of removal of glucose from the blood or by a decreased rate of entry of glucose into the blood.*

Increased Removal of Glucose from the Blood

> *When the availability of fatty acids is low, oxidation of glucose will increase.*

Clinically, the most frequent cause of hypoglycemia is a fast rate of oxidation of glucose resulting from a lack of fat-derived fuels (fatty acids) for oxidation (Table 5·1). Few fatty acids are available because an excess of insulin inhibits their release from triacylglycerols in adipose tissue (Figure 1·10, page 17). The high levels of insulin will also promote conversion of glucose to its storage forms—glycogen (in the liver) or triacylglycerols (in adipose tissue)—thereby removing glucose from the blood.

Table 5·1
Causes of low levels of fatty acids during hypoglycemia

1. Insulin
2. Hormones that act like insulin (e.g., insulin-like growth factors)
3. Lack of hormones whose actions oppose insulin (e.g., glucocorticoids)
4. Drugs that inhibit the release of fatty acids from adipocytes (e.g., niacin)

The level of insulin in the blood is normally closely controlled by the level of glucose in the blood. Excess release of insulin can be caused by an *insulinoma* (a tumor of β cells in the pancreas). The release of insulin from β cells can also be stimulated

Sympathetic nervous system:
The part of the autonomic nervous system that is concerned with the involuntary "fight or flight" response. Adrenaline and noradrenaline are the principal mediators.

Coma:
A problem in the central nervous system that compromises function so that the patient loses consciousness and does not respond to external stimuli.

Blood-brain barrier:
The cell structures lining the walls of the capillaries that separate the blood from the cerebrospinal fluid. These structures exert close control over the fluids surrounding the cells of the brain.

Insulinoma:
A tumor of the β cells of the pancreas that produces insulin in large amounts but not under the control of glucose.

Oral hypoglycemic drugs:
Drugs taken by mouth that decrease the concentration of glucose in blood.

Glucocorticoids:
Hormones (mainly cortisol) from the adrenal cortex that promote the synthesis of glucose.

Hypoxia:
A condition in which tissues receive an insufficient amount of oxygen for regeneration of ATP.

Necrotic cells:
Cells that have died because they could not obtain the fuels they needed or eliminate their waste products.

Inherited disorders of metabolism:
Defects in enzymes caused by alteration of genes.

by certain amino acids and certain *oral hypoglycemic drugs* (e.g., Chlorpropamide) used in treatment of adult-onset diabetes mellitus.

Factors that can simulate insulin include insulin-like hormones (e.g., growth factors released from some tumors), drugs that mimic the antilipolytic action of insulin (e.g., niacin), or the lack of a hormone that normally counters the actions of insulin (e.g., *glucocorticoids*).

Hypoxia in part of the body can also lead to hypoglycemia. Hypoxia causes very rapid consumption of glucose as the tissues struggle to regenerate ATP normally obtained through oxidative metabolism. *Necrotic cells* occur in large tumors or in severely infected areas and can, by swelling, make neighboring cells hypoxic.

Loss of glucose in urine or through lactation are very rare causes of hypoglycemia.

Decreased Entry of Glucose into the Blood

When glucose is not entering the blood rapidly enough, the liver is usually producing less glucose.

Glucose can be released into the blood from a number of sources (Table 5·2). The absence of glucose in the diet is very unlikely to be the sole cause of hypoglycemia since the body has extensive mechanisms to compensate for a low dietary source of glucose. Failure of these adaptive mechanisms is more likely to cause hypoglycemia. In children, defects in the release of glucose from liver glycogen may result from *inherited disorders of metabolism* and can cause hypoglycemia shortly after meals. Decreased synthesis of glucose from lactate or from amino acids will lead to hypoglycemia, especially after stores of glycogen have been exhausted. This decreased synthesis may be due to inherited deficiencies of glucogenic enzymes, or it may be caused by ingested materials such as ethanol or inhibitors of enzymes required for synthesis of glucose.

Table 5·2
Potential sources of glucose for a normal person on a normal diet

Source of glucose	Quantity of glucose mmol (g)		Normal controls
	Total available	Released per day	
Diet			
Carbohydrates	1500 (270)	*	Appetite or habit
Proteins	333 (60)	*	Amount of protein ingested
Glycogen in liver	550 (100)	0 (net)**	Fall in the level of glucose in blood (via glucagon)
Glycogen in muscle (via lactate)	2500 (450)	0 (net)**	Exercise, catecholamines
Proteins in body stores	20 000 (3600)	0 (net)**	Net concentration of insulin
Glycerol in triacylglycerol	8300 (1500)	0 (net)**	Net concentration of insulin

*Figures are not available.
**Because the person is in caloric balance, any fuels broken down during the day are replenished from dietary sources.

CLINICAL APPROACH TO A PATIENT WITH HYPOGLYCEMIA

> *Ask why the patient may be oxidizing glucose more rapidly.*
> *Ask why the supply of glucose from endogenous compounds is low.*

Assume that you are asked to treat a patient who, from signs and symptoms, appears to be hypoglycemic. Simply confirming your suspicions by administering glucose and watching the patient recover rapidly is not enough; the underlying cause needs to be understood to prevent further episodes. Before administering glucose, a blood sample should be saved for further analysis.

A logical approach to diagnosing the cause of hypoglycemia is based on three questions (Figure 5·1).

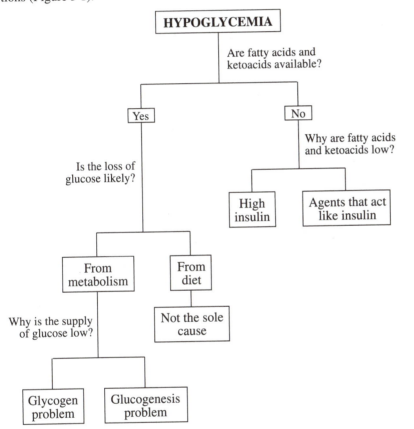

Figure 5·1

Approach to diagnosis of the cause of hypoglycemia. The most important diagnostic and therapeutic consideration in a patient with hypoglycemia is whether or not ketoacids are available for oxidation by the brain (ketoacids are the only major alternative fuel for the brain at this time).

1. Are fatty acids and ketoacids available?

   Measuring levels of fatty acids and ketoacids can indicate whether levels of insulin are consistent with the observed low levels of glucose in blood—a useful technique because it provides the answer faster than can be achieved by direct measurement of insulin levels. High levels of fatty acids indicate a normal response, via low insulin, to the hypoglycemia; if levels of ketoacids are also high, low levels of insulin have probably persisted for some time. High levels of ketoacids also offer the brain an alternative to glucose as an energy fuel, and thus a greater degree of hypoglycemia can be tolerated by the patient.

   Low levels of fatty acids and ketoacids indicate that the hypoglycemia is caused by an excess of insulin or factors simulating this condition (Table 5·1); these possibilities can be investigated through measurements on the blood sample.

If you conclude that the response of insulin to hypoglycemia is normal, consider next how the body is using glucose.

Glucosuria:
The excretion of glucose in the urine.

2. Is excess use or loss of glucose likely?

The usual reasons for a positive answer are: loss in body fluids (during *glucosuria* or lactation), which can be easily tested; excess use because part of the body is hypoxic (hypoxia should be accompanied by other signs and will cause high levels of lactate); or excess use because fatty acids, though available, are not being oxidized to yield ATP, and the body must therefore oxidize glucose for most of its energy. Excess use by muscle during prolonged exercise is another possibility (see Case 1·3, page 14).

If it is unlikely that the hypoglycemia is caused by excess use or loss of glucose, the conclusion must be that the supply of glucose is low.

3. Why is the supply of glucose low?

Hypoglycemia from a low supply of glucose indicates that the body has failed to respond normally to a lack of dietary glucose. Inadequate ability to release glucose from glycogen in the liver, usually from an inherited defect, is indicated by repeated bouts of hypoglycemia shortly after absorption of meals. Inadequate synthesis of glucose from precursors (lactate or amino acids) is indicated by hypoglycemia occurring many hours after meals (glycogen in the liver should provide the needed glucose until then).

Discussion of Case 5·1
Convulsion in a Diabetic
(Case presented on page 122)

What treatment should Michael receive?

Michael's symptoms (sympathetic overdrive followed by dysfunction of the central nervous system) should lead you to suspect hypoglycemia; therefore, immediately take a blood sample to confirm the diagnosis. Treatment consists of rapid administration of glucose. Sudden disappearance of symptoms confirms that Michael indeed had hypoglycemia. The next step is to establish the cause(s) of hypoglycemia. Refer to the laboratory results below and Figure 5·1.

	Normal levels	Michael's levels
Glucose	4-8 mmol/l (72-144 mg/dl)	1.8 mmol/l (32 mg/dl)
Ketoacids	< 0.05 mmol/l	< 0.05 mmol/l
Fatty acids	0.5 mmol/l	< 0.25 mmol/l
Insulin	Should be close to zero with hypoglycemia	Very high

Low levels of fatty acids and the absence of elevated levels of ketoacids suggest an excess of insulin. Michael's overdose of insulin may have been intentional or accidental. For additional treatment, he will require education or counseling, with careful follow-up.

Why did it take so long for Michael to develop symptoms?

The 45-minute delay before signs of hypoglycemia appeared reflects the time needed for the excess insulin to close down release of fatty acids from adipose tissue triacylglycerols (which occurs rapidly) and for the body to consume the fatty acids and glucose in the extracellular fluid (ECF).

In more detail, assume that before Michael injected the excess of insulin, his blood glucose was 9 mmol/l (162 mg/dl); this level is higher than normal for a non-diabetic, since the effects of his previous injection of insulin would have partially worn off. He thus had close to 160 mmol (30 g) of glucose in his ECF and may also have had elevated levels of fatty acids and ketoacids. The excess insulin rapidly closed down release of fatty acids from triacylglycerols and glucose from glycogen in the liver; it also stimulated deposition of glucose and amino acids into body stores and diminished synthesis of glucose from precursors. Circulating fuels became his only source of energy at that point but did not last long because his daily energy consumption of 2400 kcal/day is equivalent to 600 g of glucose per day, or 25 g/hr.

ADDITIONAL QUESTIONS FOR CASE 5·1

How much glucose should initially be given to Michael? What are the dangers of giving too much?

To bring Michael's level of glucose into the normal physiological range, raise the concentration of glucose in his blood by 5 mmol/l. Since glucose normally has access to the ECF and to tissues that are not sensitive to insulin (about 50% of total body water) and body water is about 60% of total body weight, the amount of glucose to be administered is approximately 100 mmol (18 g), assuming that Michael weighs 70 kg. Michael will need more glucose because much of it will be oxidized and some will be converted to glycogen; hence, he requires frequent monitoring of the level of glucose in his blood.

Administering too much glucose may cause the level to exceed the renal threshold. Excreting glucose in the urine causes loss of water and salts, which then have to be replaced.

How might certain brain cells detect ahead of time that the level of glucose is falling too fast or that it is declining to a level that is too low for optimum function of most brain cells?

The symptoms of sympathetic overdrive can last for 15 minutes before convulsions or coma show that the CNS is severely compromised. The signal is not specific to glucose because infusion of ketoacids, lactate, or acetate can correct the symptoms of sympathetic overdrive. Possibly, brain cells with a high rate of turnover of ATP can sense when low levels of fuels are available for the brain. For example, if these sensor cells have a large leak that allows Na^+ to enter, and these Na^+ must be pumped out rapidly using ATP in the process, a large amount of ATP would be required to maintain polarization of these cells. An insufficient availability of fuels would cause depolarization of these cells, thereby triggering sympathetic overdrive.

Discussion of Case 5·2
Hypoglycemia in a Healthy Tourist
(Case presented on page 122)

What caused Jackie's hypoglycemia?

The glucose Jackie received caused her to recover quickly, but continued monitoring of the concentration of glucose in her blood showed a subsequent decline. Following the approach summarized in Figure 5·1, the levels of fatty acids should be examined; they continued to rise while levels of ketoacids remained very low. Because levels of fatty acids are high, it appears that release of fatty acids from adipose tissue is not impaired and that the levels of insulin are low—consistent with the hypoglycemia. The very low levels of ketoacids despite high levels of fatty acids suggest a defect in oxidation of fatty acids, certainly in the liver and perhaps also in other tissues. She may therefore be consuming glucose rapidly because fat-derived fuels are not available to regenerate ATP. Since oxidation of fatty acids is an important source of energy for the liver when glucose is not provided from the diet, inhibited oxidation of fatty acids can deny the liver its supply of ATP and the appropriate signals it needs for glucogenesis.

Because Jackie had not had any previous episodes, inherited defects of metabolism are very unlikely. A toxin that inhibits oxidation of fatty acids, especially in the liver, must be suspected. Further discussion with Jackie revealed that she had picked and eaten an unripe Ackee-Ackee fruit during lunch. The ripe fruit is delicious and widely eaten, but unripe fruit contains a toxin that inactivates *carnitine,* which is essential for the transport of fatty acids into mitochondria for oxidation (Figure 5·2). Continued treatment should involve close monitoring of the glucose in her blood and administration of glucose, as needed, to avoid further hypoglycemia. Carnitine should also be given to replace that destroyed by the toxin, and she should be taught to recognize an unripe Ackee-Ackee fruit.

Carnitine:

A compound essential for the transport of activated fatty acids (fatty acyl-CoA) from the cytosol to the mitochondria, where they are oxidized. The mechanism for transport is indicated in Figure 5·2.

Figure 5·2

A deficiency of carnitine causes low oxidation of fatty acids. Carnitine is needed to transport the fatty acyl groups from the cytosol into mitochondria because CoA-derivatives do not cross the mitochondrial membrane.

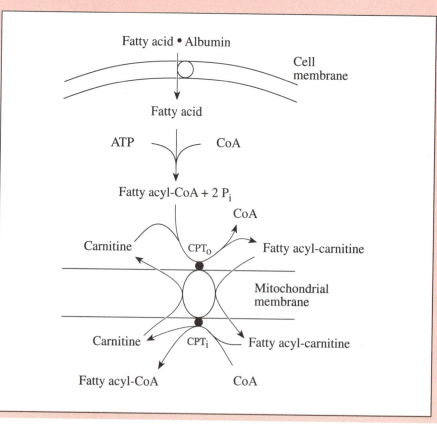

Why did the hypoglycemia occur so long (6 hours) after her last meal?

A number of factors may have contributed to Jackie's slow onset of hypoglycemia. The absorption of the toxin from the gut and the development of its effects on carnitine might have occurred slowly; also, the availability of glucose stored as glycogen in the liver could have delayed the development of hypoglycemia.

Assume that her stores of glycogen in the liver were full just after lunch. Glycogen in the liver can normally meet the brain's requirements for glucose for about 20 hours. However, her inability to oxidize fatty acids forced the rest of her body to oxidize only glucose. Since the brain normally consumes about 25% of the body's energy, and she was physically active between her lunch and the onset of convulsion, the glycogen in her liver could have been exhausted in less than six hours.

Discussion of Case 5·3
Hypoglycemia in an Alcoholic

(Case presented on page 122)

How did hypoglycemia develop?

To determine the cause of hypoglycemia, first examine the levels of fatty acids in the blood to see if hypoglycemia has produced the normal response—a decrease in the level of insulin. As expected, the levels of fatty acids and ketoacids are high and insulin is low. Because fatty acids and ketoacids could have provided alternative fuels for regeneration of ATP, excess loss of glucose by oxidation is unlikely to be the cause of the hypoglycemia. The hypoglycemia is probably due to a decreased rate of synthesis from precursors. *Ethanol* can lower this rate in two ways:

1. Ethanol is metabolized to acetaldehyde and acetic acid in the liver (Figure 5·3), increasing the level of *NADH* (or more accurately a high ratio of NADH to NAD^+; see also Chapters 6, 7, and 11). The high ratio of NADH to NAD^+ decreases the concentrations of pyruvate and *oxaloacetate*, two essential intermediates of glucogenesis (Chapter 8).

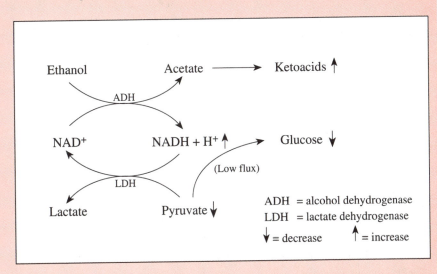

2. The formation of ATP from ethanol is very efficient (acetic acid can be converted to ketoacids or to ATP) and can fully meet the liver's requirements for ATP. Metabolism of ethanol promotes glucogenesis from lactate (the Cori cycle; see Figure 1·19, page 27) by providing the ATP needed for the process. Glucogenesis from amino acids, which is required to make new glucose for the brain, does not require ATP; instead, it produces excess NADH (and hence ATP). Ethanol and amino acids may therefore be competing for the capacity to regenerate ATP in the liver, and ethanol always wins.

Ethanol:
The alcohol in spirits, wines, etc. Ethanol is oxidized rapidly to acetaldehyde and acetate, which can enter metabolism as acetyl-CoA and can thus be used in the fat system and ATP generation system. The conversion of ethanol to acetate rapidly forms NADH (Figure 5·3; see page 292 for the structure of ethanol).

NADH and NAD^+:
One of many metabolic coenzymes involved in oxidation and reduction (removal and addition of electrons, respectively). NAD^+ accepts H atoms to form NADH, which can then pass them to another reaction. The H on NADH can also be used in the ATP generation system to yield H_2O and ATP (3 ATP per NADH used).

Oxaloacetate:
A four-carbon intermediate that is part of both glucogenesis and the tricarboxylic acid cycle of the ATP generation system.

Figure 5·3
Effect of ethanol on hepatic production of glucose. Conversion of ethanol to acetate generates NADH from NAD^+, thereby decreasing the concentration of pyruvate (which is converted to lactate rather than glucose). The acetate is rapidly converted to ketoacids.

What should be considered in his treatment?

Alcoholic hypoglycemia has to be treated with care to avoid the risk of causing permanent brain damage. Vitamin B_1 (thiamine) should be administered to protect the brain from being damaged by depletion of ATP and overproduction of H^+ (see below). Potassium should be administered to avoid the risk of cardiac arrhythmias.

WERNICKE-KORSAKOFF SYNDROME

Because alcoholics are often chronically malnourished, they may lack thiamine, which is essential for the activity of pyruvate dehydrogenase (PDH), the first enzyme in the oxidation of pyruvate to regenerate ATP. The actions of PDH are critically important for the brain, which always requires oxidation of glucose to regenerate ATP.

Because ethanol can be rapidly converted to ketoacids, alcoholics on a binge generally have ketoacidosis (the brain can use ketoacids to supply most of its requirement for ATP). As the patient "dries out" the level of ketoacids falls and glucose must be oxidized to supply an increasingly large fraction of the ATP needed. If PDH is compromised by lack of thiamine, the ATP must come from glycolysis alone; the large amounts of lactic acid released can destroy areas of the brain that have rapid rates of turnover of ATP or lower levels of thiamine. The resulting condition, which is called Wernicke-Korsakoff Syndrome, can be prevented by treatment with thiamine at the outset (see also Case 6·5, The Brain Was Pickled).

ARRHYTHMIAS

Administration of glucose to correct alcoholic hypoglycemia also runs the risk of causing abnormal heart rhythms and possibly death. Excess intake of alcohol irritates the stomach lining and may cause vomiting, which can deplete the body of K^+. The hypoinsulinemia caused by the hypoglycemia masks this deficit of K^+ by allowing K^+ to leak from the ICF to the ECF (the normal $[K^+]$ is 150 mmol/l in the ICF and 4 mmol/l in the ECF). Giving glucose will raise the concentration of insulin in the blood, providing that the ECF volume has returned to normal (lowering the level of adrenaline, an inhibitor of the release of insulin). An increased level of insulin will cause the cells to reaccumulate K^+; the resulting hypokalemia in the ECF can cause cardiac arrhythmias and even death. Administration of potassium chloride can decrease this risk (do not give too much or administer too quickly because an excess is as lethal as a deficiency).

Why did the hypoglycemia take so long to develop?

Hypoglycemia will develop when the liver cannot supply glucose fast enough to match the rate of oxidation of glucose. Ed has high levels of ketoacids, so he will oxidize glucose at a slow rate. For hypoglycemia to have developed, he must have been producing glucose very slowly. Since glucose can be made from glycogen (quickly) and glucogenesis (less quickly), glycogen in his liver must have been depleted before hypoglycemia developed (i.e., he must not have consumed carbohydrates for the past two days). In addition, since metabolism of ethanol leads to the production of high levels of NADH, the rate of glucogenesis in his liver was reduced (Figure 5·3) from consumption of alcohol.

Questions

(Discussions on pages 265–66)

5·1 How much glucose can be synthesized from one kilogram of muscle? (Assume muscle is 80% water and 20% proteins).

5·2 How much would the concentration of lactate in the blood rise if all the glucose in the body fluids were converted to lactate?

5·3 How would a rapid infusion of 1 liter of D_5W affect the concentration of glucose in the blood?

D_5W:
Dextrose (glucose) 5 g/100 ml water.

5·4 If the rate of filtration in the kidney (the glomerular filtration rate, GFR) is 180 liters per day in a normal adult, and a normal kidney can reabsorb 1800 mmol (324 g) of glucose per day, how high must the concentration of glucose be in the blood before glucose will appear in the urine?

5·5 How long will it take to develop symptoms of hypoglycemia after an overdose of insulin?

5·6 Will glycogen in muscle be converted into a fuel for the brain during hypoglycemia?

5·7 What essential differences between Cases 5·1, 5·2, and 5·3 affect the time taken for signs of hypoglycemia to develop?

PART B
CASES FOR REVIEW

Case 5·4
Living with Chronic Hyperinsulinemia
(Case discussed on page 134)

Bertha has an insulinoma, which can release surges of insulin that are larger than that injected by Michael in Case 5·1, but she shows few symptoms of hypoglycemia. Explain her relative lack of symptoms.

Bertha, who has always been athletic, can run to catch a train (a few minutes of aerobic exercise) without symptoms of hypoglycemia, despite her insulinoma. What can you conclude from this ability?

Case 5·5
Hypoglycemia Three to Four Hours After Meals
(Case discussed on page 135)

One year ago, Peter had disseminated tuberculosis. Even though antibiotic therapy cured the tuberculosis, he never recovered his strength. His skin developed dark pigmentation (this sign suggests destruction of his *adrenal glands*). Over the past few weeks, Peter has had symptoms of hypoglycemia three hours after eating a meal; these symptoms disappear after he drinks sweetened orange juice. The levels of fatty acids in his plasma are not elevated when he is hypoglycemic. What do you conclude from the timing of the onset of the hypoglycemia? What might cause the hypoglycemia? How would you treat Peter?

Adrenal glands:
Small glands situated above the kidneys. The outer part (cortex) releases glucocorticoids and mineralocorticoids. The central part (medulla) is stimulated by the sympathetic nervous system to release noradrenaline and adrenaline.

Case 5·6
Frequent Hypoglycemia in an Infant
(Case discussed on page 136)

Baby Alice was well until six weeks of age but has not been thriving since her parents lengthened the time between feedings; she wakes frequently and is cranky but calms down with feeding. Her tummy is protruding more and more. What could be wrong? What is the most likely reason for her hypoglycemia?

Case 5·7
Nocturnal Hypoglycemia in a Child
(Case discussed on page 136)

Desmond, a two-year-old, is fine during the day but develops signs of hypoglycemia in the early morning (before the household normally gets up) unless he is fed during the night. You suspect an inherited defect in hepatic glucogenesis. What levels of energy fuels (fatty acids, ketoacids, lactate, glucose) would you expect to find in blood samples taken long after a meal or at the end of the absorptive phase?

Case 5·8
An Inherited Carnitine Deficiency
(Case discussed on page 137)

Bonnie, seven months old, has had multiple admissions to the hospital for cardiac failure, dysfunction of the CNS, and an enlarged liver. Because an "older" brother died at three months after a similar history, her parents insisted that careful studies be carried out on Bonnie.

Hypoglycemia, as low as 1 mmol of glucose per liter, was noted on each admission, and dietary studies showed that it developed after 10 hours of fasting. Physical examination showed general weakness and reduced muscle tone. Laboratory findings on blood samples revealed hypoglycemia, high levels of fatty acids, zero concentration of ketoacids (even after fasting), ammonium levels four times higher than normal, normal levels of carnitine, and mild increases in the activities of enzymes normally associated with the liver and muscle. Biopsies of the liver and muscle showed accumulation of triacylglycerols and markedly decreased levels of carnitine in both organs. Treatment with carnitine caused the signs and symptoms to disappear.

Explain each of the observed signs and symptoms in terms of the successful treatment with carnitine.

Case 5·9
Hypoglycemia in a Cancer Patient
(Case discussed on page 138)

Anita developed hypoglycemia. She had a large sarcoma, a connective tissue malignancy that can grow rapidly to an immense size and usually has a large necrotic center. Three causes for the hypoglycemia can be suggested: the tumor secretes insulin or an insulin-like agent; the tumor consumes a lot of glucose; the tumor secretes products (e.g., tryptophan) that inhibit hepatic glucogenesis. How would you distinguish between these possibilities?

Discussion of Case 5·4
Living with Chronic Hyperinsulinemia
(Case presented on page 132)

RELATIVE LACK OF SYMPTOMS DESPITE AN INSULINOMA

Hormones generally act on cells by binding to specific receptors on the cell surface. The hormone-receptor complex tells the cell that the hormone is trying to exert its influence. The binding follows the same general equilibrium reactions as described in Chapter 2 for buffers.

The strength of the signal to the cell depends on the number of hormone-receptor complexes on the surface of the cell; this number can be influenced by the concentration of the hormone, the number of receptors per cell, or the equilibrium constant of the equation (which might be changed by chemical modification of the receptor).

Homeostatic controls generally minimize the effects of untoward events. Chronic hyperinsulinemia such as Bertha's leads to a chronic excess of *insulin receptors* on cells responding to insulin. Homeostatic controls might therefore be expected to decrease either the number of receptors or the affinity of those receptors for insulin, thus minimizing the effects of surges of insulin. However, because the concentration of insulin rises to a much greater degree than the concentration of receptors falls, signs and symptoms of excess insulin still occur, but at higher levels of this hormone.

With chronic hypoglycemia and the adrenergic response, lipolysis is activated and more fatty acids are released. Patients with an insulinoma therefore usually have high enough levels of these alternate fuels to prevent hypoglycemia from excessive oxidation of glucose (at rest).

VIGOROUS EXERCISE AND CHRONIC HYPERINSULINEMIA

At normal levels of glucose, Bertha's extracellular fluid would contain about 15 g glucose (5.5 mmol/l, or 100 mg/dl, in 15 liters of extracellular fluid). Vigorous exercise, during which muscle (24 kg) can oxidize 0.2 g of glucose per kilogram, could deplete this circulating glucose in a few minutes. Bertha therefore relies on other fuels besides circulating glucose to regenerate ATP in her muscle during vigorous exercise.

As with a healthy athlete, Bertha's muscle can use endogenous glycogen (present at about 20 g/kg) as a major source of glucose. Although she has an insulinoma, she does not have low circulating levels of fatty acids, despite the expected effects of insulin on release of fatty acids from triacylglycerols. This lack of effect could be related to a) homeostatic adaptations to the chronic hyperinsulinism, such as those referred to above; b) chronically high levels of adrenaline, which stimulates lipolysis in adipose tissue; c) chronic hypoglycemia, which limits the availability of glucose for deposition of fatty acids in adipose tissue (glucose provides the glycerol backbone of triacylglycerols; see Chapter 11, page 228). As a result of such mechanisms, fatty acids may be more readily available to her than to a normal athlete; oxidation of fatty acids to regenerate ATP for muscle could spare glucose by mechanisms described in Chapters 8 and 9.

Note that patients with an insulinoma will become hypoglycemic when they exercise, but more slowly than expected because glucose is not the only fuel that their muscles will use.

Insulin receptors:

Specific chemical structures on the surface of the cell (part facing the extracellular fluid and part facing the cytosol). The part facing the extracellular fluid can bind insulin with a very high specificity. Once insulin is bound, the cell receives a signal (see Chapter 7, pages 183–84).

Discussion of Case 5·5
Hypoglycemia Three to Four Hours After Meals
(Case presented on page 132)

What do you conclude from the timing of onset of the hypoglycemia?

The body absorbs the nutrients in a normal meal in about three hours. Peter's hypoglycemia starts shortly thereafter, which suggests that the release of glucose from glycogen in his liver is not adequate. In addition, glucogenesis from amino acids may not occur at an appropriate rate. Thus, signals from either too much insulin or not enough counter-insulin hormones seem to be the cause of the problem.

What might cause the hypoglycemia?

Insulin, which signals storage of glucose, is normally balanced by the counter-insulin hormones—glucagon, glucocorticoids, adrenaline, and *thyroid hormone* (Table 5·3). Glucocorticoids, produced by the adrenal cortex, permit more breakdown of glycogen and more synthesis of glucose from precursors, thereby helping the liver to release glucose. Absence of glucocorticoids impairs these signals (Chapter 8), diminishes catabolism of proteins (and thus mobilization of precursors for glucogenesis), and prevents induction of key glucogenic enzymes (thereby slowing the rate of glucogenesis).

Thyroid hormone:
A hormone from the thyroid gland that increases the basal metabolic rate in all organs except the brain.

How would you treat Peter?

Peter should immediately be given oral glucose in small, frequent feedings. He should be given glucocorticoids and *mineralocorticoids* (also produced by the adrenal cortex) as soon as possible as replacement therapy for his inactive adrenal cortex.

Mineralocorticoids:
Hormones, such as aldosterone, that promote the reabsorption of Na^+ and the secretion of K^+ by the kidneys.

Table 5·3
Actions of counter-insulin hormones

Hormone	Actions
Glucagon	Promotes breakdown of glycogen in the liver
Glucocorticoids	Promotes breakdown of glycogen and proteins and stimulates glucogenesis
Adrenaline	Promotes breakdown of glycogen (primarily in muscle) and activates hormone-sensitive lipase in adipose tissue
Thyroid hormone	Potentiates actions of adrenaline
Pituitary growth hormone	Promotes release of glucagon and stimulates growth in many tissues

⌐ Discussion of Case 5·6
Frequent Hypoglycemia in an Infant
(Case presented on page 132)

The frequent appearance of symptoms and their relief after a feeding indicate that hypoglycemia might be the problem. The cause of hypoglycemia is likely to be an inadequate production of glucose. As with Peter (Case 5·5), Alice seems troubled at the time when her body has finished absorbing dietary glucose—probably the result of an insufficient release of glucose from her liver.

The problem, which became evident when Alice's parents tried to lengthen the period between feedings, may have been present from birth but was not noticed because of the frequency of feedings. An inherited enzyme deficiency is possible; glucose 6-phosphatase (Chapter 8) is the enzyme common to both glucogenesis and breakdown of glycogen. A biopsy of the liver confirmed the absence of this enzyme.

To avoid the hypoglycemia, Alice must be given glucose frequently. Because her liver cannot readily convert its store of glycogen to glucose (though glycogen can be broken down through glycolysis), her liver will become very enlarged, "stuffed" with glycogen. At present, this disease cannot be treated; in the future, insertion of this enzyme into the liver may become possible.

⌐ Discussion of Case 5·7
Nocturnal Hypoglycemia in a Child
(Case presented on page 132)

If hepatic glucogenesis is not functioning properly, Desmond's exchanges between glucose and glycogen should not be the major problem (in contrast to Alice, Case 5·6), but he will be unable to synthesize glucose from lactate and amino acids at all times. Hypoglycemia can thus be expected to develop many hours after a meal (after dietary glucose has been metabolized). This hypoglycemia will cause low levels of insulin, leading to the release of fatty acids from triacylglycerols and to the formation of ketoacids. The release of insulin in response to dietary glucose will inhibit release of fatty acids so that the levels of fatty acids and ketoacids, which were high just before meals, will fall shortly after intake of food. The inactive glucogenesis will prevent use of precursors of glucose (lactate, alanine), which will be found in above normal concentrations at all times.

Question

(Discussion on page 267)

5·8   What are the essential metabolic differences between Cases 5·6 and 5·7 concerning the timing of hypoglycemia and the anticipated levels of fuels in the blood?

Discussion of Case 5·8
An Inherited Carnitine Deficiency
(Case presented on page 133)

If carnitine, which is essential for oxidation of fatty acids, is lacking in organs such as muscle and the liver (in which oxidation of fatty acids is important), these organs will become depleted of ATP and thus will be damaged. Note Bonnie's general weakness and reduced muscle tone and the presence in blood of enzymes normally found in muscle and the liver—all symptoms of a lack of carnitine.

Although Bonnie has high levels of fatty acids in her blood, her liver is not oxidizing them. Her levels of ketoacids are therefore low, and her liver is unable to regenerate the ATP it needs to synthesize glucose, thus explaining the hypoglycemia. High levels of fatty acyl-CoA in the cytosol in her liver may compromise the adenine nucleotide translocase (Chapter 7, page 182, and Chapter 10), thereby affecting other metabolic events in the liver, such as the synthesis of urea (Chapter 12). As a result, hyperammonemia may ensue.

Hypoglycemia causes the levels of insulin to fall, triggering an increased release of fatty acids from adipose tissue. Lower levels of insulin should promote the oxidation of fatty acids, but, because of low levels of carnitine in tissues, fatty acids have accumulated. When fatty acids accumulate in plasma, organs store them as triacylglycerols.

Bonnie's cardiac problems stem from a lack of carnitine; heart muscle, like other types of muscle, relies heavily on oxidation of fatty acids. Her brain dysfunction results from hypoglycemic episodes, which have more severe effects on the brain because of the absence of ketoacids.

Similar findings in a sibling suggest a hereditary problem. The normal levels of carnitine in blood with low levels in tissues suggest that the specific problem might be in the uptake of carnitine from the circulation. Bonnie will require dietary supplements of carnitine for the rest of her life. Any permanent damage that she may have suffered before the causes of hypoglycemia were recognized will not be repairable, but carnitine therapy should minimize further problems.

Discussion of Case 5·9
Hypoglycemia in a Cancer Patient
(Case presented on page 133)

If the tumor is releasing insulin-like factors, the circulating levels of fatty acids and ketoacids will be low, and hypoglycemia will result from the need for glucose to provide energy for the whole body (Figure 5·4; see also Cases 5·1 and 5·4).

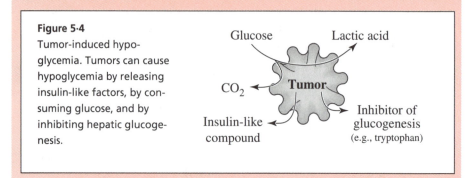

Figure 5·4
Tumor-induced hypoglycemia. Tumors can cause hypoglycemia by releasing insulin-like factors, by consuming glucose, and by inhibiting hepatic glucogenesis.

If the tumor is consuming excess glucose, the resulting low levels of insulin will cause the concentrations of fatty acids and ketoacids to be high, especially between meals. Such tumors often produce large amounts of lactate (for unknown reasons) even if the cells in the tumor are not hypoxic; the increase in production of lactic acid will decrease the concentration of HCO_3^- in the blood. In such cases, the tumor can consume very large amounts of glucose, so exogenous glucose must be provided in abundance, and lactic acidosis must be controlled.

If the tumor is causing hypoglycemia by inhibiting hepatic glucogenesis, the low levels of insulin will lead to high concentrations of fatty acids and ketoacids, and the precursors of glucogenesis (lactate, alanine) will accumulate in the blood. If the metabolism of glycogen is unaffected, glycogen in the liver should delay the development of hypoglycemia after meals, but low glucogenesis may limit the degree of accumulation of glycogen during meals. The amounts of glucose that must be administered to maintain normal levels in blood will be much less than if the tumor is rapidly consuming glucose.

Question

(Discussion on page 267)

5·9 If a tumor converts half of the glucose that it consumes to lactic acid rather than carbon dioxide, what fraction of the tumor's ATP would be made anaerobically? Assume that there is no change in the rate of turnover of ATP in the tumor and that all the generation of ATP in the tumor is by aerobic or anaerobic metabolism of glucose.

CHAPTER 6
METABOLIC ACIDOSIS

PART A
LACTIC ACIDOSIS

Lactic acid accumulates in blood when hypoxia increases the production of lactic acid or when inhibited glucogenesis decreases its use. Accumulation of lactic acid as a result of hypoxia can rapidly become life-threatening.

PART B
KETOACIDOSIS

Ketoacids are synthesized when the concentration of fatty acids in blood rises as a result of enhanced lipolysis (breakdown of triacylglycerols in adipose tissue). Ketoacids can also be formed from ethanol. Ketoacidosis becomes life-threatening when the rate of oxidation of ketoacids in the brain and kidneys is markedly reduced.

PART C
CASES FOR REVIEW

PART A
LACTIC ACIDOSIS

Case 6·1
Hugh Was Found Unconscious in the Garage
(Case discussed on page 148)

Hugh's mother opened the garage door and found Hugh unconscious in the family car; the engine was running. She rushed him to the hospital. Laboratory data (Table 6·1) show a large anion gap (Chapter 2, pages 56–57) and high levels of lactate in his plasma. Why did these abnormalities occur? Which of Hugh's organs is most at risk? What treatment should Hugh receive? Will his blood be blue or red in color?

Table 6·1
Hugh's laboratory values on admission

Arterial blood		Normal	Hugh
H$^+$	nmol/l	40	50
pH		7.40	7.30
HCO$_3^-$	mmol/l	25	15
Pco$_2$	mm Hg	40	30
Po$_2$	mm Hg	100	105
Anion gap	mEq/l	12	22
Hemoglobin	g/l	140	140
Lactate	mmol/l	1	10

Case 6·2
Greta Has Cancer with Jaundice
(Case discussed on page 149)

Greta, a patient with cancer of the colon, develops severe jaundice, loses a lot of weight, and lacks energy. Blood tests during her most recent monthly visit to the hospital for chemotherapy show metabolic acidosis with a large increase in the anion gap (Table 6·2). She has no apparent circulatory problems. Is the problem due to a low availability of O$_2$? Is flux through pyruvate dehydrogenase (Chapter 9) compromised? Is her liver removing lactic acid too slowly? How can the oxidation of lactate be increased?

Table 6·2
Greta's laboratory values on admission

Arterial blood		Normal	Greta
H$^+$	nmol/l	40	60
pH		7.40	7.22
HCO$_3^-$	mmol/l	25	10
Pco$_2$	mm Hg	40	25
Po$_2$	mm Hg	100	105
Anion gap	mEq/l	12	27
Hemoglobin	g/l	140	120
Glucose	mmol/l (mg/dl)	5 (90)	4 (72)
Ketoacids	mmol/l	0.05	0.05
Creatinine	μmol/ (mg/dl)	113 (1.0)	113 (1.0)
Lactate	mmol/l	1	15

Po$_2$:
The partial pressure of oxygen in solution. Po$_2$ reflects dissolved O$_2$, not O$_2$ carried in hemoglobin.

Case 6·3
Garth Has a Heart Attack
(Case discussed on page 150)

Garth arrived at the hospital complaining of crushing anterior chest pain. His blood pressure is low and his pulses are difficult to palpate. Laboratory tests indicate normal values for the bicarbonate buffer system, blood pH, anion gap, and hemoglobin. Do you expect lactic acidemia? Should you give Garth oxygen? How would your answer change if the anion gap were significantly elevated?

Case 6·4
Archie Has Severe Hypoxia but Feels Fine
(Case discussed on page 150)

A blood sample from Archie, a patient with *leukemia*, was taken with a number of other blood samples to the laboratory for a range of assays. The value obtained for the oxygen content was so low that the physician rushed him to the ICU. However, Archie felt no worse than usual. What happened?

Leukemia:
A cancer involving white blood cells, which may accumulate to extremely high levels in the blood.

Case 6·5
Julian's Brain Was Pickled
(Case discussed on pages 151–52)

Julian, a chronic alcoholic, came to the hospital for care. He was obviously malnourished. While he could carry on a conversation, Julian was obviously confabulating (telling made-up stories lacking a basis in reality). Julian had been to the hospital on previous occasions with ketoacidosis resulting from abuse of ethanol. He was given intravenous solutions of glucose on these occasions. Why was this treatment dangerous? What precautions should have been taken beforehand?

Questions to Consider

What factors can increase the rate of formation of lactic acid?

What factors can decrease the rate of consumption of lactic acid?

What are the relative rates at which lactic acid accumulates in hypoxia and is used when supplies of oxygen return?

What can cause the delivery of O_2 to tissues to be compromised?

BACKGROUND

LEVELS OF LACTIC ACID

In a normal fed person, levels of lactic acid in plasma are low (less than 1 mmol/l), but they may rise to 10–15 mmol/l during vigorous anaerobic exercise. In pathological conditions, lactic acid may accumulate and cause metabolic acidosis. Table 6·3 (page 144) contains a list of the causes of lactic acidosis.

Chapter 2 provides information on the physiological systems that control the [H$^+$] in blood and tissues.

LACTIC ACID METABOLISM

> *Lactic acid is a metabolic cul-de-sac; it is only made from pyruvic acid, and can only be converted to pyruvic acid. This interconversion of intermediates is in equilibrium with the ratio of NADH to NAD$^+$.*

Dehydrogenase:

A common group of enzymes that catalyze oxidation (removal of H or addition of O) or reduction (addition of H or removal of O). The enzymes always require a cofactor, usually NAD$^+$ or NADP$^+$ (see pages 186–87) to accept the H (forming NADH or NADPH) or to provide it.

Cofactors:

Compounds that generally participate in enzyme-catalyzed reactions by carrying chemical groups or energy. For example, the NADH-NAD$^+$ system carries H, the ATP-ADP-AMP system carries phosphate, and coenzyme A carries acetyl or other acyl groups.

All tissues can make lactic and pyruvic acids from glucose. Red blood cells make lactic acid as the byproduct from regeneration of ATP during anaerobic glycolysis but cannot use lactic acid. In virtually all other tissues, lactic acid can be both made and used; it is made from glucose (the main source) or amino acids (a minor source) and can be used to make acetyl-CoA via pyruvate dehydrogenase (PDH) in all tissues except red blood cells. It is also used in glucogenesis (in the liver or kidneys) and in synthesis of glycogen (in muscle).

Lactate dehydrogenase (LDH), an enzyme that is located in the cytosol, catalyzes the interconversion between lactate and pyruvate; lactate can be oxidized to pyruvate, and pyruvate is reduced to lactate (equation 1). The NADH-NAD$^+$ cofactor system exchanges the H atoms released or consumed.

$$\text{Lactate}^- + \text{NAD}^+ \xrightleftharpoons{\text{LDH}} \text{Pyruvate}^- + \text{H}^+ + \text{NADH} \qquad (1)$$

Because the reaction is always at chemical equilibrium, the ratio of lactate to pyruvate is always proportional to [NADH]/[NAD$^+$] in the cytosol (equation 2).

$$\frac{[\text{Lactate}^-]}{[\text{Pyruvate}^-]} = K \, \frac{[\text{NADH}]}{[\text{NAD}^+]} \qquad (2)$$

(Note: K contains the value for the [H$^+$], which, in the absence of hypoxia, remains essentially constant inside cells.)

In plasma, the normal ratio of lactate to pyruvate ranges from approximately 4:1 to 10:1. Hence, though pyruvate and lactate anions both enter plasma, the lactate anion is usually regarded as the principal circulating fuel. The concentration of lactate in blood is normally 0.5–1.0 mmol/l; in severe lactic acidosis, it can rise to more than 20 mmol/l.

FACTORS CAUSING CHANGES IN THE CONCENTRATION OF LACTIC ACID OR LACTATE ANIONS

A high concentration of lactate is due to a high concentration of pyruvate or a high level of NADH in the cytosol.

High NADH/NAD+

The ratio of lactate to pyruvate varies with the ratio of NADH to NAD^+ (Figure 6·1). A high ratio is most commonly the result of underutilization of NADH (hypoxia) and is sometimes due to overproduction of NADH (hypoxia, oxidation of ethanol in the liver).

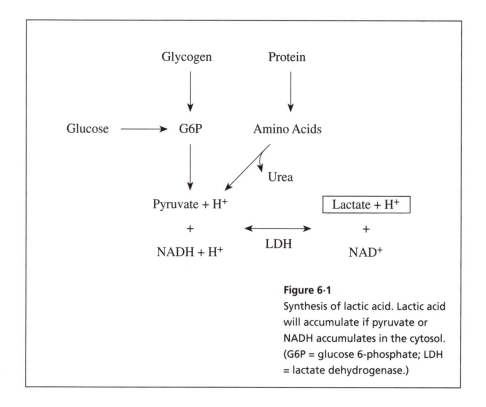

Figure 6·1
Synthesis of lactic acid. Lactic acid will accumulate if pyruvate or NADH accumulates in the cytosol. (G6P = glucose 6-phosphate; LDH = lactate dehydrogenase.)

High Level of Pyruvate

The concentration of lactate will increase if that of pyruvate increases. The concentration of pyruvate (and hence of lactate), will increase if its rate of formation increases or if it is used less rapidly. The most important influence on the rate of formation of pyruvate is its formation from glucose or glycogen. This rate can increase by 50-fold if either glucose or glycogen is required to regenerate ATP in the absence of oxygen.

Pyruvate can also be converted into glucose or glycogen; thus, glucogenesis in the liver and synthesis of glycogen in muscle are powerful means of using pyruvate (and lactate). However, the most important control over the rate of removal of pyruvate is the activity of PDH. This enzyme catalyzes the oxidation of pyruvate to acetyl-CoA, which, as part of the fat system, cannot be made into glucose. PDH is inhibited when fatty acids or ketoacids are being oxidized.

CAUSES OF LACTIC ACIDOSIS

> *Fast lactic acidosis occurs when lactic acid is produced at a rapid rate (as in hypoxia). Slow lactic acidosis occurs when the rate of removal of lactic acid is compromised (usually a liver problem).*

Hypoxia and inhibition of glucogenesis are the two common causes of lactic acidosis (Table 6·3). Hypoxia causes a rapid formation of lactic acid because it prevents the oxidation of NADH to yield ATP. Hypoxia therefore forces tissues to regenerate all their ATP from anaerobic glycolysis (the conversion of glucose to lactic acid). Hence, flux through PDH stops and glycolysis goes on very rapidly (Figure 6·2). Anaerobic glycolysis yields two molecules of ATP (plus two molecules of lactic acid) per molecule of glucose consumed. Complete oxidation of glucose to $CO_2 + H_2O$ via glycolysis, PDH, and the ATP generation system yields 36–40 ATP per glucose. Anaerobic glycolysis must therefore use glucose 18–20 times faster than complete oxidation to meet the normal demands of a tissue for regeneration of ATP. Hence, this pathway can produce H^+ very rapidly.

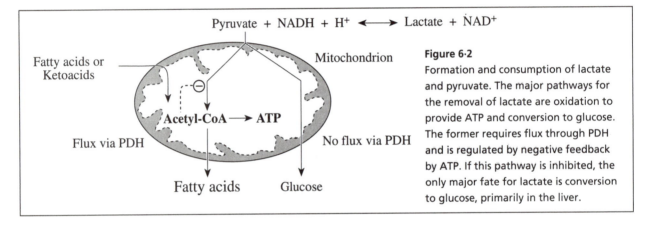

Figure 6·2

Formation and consumption of lactate and pyruvate. The major pathways for the removal of lactate are oxidation to provide ATP and conversion to glucose. The former requires flux through PDH and is regulated by negative feedback by ATP. If this pathway is inhibited, the only major fate for lactate is conversion to glucose, primarily in the liver.

Glucogenesis from lactate in the liver or kidneys completes the Cori cycle (Figure 1·19, page 27). The liver's normal capacity for glucogenesis is approximately five times the normal rate at which tissues such as red blood cells and the brain form lactate, but this capacity is inadequate to meet the load of lactate when it is generated very rapidly (e.g., during hypoxia and vigorous muscular activity).

Table 6·3
Causes of lactic acidosis

Deficit of oxygen
- Lung problem (low Po_2)
- Circulatory problem (poor delivery of O_2)
- Hemoglobin problem (low capacity of blood to carry O_2)

Compromised metabolism of lactate without hypoxia
- Excessive formation of lactic acid (increased glycolysis as a result of low ATP, e.g., from exercise or from the presence of agents that uncouple oxidative phosphorylation)
- Insufficient utilization of lactic acid
  - PDH problem (from a deficiency of vitamin B_1 or an inborn error)
  - Increased availability of other fuels (fatty acids)
  - Low flux through the ATP generation system (less biological work)
- Decreased conversion of lactate to glucose, a liver problem
  - Destruction or replacement of cells in the liver
  - Defect in glucogenesis (from an inborn error or inhibitors of glucogenesis, such as drugs, ethanol, tryptophan)

Questions

(Discussions on pages 267–68)

6·1 How rapidly can the liver synthesize glucose from the lactate produced in muscle during a 10-second sprint?

6·2 Assuming that the body uses 12 mmol of O_2 per minute and that the total amount of glucose in solution in the body is 100 mmol (5 mmol/l $\times$ 20 liters), how long can free glucose meet normal bodily demands for ATP through anaerobic glycolysis if all the glucose is used?

What will be the final concentration of lactate?

What might happen to the pH of the blood?

When oxygen returns, what is the maximal theoretical rate at which the lactate can be oxidized to CO_2 and H_2O?

DIAGNOSTIC APPROACH TO THE PATIENT WITH LACTIC ACIDOSIS

> *The important fact to establish is whether or not the rate of formation of lactic acid is rapid.*

Laboratory results (anion gap and direct measurement of lactate) tell you that your patient has high levels of lactate in blood. Your problem now is to find out why (refer to Figure 6.3).

Figure 6.3

Daignostic approach to the patient with lactic acidosis.

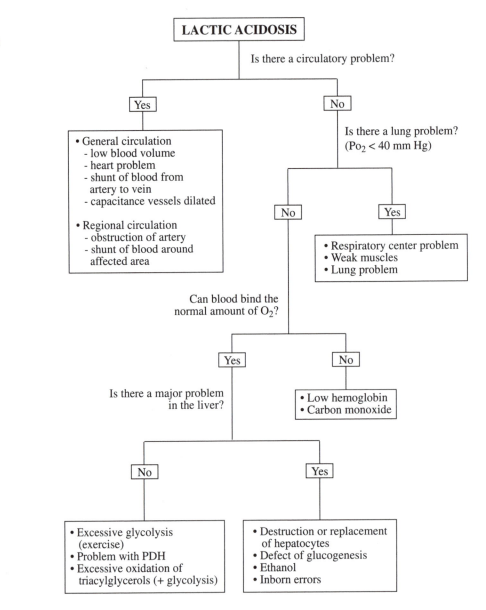

Your first concern is whether the lactic acidosis is due to hypoxia or anoxia in part of the body, since very high rates of formation of lactic acid could be lethal or cause permanent damage. Hypoxia can be due to poor circulation, inadequate functioning of the lungs, or decreased oxygen-carrying capacity of the blood.

1. Poor circulation throughout the body can result either from a decrease in blood volume, which will cause low arterial and venous blood pressure, or from heart problems that reduce the volume of blood pumped and thereby cause low arterial pressure but high venous pressure. Parts of the body can also suffer from poor circulation; examples include arterial or venous occlusion and sepsis, which can cause shunting of blood around areas needing a supply of oxygen.

2. Inadequate lung function can be caused by a condition such as emphysema. The Po_2 of arterial blood is an indirect measure of the availability of oxygen in the blood.

3. The ability of the blood to carry O_2 depends upon the concentration of hemoglobin (determined by hematocrit or direct measurement) and whether or not the hemoglobin has been poisoned (e.g., by carbon monoxide; refer to Case 6·1).

As would be expected from the body's ability to perform prolonged vigorous exercise, the *cardiorespiratory system* has considerable reserve capacity to deliver oxygen to tissues.

> **Cardiorespiratory system:**
> The entire system that delivers oxygen to tissues (and removes CO_2), including the heart, lungs, and blood system.

- With normal cardiac output (5 liters/min) and normal oxygen content in arterial blood (8–9 mmol/l), the blood can deliver 40–45 mmol of O_2 per minute.

- At rest, demand for oxygen is 12 mmol of O_2 per minute. Thus, the tissues consume less than one-third of the O_2 delivered to them; mixed venous blood normally contains 60–70% as much oxygen as arterial blood.

- If demands for oxygen increase, a greater fraction of oxygen can be extracted from blood, the heart can quadruple the volume of blood it delivers to the periphery, and increased rate and depth of respiration can ensure that the hemoglobin in arterial blood remains almost 100% saturated with oxygen.

- Overall, the excess capacity of the cardiorespiratory system means that only very major deficiencies of heart function, pulmonary function, or oxygen-carrying capacity of the blood will be likely to lead to hypoxia in a resting patient.

- Note, however, that in disease states, the ability of the lungs to excrete CO_2 may be compromised before delivery of O_2 to tissues becomes inadequate to meet their needs for O_2 (see Cases 2·2 and 2·3).

CLINICAL APPROACH TO A PATIENT WITH LACTIC ACIDOSIS

The first three questions in Figure 6·3 address the potential causes of hypoxia. If you conclude that hypoxia has not caused the observed lactic acidosis, it is probably due to decreased rates of use of lactic acid, either through glucogenesis or through PDH. Among the possible causes of inhibited glucogenesis are destruction of the liver (indicated by signs such as jaundice), severe intoxication from ethanol, use of drugs that inhibit glucogenesis (e.g., certain oral hypoglycemic agents), ingestion of toxins affecting glucogenesis, and inherited disorders of metabolism in the glucogenic pathway. Inhibition of PDH arises primarily from inherited disorders and from a deficiency of vitamin B_1 (thiamine).

Question

(Discussion on page 268)

6·3 Would a patient with a level of hemoglobin in blood that is 25% of normal be expected to develop lactic acidosis either at rest or during exercise that triples the overall consumption of O_2? Assume that the normal concentration of hemoglobin is 2.25 mmol/l (140 g/l), the normal cardiac output at rest is 5 liters/min (which can rise to 15 liters/min when needed), and the normal consumption of oxygen at rest is 12 mmol/min. Given: 2 mmol of hemoglobin per liter can carry 8 mmol of O_2 per liter.

Discussion of Case 6·1
Hugh Was Found Unconscious in the Garage
(Case presented on page 140)

Why did these abnormalities occur?

The increased anion gap (which matched the fall in the $[HCO_3^-]$ in plasma) indicates metabolic acidosis. The case history is consistent with lactic acidosis from poisoning by carbon monoxide, which binds avidly to hemoglobin. Such bonds essentially irreversibly decrease the ability of the blood to carry O_2 from the lungs. Hugh was unconscious because of the low delivery of oxygen to his brain.

Which of Hugh's organs is most at risk?

Because of its very high demand for regenerating ATP and its limited capacity to buffer H^+, the brain is the organ most susceptible to permanent damage from carbon monoxide poisoning. Treatment, therefore, should pay special attention to saving the brain while addressing the overall problem.

What treatment should Hugh receive?

To decrease the brain's need for O_2 to regenerate ATP, administer anesthesia or initiate hypothermia. Exchange transfusion (replacing the red cells containing poisoned hemoglobin with normal cells) can help in the delivery of oxygen to tissues since the hemoglobin is normal in the transfused cells. Treatment with hyperbaric oxygen (oxygen at a pressure higher than atmospheric, administered by placing the patient in a pressure chamber) is also effective because the high P_{O_2} enables oxygen to displace carbon monoxide from hemoglobin. This treatment will usually require considerable time, so the above measures should be initiated immediately if the brain appears to be compromised.

Will his blood be blue or red in color?

If blood has a very low content of oxygen, it will be blue (the clinical term is "cyanosis"). Hugh's blood was red despite a low content of oxygen because hemoglobin is red when carbon monoxide binds to it.

Discussion of Case 6·2
Greta Has Cancer with Jaundice
(Case presented on page 140)

Is the problem due to a low availability of O_2?

Greta's high Po_2, the almost normal level of hemoglobin in her blood, and lack of evidence of a problem in her circulation indicate that the high anion gap is not due to hypoxia. However, laboratory values show lactic acidosis and normal levels of ketoacids.

Is flux through PDH compromised?

Tumors can deplete the rest of the body of vitamins. A deficiency of vitamin B_1 can be tested by administering thiamine; if the level of lactate does not start to fall soon, the lactic acidosis is not due to a deficiency of vitamin B_1. Thiamine had no effect on Greta. Hence, severe liver damage, consistent with the observed jaundice, is the most likely cause of lactic acidosis (see Table 6·3).

Is her liver removing lactic acid too slowly?

The jaundice indicates that Greta's liver is severely damaged, probably by the spread of cancer cells from her colon. The removal of lactate by glucogenesis will therefore be ineffective. Also, efforts to limit the use of glucose could have undesirable side effects on both the brain and red blood cells, which depend on metabolism of glucose to regenerate their ATP (note the normal low levels of ketoacids, the only alternative brain fuel). Hence, stimulating the oxidation of lactate (pyruvate) through PDH seems to be the only choice.

How can the oxidation of lactate be increased?

Dichloroacetate stimulates PDH and can increase the rate of consumption of pyruvate either when a need exists for more regeneration of ATP or when the rate at which other fuels are oxidized to regenerate ATP decreases (Figure 6·4). Increased oxidation of pyruvate could decrease the need for acetyl-CoA from fatty acids. However, treatment with dichloroacetic acid must be monitored closely since removing lactate will lessen flux through the small remaining capacity for hepatic glucogenesis and could possibly lead to hypoglycemia; as the level of lactate falls, administration of glucose may be necessary.

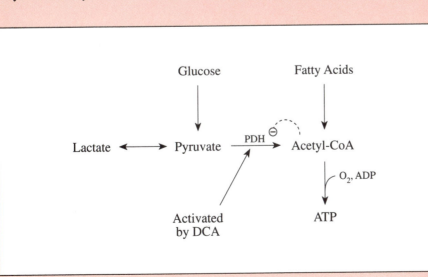

Figure 6·4

Site of action of dichloroacetate in lactic acidosis. Dichloroacetate (DCA) activates PDH, thus favoring the oxidation of lactate despite the presence of fatty acids. Oxygen must be present for lactate to be oxidized at a high rate.

Discussion of Case 6·3
Garth Has a Heart Attack
(Case presented on page 141)

Although Garth has a problem with the amount of blood that his heart is pumping, he shows no sign of hypoxia or acidosis now. Hence, you can focus completely on improving his cardiac output, without being distracted by the need to provide oxygen masks or to control acidosis. Even if he has acidosis, his hemoglobin should be 100% saturated with O_2 as long as he is breathing fairly normally. He will need O_2 only if his respiration rate falls to a point at which the saturation of arterial hemoglobin with O_2 becomes compromised.

Discussion of Case 6·4
Archie Has Severe Hypoxia but Feels Fine
(Case presented on page 141)

Blood samples from healthy patients can sit for up to one hour without undergoing a change in content of oxygen because the red blood cells do not consume oxygen and the very few white blood cells consume oxygen at a slow rate. A patient with leukemia, however, has a high concentration of white blood cells, which consume oxygen rapidly. Archie's leukemic cells caused the appearance of hypoxia but did not actually cause Archie to become hypoxic. Given the large capacity of blood to carry O_2, his leukemic cells would not be able to remove a sufficient amount of oxygen to cause hypoxia in the time that it takes each liter of blood to circulate through his body (1 minute).

To obtain accurate readings of the levels of oxygen in Archie's blood, use an ear oximeter (which measures the content of O_2 in blood), perform the assay for oxygen very rapidly, or inhibit aerobic metabolism in the blood samples by cooling them immediately (usually not adequate) or by using poisons (e.g., cyanide).

Discussion of Case 6·5
Julian's Brain Was Pickled
(Case presented on page 141)

Why was treatment with glucose dangerous?

A DEFICIENCY OF VITAMIN B$_1$

Like Julian, many alcoholics are chronically malnourished; they obtain their energy from oxidizing ethanol, fatty acids, and ketoacids. Their inadequate diet causes vitamin deficiencies. A deficiency of thiamine (vitamin B$_1$) is usually the first to become evident.

Thiamine is an essential cofactor for the activity of PDH and 2-oxoglutarate dehydrogenase, an enzyme of the TCA cycle (Chapter 10). Since a deficiency of thiamine affects PDH first, the body's ability to regenerate ATP primarily from glucose will be compromised because PDH is not required to regenerate ATP from fatty acids or ketoacids.

There are two hypotheses to explain why a patient such as Julian may suffer from brain damage after receiving glucose without thiamine. First, with a low activity of PDH and low levels of ketoacids, ATP must be regenerated by anaerobic glycolysis, yielding lactic acid and leading to local damage from a high [H$^+$]. Second, by inhibiting glycolysis, the high [H$^+$] can lead to an inadequate rate of regeneration of ATP. The name given to this type of brain damage is the Wernicke-Korsakoff syndrome. The areas of the brain that are affected are those with a lower level of thiamine or a higher metabolic rate.

A RELATIVE LACK OF INSULIN

A relative lack of insulin may result from either a low level of glucose, which stimulates the release of insulin from β cells of the islets of Langerhans in the pancreas, or from the presence of an inhibitor (α-adrenergics) for those cells. The level of α-adrenergics increases in response to a contracted ECF volume (usually the result of vomiting).

What precautions should have been taken beforehand?

The aim of therapy is to correct the deficit of thiamine before the level of ketoacids declines. The danger of therapy is that the administration of glucose and saline (which reexpands the ECF volume) leads to the release of insulin. Higher levels of insulin will decrease ketogenesis.

Ketoacidosis allows the brain to regenerate up to 80% of its ATP from a source other than glucose, but when the degree of ketoacidosis diminishes, the brain becomes fully dependent on glucose and PDH to regenerate the ATP it needs. To avoid an energy crisis, administer thiamine via the intravenous route before much glucose and saline are given. This treatment should provide sufficient time to restore the activity of PDH before ketoacid levels fall significantly.

ADDITIONAL QUESTION FOR CASE 6·5

What will the concentration of ketoacids in Julian's blood be one hour from now if ketogenesis is completely stopped? The level of ketoacids in his blood is now 10 mmol/l. These ketoacids are distributed in 30 liters, two-thirds of his total body water. Assume that the only pathway for removal of ketoacids is oxidation in the brain and kidneys.

To answer this question, we need to know the quantity of ketoacids available for oxidation and how fast such oxidation can occur.

Quantity

Since ketoacids are distributed in 30 liters of his body water, he has a total of 300 mmol of ketoacids.

Rate of Oxidation

Brain. The brain can oxidize ketoacids to regenerate close to 80% of the ATP it needs to perform its biological work—about 0.5 mmol/min, or 750 mmol/day. In one hour, the brain can oxidize 30 mmol of ketoacids.

Kidneys. The kidneys oxidize ketoacids and excrete ketoacid anions with NH_4^+ at about half the rate at which ketoacids are removed by the brain. Thus, the kidneys can remove another 15 mmol in an hour.

Other Organs. The oxidation of ketoacids in other organs does not occur rapidly. The conversion of acetoacetate to acetone also occurs slowly. A reasonable estimate of the removal of ketoacids in other organs is 10–20% of the total rate of disappearance.

Summary. The rate of removal of ketoacids should be about 60 mmol in the next hour. This amount is about 20% of the total (300 mmol), so the decline in concentration of ketoacids should be 2 mmol/l (20% of 10 mmol/l) if no ketoacids are formed in this time period. Thus, the estimated concentration after one hour is 8 mmol/l if no other fuels are oxidized.

PART B
KETOACIDOSIS

Case 6·6
Wendy Became Confused and Then Comatose
(Case discussed on pages 159–60)

Recently, Wendy, aged 12 years, became insatiably hungry and thirsty, urinated more often, and lost weight. This morning Wendy was confused, and she became comatose in the afternoon. Her parents called Linda, the family doctor, who immediately ordered blood tests (Table 6·4). Linda noted a rapid, weak pulse and low blood pressure (indicating contraction of ECF volume). She smelled acetone on Wendy's breath, and noted a large amount of glucose in her urine. What are the threats to Wendy's life? What should Linda do immediately? Will the degree of ketoacidosis be affected by Wendy's comatose state? Why will the initial treatment cause the level of glucose in her blood to fall?

Table 6·4
Wendy's laboratory values on admission

Arterial blood		Normal	Wendy
H^+	nmol/l	40	62
pH		7.40	7.22
HCO_3^-	mmol/l	25	10
Pco_2	mm Hg	40	25
Po_2	mm Hg	100	110
Anion gap	mEq/l	12	30
Hemoglobin	g/l	140	150
Glucose	mmol/l (mg/dl)	5 (90)	50 (900)
Urea	mmol/l (mg/dl)	4 (11)	20 (56)
Ketoacids	mmol/l	0.05	17
Creatinine	μmol/ (mg/dl)	113 (1.0)	385 (3.5)
Osmolality	mosmol/kg H_2O	295	355

Case 6·7
Bud Is an Alcoholic
(Case discussed on page 161)

Once again, Bud went on a spree. Friends brought him to the emergency room because he became comatose after vomiting profusely; they swore he had drunk only "store-bought" liquor. He responded only to painful stimuli and had marked contraction of ECF volume and a normal rate and depth of respiration. Laboratory values on a blood sample showed hyperglycemia and ketonemia but normal levels of the components of the bicarbonate buffer system in his plasma (Table 6·5). The hospital records showed no evidence of diabetes mellitus. How can Bud have ketoacidosis and hyperglycemia without diabetes mellitus? How can his pH, $[HCO_3^-]$, and Pco_2 be normal with a high level of ketoacids? What treatment should he receive?

Table 6·5
Bud's laboratory values on admission

Arterial blood		Normal	Bud
Ketoacids	mmol/l	<1	8
Glucose	mmol/l (mg/dl)	3.3–5 (60–90)	20 (360)
H+	nmol/l	40	40
pH		7.40	7.40
Pco$_2$	mm Hg	40	40
HCO$_3$–	mmol/l	24	24

Case 6·8
Collin Cannot Remember the Party
(Case discussed on pages 162–63)

Collin, a first year medical student, much too enthusiastically celebrated finishing final exams. He arrived at the emergency room vomiting profusely, reeking of alcohol, confused, and with marked contraction of his ECF volume. The quick screening test for ketoacids is negative. Do you trust the results of the test?

Questions to Consider

What factors can increase the rate of formation of ketoacids?

How fast can ketoacids be synthesized?

In what organs are ketoacids oxidized?

What factors can decrease the rate of consumption of ketoacids?

How might the levels of glucose in blood indicate the cause of ketoacidosis?

What problems are caused when ketoacid anions are excreted in the urine?

Why does coma aggravate ketoacidosis?

Why would a clinical screening test for ketoacids in blood be falsely negative?

What significance do you attribute to hypoglycemia during ketoacidosis?

BACKGROUND

KETOACID METABOLISM

Ketoacidosis results from a relative lack of insulin.

Ketogenesis

Ketoacids are produced almost exclusively in the liver. Concentrations of ketoacids in blood rise when levels of insulin fall markedly, partly because low concentrations of insulin activate lipolysis in adipose tissue. The low levels of circulating insulin and high levels of glucagon act on the liver to steer fatty acids towards oxidation (and thus ketogenesis) and away from synthesis of triacylglycerols (see Chapter 11). Ketoacids, although normally derived from fatty acids, can also be made from ethanol.

Concentrations of ketoacids in normal fed humans are usually very low (< 0.05 mmol/l). During fasting, the levels rise over 2–5 days to a stable level of approximately 4–6 mmol/l (with some loss in urine). In uncontrolled diabetes mellitus (diabetic ketoacidosis) the concentration of ketoacids can rise to more than 20 mmol/l, and large quantities of ketoacid anions are excreted in the urine (200–400 mmol/day).

Chemistry

Ketoacids (a chemical misnomer) consist of two compounds: acetoacetic acid, a true ketoacid that is both formed from acetyl-CoA and reconverted back to it, and β-hydroxybutyric acid, which, like lactic acid, is a hydroxy acid and a metabolic cul-de-sac. Acetoacetate and β-hydroxybutyrate anions can be interconverted through an NADH/NAD$^+$-linked dehydrogenase, which is always at equilibrium (see the discussion of lactate dehydrogenase on page 142; β-hydroxybutyrate dehydrogenase (β-HBDH) has a similar function, but it is located in mitochondria rather than in the cytosol). The usual ratio of β-hydroxybutyrate to acetoacetate is approximately 2:1 during the ketoacidosis of chronic fasting.

$$\text{Acetoacetate}^- + \text{NADH} + \text{H}^+ \xleftrightarrow{\text{β-HBDH}} \text{β-hydroxybutrate}^- + \text{NAD}^+$$

FACTORS CAUSING CHANGES IN THE CONCENTRATION OF KETOACIDS

> *The maximal rate of formation of ketoacids is relatively slow (1 mmol/min, or 1500 mmol/day).*
>
> *Most of the ketoacids produced are oxidized by the brain (750 mmol/day) and kidneys (250 mmol/day).*

There are several factors to consider in the control of ketogenesis. Ketogenesis can be influenced by the supply of substrate (fatty acids), by the activity of the enzymes in this pathway, or by the rate at which the liver removes ATP, one of the ultimate products of this pathway. The supply of substrate is rarely an important consideration (ketogenesis may respond to increased availability of fatty acids, but only up to a maximum set by the ability of the liver to utilize ATP). The activity of enzymes is also unimportant in regulating ketogenesis. In insulin deficiency, the rate of utilization of ATP in hepatocytes is important in determining the maximum rate of synthesis of ketoacids. The conversion of fatty acids to ketoacids yields NADH, thereby regenerating ATP; hence, the theoretical maximal rate of ketogenesis is determined by the liver's overall need for ATP and is unlikely to exceed 2000 mmol/day.

Removal of Ketoacids

Ketoacids are removed from the circulation by spontaneous conversion to acetone and by oxidation (Figure 6·5).

Conversion to Acetone. Spontaneous conversion to acetone plus CO_2 (both uncharged) occurs without enzymes and consumes the H^+ formed with the acetoacetate. Approximately 10% of the maximal production of ketoacids (150 mmol/day) is converted to acetone and is excreted through the lungs (acetone is very volatile and has the fruity odor that clinicians detect when diagnosing diabetic ketoacidosis).

Figure 6·5

Biochemistry of ketoacid metabolism. The liver forms ketoacids from fatty acids derived from adipose tissue triacylglycerols (TG). A lack of insulin is required for usual flux through this pathway. The brain and kidneys remove 1150 mmol of ketoacids per day. Acetone in breath removes another 150 mmol/day. Other organs, such as muscle and the intestines (not shown in the figure), consume the remaining 200 mmol of ketoacids per day.

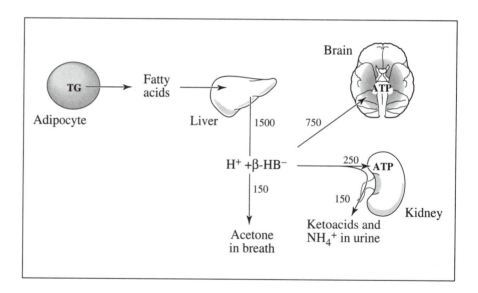

Oxidation. The brain and the kidneys are the major organs that oxidize ketoacids. Ketoacids can meet 80% of the brain's energy requirements (approximately 500 kcal/day) when ketoacid levels in blood are 5 mmol/l or more—equivalent to about half of the daily production of ketoacids in normal prolonged fasting or during diabetic ketoacidosis (approximately 750 mmol/day). The kidneys both excrete and oxidize ketoacids. In normal prolonged fasting, the kidneys excrete approximately 150 mmol of ketoacids per day and oxidize approximately 250 mmol/day.

Muscle does not consume an appreciable amount of ketoacids during chronic fasting, possibly to help in conserving fuel for the brain. Muscle does, however, oxidize fatty acids to regenerate ATP at this time.

DIAGNOSIS OF THE CAUSE OF KETOACIDOSIS

> *Look for the cause of the low level of insulin.*

Ketoacidosis can occur with high, normal, or low concentrations of glucose in blood. The first step in diagnosing the cause of ketoacidosis is therefore to determine the level of glucose in the patient's blood (Figure 6·6). A high level of glucose with ketoacidosis indicates that diabetes mellitus may have caused the lack of insulin (through selective destruction of the β cells of the pancreas) or that the patient has suffered from general damage to the pancreas. A normal level of glucose with ketoacidosis may be consistent with diabetic ketoacidosis if an inhibitor of the production of glucose is also present (e.g., ethanol). Ketoacidosis with a low level of glucose in blood implies a low flow through glycogenolysis or glucogenesis. Chronic fasting could be the cause, as could an inherited defect in synthesis or release of glucose. Ethanol or other drugs may also inhibit these pathways and contribute to the ketoacidosis.

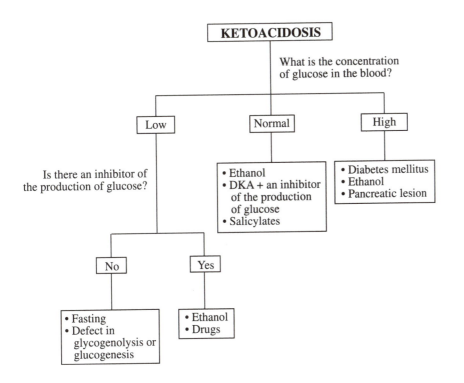

Figure 6·6
Clinical classification of ketoacidosis. The critical step is to assess whether the patient has a low, normal, or high concentration of glucose in the blood. (DKA = diabetic ketoacidosis.)

Questions

(Discussions on page 269)

6·4 Lactate dehydrogenase occurs only in the cytosol of cells, whereas β-hydroxy-butyrate dehydrogenase occurs only in the mitochondria. How can you calculate the ratio of NADH to NAD^+ in both cytosol and in mitochondria from measurements made on the venous blood leaving an organ such as the liver?

What assumption must be made to perform this calculation?

6·5 The liver normally uses 3 mol of O_2 per day. Metabolism of palmitic acid (16 carbons) to ketoacids can be described by the following equations:

$$C_{16}\ \text{Fatty acid} + 4\ NAD^+ + 7\ FAD \longrightarrow \text{Acetoacetate} + 3\ \beta\text{-hydroxybutyrate} + 4\ NADH + 7\ FADH_2$$

$$NADH + H^+ + \tfrac{1}{2}O_2 \longrightarrow H_2O + 3\ ATP$$

$$FADH_2 + \tfrac{1}{2}O_2 \longrightarrow H_2O + 2\ ATP$$

How fast can the liver make ketoacids (theoretically)?

FADH₂:

A cofactor for carrying H. In contrast to NAD^+-NADH, FAD-$FADH_2$ is always bound to proteins and often feeds directly into the ATP generation system.

Discussion of Case 6·6
Wendy Became Confused and Then Comatose
(Case presented on page 153)

Wendy's clinical picture—acetone on the breath, hyperglycemia, and contraction of ECF volume—points immediately to diabetic ketoacidosis. The laboratory data (Table 6·4, page 153) confirm the diagnosis; she has metabolic acidosis with an elevated anion gap in plasma. The high creatinine level indicates poor renal function from contraction of ECF volume. The high osmolality in plasma reflects the hyperglycemia and the high concentration of urea.

What are the threats to Wendy's life?

The threats to Wendy's life include:

1. poor circulation from contraction of ECF volume (due, in turn, to excretion of sodium chloride and potassium caused by rapid loss of glucose and ketoacid anions in urine);
2. acidosis resulting from the level of ketoacids;
3. cardiac arrhythmias resulting from low serum potassium 1–2 hours after insulin is given;
4. hypoglycemia 6–8 hours after insulin is given;
5. the underlying cause of the relative lack of insulin.

Treatment over the next 48 hours must address all these factors, essentially titrating the amount of insulin given with the level of glucose in blood until Wendy's condition stabilizes.

What should Linda do immediately?

Linda should start a rapid infusion of isotonic saline to reexpand Wendy's ECF volume and administer insulin to stop the formation of ketoacids. Later (usually one or two hours, depending on the degree of hyperkalemia), Wendy should receive KCl (her need to excrete glucose and ketoacids has resulted in polyuria, which caused a deficit of K^+).

Will the degree of ketoacidosis be affected by Wendy's comatose state?

Wendy's comatose state will enhance her ketoacidosis. Coma decreases the brain's use of ATP, thus decreasing its demand for regenerating ATP. Since the brain consumes half of the ketoacids produced in prolonged fasting, coma can markedly increase the degree of ketoacidosis.

Why will the initial treatment cause the level of glucose in her blood to fall?

The treatment that Linda started will cause the concentration of glucose in Wendy's blood to fall through dilution, excretion, and, lastly, metabolism (see Case 4·3, pages 112–13). The intravenous infusion of saline will dilute the glucose by expanding the ECF volume. The increased ECF volume will improve renal function, causing more filtration of glucose. Because the level of glucose in blood is well above the renal threshold, glucose will be excreted. Since the synthesis of glycogen will not occur at a rapid rate initially, metabolism will start to consume glucose only when the concentrations of fatty acids and ketoacids in blood have fallen to low levels and thus no longer inhibit PDH; this process can take several hours.

ADDITIONAL QUESTION FOR CASE 6·6

Approximately how long after injection of insulin will it take for Wendy's body to start to oxidizing glucose at the maximal rate permitted by her consumption of oxygen? Assume that she is burning 12 mmol of O_2 per minute while receiving initial therapy, her serum concentrations of ketoacids and fatty acids are 12 and 1.5 mmol/l, respectively, her blood volume is 5 liters, and her ECF volume is 30 liters.

To oxidize glucose to meet all the body's needs for ATP, the concentrations of ketoacids and fatty acids must be very low in the circulation, given the hierarchy of fuels described in Table 1·4, page 9. Hence, to answer this question, the time taken to oxidize the fatty acids and ketoacids in blood must be calculated.

Ketoacids

• The total quantity is 12 mmol/l × 30 liters, or 360 mmol.

• If a person consumes oxygen at a normal rate, the maximum rate at which he or she can remove ketoacids is close to 1 mmol/min, which is also the rate of production of ketoacids (page 156).

• Therefore, ketoacids will be oxidized in six hours if no more are produced.

Fatty Acids

• The total quantity in the circulation is 1.5 mmol × a plasma volume of 3 liters (remember that 40% of her blood volume is red blood cells), or 4.5 mmol of fatty acids.

• The quantity of albumin outside her circulation is close to the quantity in plasma, so add another 4.5 mmol of fatty acids to the total.

• Because fatty acids are sparingly soluble in water, their concentrations in aqueous solutions will not be appreciable.

• Fatty acids will be bound to other proteins as well as albumin, but, for simplicity, we shall ignore these.

• The stoichiometry for oxidation of fatty acids is 24 mmol of O_2 per mmol of palmitate. Thus, 0.5 mmol of palmitate can be oxidized per minute if oxygen is consumed at a rate of 12 mmol/min. If palmitate were the only fuel oxidized, it would take 18 minutes to oxidize the 9 mmol present. Thus, if palmitate were used as a fuel for only half the organs (not the brain or kidneys), fatty acids would disappear much more quickly than would ketoacids.

Glucose will be oxidized at rapid rates six hours after ketogenesis is arrested. During the treatment of diabetic ketoacidosis, glucose is typically administered when its concentration falls to 15 mmol/l (270 mg/dl), usually about six hours after insulin is given.

Discussion of Case 6·7
Bud Is an Alcoholic
(Case presented on pages 153–54)

How can Bud have ketoacidosis and hyperglycemia without diabetes mellitus?

Bud's ketonemia and hyperglycemia would suggest diabetes mellitus, but his previous hospital records indicate that he does not have this problem. However, a relative lack of circulating insulin could be caused by adrenaline, which is released as a result of a contracted ECF volume and, in turn, inhibits the release of insulin from the pancreas.

Even though Bud has not eaten for several days, fasting is not the cause of his ketoacidosis because he has hyperglycemia, not hypoglycemia.

How can his pH, $[HCO_3^-]$, and Pco_2 be normal with a high level of ketoacids?

Metabolic acidosis usually results in an increase in the $[H^+]$ and a decrease in the $[HCO_3^-]$ that is comparable to the concentration of metabolic anions. The normal values in Table 6·5 (page 154) can be explained by Bud's history of vomiting, which causes loss of stomach acid; in his case, the loss of H^+ through vomiting matched the gain of H^+ caused by the ketoacidosis.

What treatment should he receive?

Bud's most immediate problem is his low ECF volume, which should be treated with a rapid, large intravenous infusion of saline. No insulin is necessary now because he is not currently acidemic. The $[K^+]$ in his plasma should be monitored, since vomiting probably caused an excessive loss of K^+. The presence of hypokalemia would be good reason to delay or avoid the administration of insulin.

Discussion of Case 6·8
Collin Cannot Remember the Party
(Case presented on page 154)

Collin's high anion gap and case history strongly suggest ketoacidosis. The negative screening test should be discounted in favor of clinical judgment pending laboratory analysis. The quick test detects only acetoacetate, not β-hydroxybutyrate, but the high concentration of ethanol (case history and smell on his breath) will shift the equilibrium toward β-hydroxybutyrate because of effects on the ratio of [NADH] to [NAD⁺] (Figure 6·7). Direct measurement confirms this possibility.

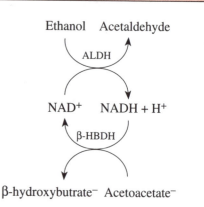

Figure 6·7
Effect of ethanol on the level of acetoacetate. When ethanol is oxidized, NADH is formed in the liver. A high level of NADH drives acetoacetate to β-hydroxybutyrate. (ALDH = alcohol dehydrogenase; β-HBDH = β-hydroxybutyrate dehydrogenase.)

ADDITIONAL QUESTIONS FOR CASE 6·8

Why isn't the pH of his blood low in the presence of ketoacidosis?

The [H⁺] in Collin's blood is lower than expected (alkalosis) given the higher-than-normal anion gap in his plasma (Table 6·6). The alkalosis is consistent with loss of H⁺ from vomiting (which persisted after he arrived at the hospital). The large anion gap indicates the presence of unidentified anions in his blood. Thus, he has both lost and gained acids.

Table 6·6
Collin's laboratory values on admission

Arterial blood		Normal	Collin
H⁺	nmol/l	40	30
pH		7.40	7.50
HCO₃⁻	mmol/l	25	32
Pco₂	mm Hg	40	40
Po₂	mm Hg	100	100
Anion gap	mEq/l	12	32
Glucose	mmol/l (mg/dl)	5 (90)	3.3 (60)
Ketoacids	mmol/l	0.05	5
Creatinine	μmol/l (mg/dl)	68 (0.6)	226 (2.0)
Osmolality	mosmol/kg H₂O	295	385

Why is the osmolality of his plasma so high?

The high osmolality is not due to glucose (he is hypoglycemic) or to the elevated level of creatinine (the concentration is measured in $\mu mol/l$ not $mmol/l$); it is consistent with the level of ethanol in his blood (70 mmol/l, or 320 mg/dl, from elevation in osmolality—four times the conviction limit for driving).

Question

(Discussion on page 270)

6·6 How much ethanol can be metabolized by the liver per day if the product is a) CO_2; b) acetate; c) β-hydroxybutyrate? Assume that the liver consumes 3 mol of O_2 per day and that conversion of ethanol to CO_2, to β-hydroxybutyrate, and to acetate produces 16, 2.5, and 6 mmol of ATP per mmol of ethanol, respectively.

PART C
CASES FOR REVIEW

Case 6·9
Steve Drank Until He Became Sick
(Case discussed on pages 165–66)

Steve, just back from a posting on the oil wells in Saudi Arabia, tried to make up for lost drinking time. He had been partying for a week, vomiting on many occasions, but had had no alcohol for 30 hours. He was brought to the hospital. You find low arterial and venous blood pressure and a rapid pulse rate. Table 6·7 shows data from laboratory analysis of a blood sample. What caused the acidosis? Is the hypoglycemia of immediate concern? What are the most important considerations for therapy?

Table 6·7
Steve's laboratory values on admission

Arterial blood		Normal	Steve
H^+	nmol/l	40	45
pH		7.40	7.35
HCO_3^-	mmol/l	25	20
Pco_2	mm Hg	40	35
Po_2	mm Hg	100	100
Anion gap	mEq/l	12	32
Hemoglobin	g/l	140	150
Glucose	mmol/l (mg/dl)	5 (90)	3 (54)
Ketoacids	mmol/l	0.05	6
Lactic acid	mmol/l	0.75	12
K^+	mmol/l	4.0	3.0
Creatinine	μmol/l (mg/dl)	113 (1.0)	225 (2.0)
Osmolality	mosmol/kg H_2O	295	295

Case 6·10
A Type II Diabetic Gets Gangrene
(Case discussed on page 167)

Gangrene:
Death of tissue caused by persistent anoxia. Gangrene leads to infection via anaerobic bacteria.

Joan, a type II diabetic for 10 years, has been treated with two oral hypoglycemic drugs: one that inhibits glucogenesis and the other that stimulates the release of insulin. In the past 10 days, poor delivery of blood to her legs has caused *gangrene* in her toes and infection of the surrounding tissues; for seven days, she has been on antibiotics, which have partially controlled the infection. On arrival in the emergency room, her diabetes is in poor control (severe hyperglycemia, a large anion gap, a low [HCO_3^-], and a low pH, as shown in Table 6·8). Physical examination shows a rapid respiratory rate, mild fever, and a slight contraction of ECF volume. Bowel sounds are normal, and renal function is not markedly reduced. What is the most likely cause for the acidosis?

Table 6·8

Joan's laboratory values on admission

		Arterial blood	Urine
Glucose	mmol/l (mg/dl)	40 (720)	300 (5400)
Urea	mmol/l (mg/dl)	11 (31)	No data
Creatinine	μmol/l (mg/dl)	126 (1.4)	12000 (140)
Na^+	mmol/l	138	23
K^+	mmol/l	4.2	24
Cl^-	mmol/l	97	14
HCO_3^-	mmol/l	10	0
pH		7.20	5.4
Pco_2	mm Hg	26	No data
Po_2	mm Hg	105	No data
Albumin	g/l	40	No data
Osmolality	mosmol/kg H_2O	350	435

Veterinary Case 6·1
Preparing to Wrestle with a Crocodile

(Case discussed on page 168)

You are asked to help a novice crocodile wrestler subdue a crocodile, an animal that uses brisk repetitive bursts of energy when defending itself. From your knowledge of energy metabolism associated with the sprint (Chapter 3), what advice would you give him? Do not use drugs or anything that might harm the crocodile in the long term.

Veterinary Case 6·2
Why Was the Goldfish Blue?

(Case discussed on page 168)

Madison found her pet goldfish on the floor beside the fishbowl; it had been out of the bowl for many hours and appeared to be dead. Her grandmother, a very practical lady, put the fish back in its bowl and it survived! Madison asked her science teacher how a fish could live without oxygen for so long (no water passed through its gills). Can you help the teacher answer this question?

Discussion of Case 6·9
Steve Drank Until He Became Sick

(Case presented on page 164)

What caused the acidosis?

The data (Table 6·7) show metabolic acidosis (high $[H^+]$ and low $[HCO_3^-]$) with an increased value for the anion gap in plasma. However, the anion gap is increased much more than the fall in $[HCO_3^-]$, probably reflecting the vomiting that preceded the admission. The cause of the metabolic acidosis may be lactic acidosis, ketoacidosis, or the ingestion of methanol, but not renal failure (creatinine is not elevated enough). Since the normal osmolal gap in plasma indicates an absence of alcohol in the blood, we shall focus on lactic acidosis and ketoacidosis.

LACTIC ACIDOSIS

The level of lactate in Steve's blood has increased considerably to 12 mmol/l (the normal value is 1 mmol/l). Because there is no evidence to support an inadequate level of O_2 in the blood or a problem with delivery of O_2 to tissues, slower removal of lactic acid by the liver is likely. The pathways for removal of lactic acid via glucogenesis and oxidation are relatively slow, so it is reasonable to suspect that lactic acid was produced at a rapid rate recently. The absence of ethanol means that ethanol did not cause lactic acidosis. Convulsion and alcohol-withdrawal symptoms may have occurred recently, and both lead to rapid production of lactic acid.

Although Steve's medical history does not suggest a deficiency of thiamine (vitamin B_1), an injection would do no harm and might be very valuable.

KETOACIDOSIS

The elevated level of β-hydroxybutyrate (6 mmol/l, normal < 0.5 mmol/l), indicates the presence of ketoacidosis. A high rate of production of ketoacids requires a relative lack of insulin. Ethanol can indirectly cause a lack of insulin by irritating the stomach and causing excessive vomiting, which results in contraction of the ECF volume. A low ECF volume stimulates a large release of adrenaline; the α-adrenergic actions of adrenaline inhibit the release of insulin from β cells of the pancreas even if hyperglycemia is present. Since Steve's level of blood sugar was low (3 mmol/l, 54 mg/dl), diabetic ketoacidosis is unlikely, especially since ethanol is no longer inhibiting gluconeogenesis.

Is the hypoglycemia of immediate concern?

Probably not. Although glucose is usually the primary fuel for the brain, Steve has a large supply of alternate fuels (ketoacids). However, once the effects of insulin cause ketoacids to decline, glucose will be needed for his brain. Therefore, it seems prudent to give a small quantity of glucose (enough to raise his concentration of glucose by 5 mmol/l). Quantitatively, since glucose distributes in half of body water (about 20 liters in a 70 kg adult), give Steve a minimum of 100 mmol of glucose (a third of a liter of D_5W).

What are the most important considerations for therapy?

The biggest immediate threats to Steve's life are marked contraction of his ECF volume (give several liters of isotonic saline), a serious degree of hypokalemia once insulin takes action (give KCl slowly at first, but more rapidly later), hypoglycemia (give glucose as described above), and the possibility of withdrawal from alcohol (delirium tremens). Hence, he should be observed carefully over the next 24 to 48 hours.

Discussion of Case 6·10
A Type II Diabetic Gets Gangrene
(Case presented on pages 164–65)

The acidosis evident from both the [H^+] in blood and the anion gap in plasma seems unrelated to poor renal function since the level of creatinine in blood is close to normal. Therefore, consider the possibility of lactic acidosis and ketoacidosis.

LACTIC ACIDOSIS

The normal Po_2 and blood pressure indicate adequate cardiorespiratory function, and there is no evidence of inadequate liver function or ingestion of drugs or toxins. Nevertheless, the acidosis could be due to accumulation of lactate caused by hypoxia in her feet (gangrene causes shunting of the blood around the affected area).

KETOACIDOSIS

Though noninsulin-dependent diabetes mellitus usually does not cause ketoacidosis, the severe hyperglycemia suggests that the oral hypoglycemic drugs are no longer effective and that she requires insulin. Hence, the lactic acidosis could be mixed with ketoacidosis resulting from a relative lack of insulin.

LABORATORY DATA FOR A DEFINITIVE DIAGNOSIS

Analysis for organic anions reveals both lactic acidosis and ketoacidosis (6 mmol/l each).

Question

(Discussion on page 270)

6·7 Can inhibiting the oxidation of fatty acids help decrease the severity of acidosis caused by overproduction of acids from the gastrointestinal tract?

Discussion of Veterinary Case 6·1
Preparing to Wrestle with a Crocodile
(Case presented on page 165)

The cardinal metabolic feature of the type of exercise used in a sprint is an exceedingly rapid rate of regeneration of ATP from anaerobic glycolysis; recall that the price to pay for this regeneration of ATP is a very large load of H^+.

Fatigue during a sprint seems to be related to the degree of accumulation of H^+ in muscle cells. Since H^+ are formed very rapidly and in large quantities with each burst of muscular activity, advise the wrestler to encourage (provoke) the crocodile to move rapidly and often, with sudden bursts of activity. Lactate anions plus H^+ are removed at a very slow rate compared with the rate at which protons are formed, so that it should not take long for a considerable quantity of H^+ to accumulate in the crocodile's muscle and induce fatigue (preventing it from performing further vigorous exercise).

Discussion of Veterinary Case 6·2
Why Was the Goldfish Blue?
(Case presented on page 165)

In humans and goldfish, the vast majority of ATP is regenerated when oxygen is consumed in the electron transport pathway (Chapter 10). Although humans can regenerate ATP rapidly via anaerobic metabolism, an enormous load of H^+ accumulates. Madison's goldfish did not seem to have this problem. Why?

A DIMINISHED NEED TO REGENERATE ATP

The fish diminishes its need for ATP by doing less metabolic and mechanical work. In humans, the brain consumes 25–30% of the oxygen used at rest. In fish, the brain is relatively small and may consume proportionately less ATP; it may also reduce its metabolic work by decreasing the permeability of brain cells for Na^+ (pumping this ion accounts for 50–60% of the daily work in brain cells). The mechanism seems to involve the release of adenosine, a product of the breakdown of ATP; adenosine is formed when anoxia causes levels of ATP to fall. The binding of adenosine to its receptor closes down channels for Na^+ in the plasma membrane.

When exposed to hypoxia, fish can also reduce the work of the heart and other organs (see Veterinary Case 2·3, Neptune and the Seal, pages 75–76).

LACK OF AN ANION (LACTATE) AS A METABOLIC END PRODUCT OF ANAEROBIC METABOLISM

Because the fish can form ethanol (a neutral compound) from anaerobic metabolism instead of lactate (an anion), H^+ do not accumulate. The enzymes involved include PDH, alcohol dehydrogenase, and aldehyde dehydrogenase; the NADH formed in both glycolysis and PDH is oxidized to NAD^+ in the latter two dehydrogenases.

Another Application

When Madison's uncle heard about this metabolic strategy, he recognized that the fish would release ethanol via its gills into the water in the fishbowl. He thought that plastic wrap across the top of the fishbowl might reduce the supply of oxygen to the goldfish and result in a drink that would put fun back into science.

SECTION THREE

Metabolic Function and Control

CHAPTER 7

INTRODUCTION TO METABOLIC PATHWAYS AND CONTROL MECHANISMS

PART A
OVERVIEW OF CONTROLS IN METABOLIC PATHWAYS

Metabolic pathways are sequences of enzyme-catalyzed reactions. Some of these reactions function in one direction and are the sites for metabolic control; these reactions normally occur at the beginning of each pathway and just after branch points. Product-inhibition and substrate-activation control mechanisms are consistent with homeostasis.

PART B
PROPERTIES OF ENZYMES

Enzyme-catalyzed reactions can be subject to a range of controls. The simplest is substrate activation; the rate of a reaction generally increases with the concentration of substrate ([S]) until the maximal velocity (V_{max}) is reached. Enzymes can be inhibited or activated by altering their V_{max} or their K_m (usually numerically equal to the value of the [S] when $v = \frac{1}{2} V_{max}$). Activation or inhibition can occur with noncovalent binding of activators or inhibitors, with covalent modification of the enzyme protein (usually by addition or removal of phosphate), or with a change in the amount of enzyme protein.

PART C
MEMBRANES AND METABOLIC CONTROL

Membranes are both barriers and routes for communication between compartments. Because membranes act as partitions, cells and organelles can have very different compositions from their surroundings; some membranes separate cells from the extracellular fluid and others separate intracellular compartments from each other. Membranes also contain transporters, pores, and channels that allow selected molecules or ions to cross them, either down electrical and concentration gradients, or, with the consumption of energy, against chemical or electrochemical gradients. Membranes also allow transmission of endocrine signals through second-messenger systems.

Case 7·1   Treatment of a Patient with Cholera   page 180

PART D
COENZYMES, ENZYMES, AND CONTROLS

Energy metabolism depends heavily on the major coenzymes ATP, NAD(H), NADP(H), and coenzyme A (CoA). ATP, the currency of energy in cells, transfers phosphate groups (mainly in the carbohydrate system) and provides energy (throughout metabolism); ATP controls its own formation through product inhibition. NADH and NADPH transfer H between reactions. Whereas NADPH is used mainly in anabolism, NADH is used mainly in catabolism. CoA is used to activate acyl groups in the fat system.

PART A
OVERVIEW OF CONTROLS IN METABOLIC PATHWAYS

> *Metabolic pathways are functional units of a metabolic process.*

The metabolic systems and processes that regenerate ATP and interconvert circulating fuels and energy stores consist of a number of metabolic pathways, which are described in Chapters 8–12. The rates of each of these pathways are controlled by means that reflect their physiological functions.

In Chapter 1 (pages 16–17), substrate activation and product inhibition are described very generally to indicate the overall controls but not to define precise mechanisms. For example, oxidation of glucose to regenerate ATP must be controlled through product inhibition by the concentration of ATP (or related compounds); the formation of triacylglycerols from glucose is controlled through substrate activation by glucose and related compounds—a more logical control because the process functions to remove surplus glucose.

Those who wish a more detailed understanding of physiological controls over metabolic systems may be helped by further background in biochemistry. This chapter provides only the core information that we feel necessary to understand these control mechanisms. Chapters 8–12 provide a somewhat closer look at the biochemistry of metabolic systems and processes, while examining the physiological responses of energy metabolism.

Questions to Consider

Why is it helpful to know the function of a metabolic pathway to understand its control?

What is the significance of the degree to which an enzyme-catalyzed reaction is displaced from its thermodynamic equilibrium?

Which sites in a metabolic pathway are the most likely points of control?

What controls the rate of an enzyme-catalyzed reaction?
How quickly can such controls respond?

How can controls be overridden?

What roles do membranes play in metabolic regulation?

SITES OF CONTROL IN METABOLIC PATHWAYS

> *Function and control are tightly linked.*
> *Pathways operate in one direction.*
> *Sites of control generally occur at the beginning and end of the pathway and after branch points.*
> *Controls are exerted at nonequilibrium reactions.*

Metabolic systems and processes are made up of metabolic pathways, which are the functional units of metabolic processes. Each pathway generally has one clearly identifiable physiological function that contributes to the needs of the body as a whole;

some pathways have more than one role. Pathways, like systems and processes, must be controlled. The principles of homeostasis, including product inhibition and substrate activation, apply broadly to physiological control mechanisms.

Metabolic pathways are made up of sequences of reactions, each catalyzed by its own specific enzyme. Metabolic pathways often branch from other pathways or join into them. Figure 7·1 illustrates a group of interconnecting metabolic pathways. A precious fuel, A, in short supply, can be converted to M, an essential product that is required for many cellular functions. M can also be synthesized from J, an abundant fuel. A can be stored as D, in limited amounts, and D can also be converted to M. Those familiar with Chapters 1 and 3 will readily identify probable candidates for A (glucose), D (glycogen), J (fatty acids), and M (ATP), and the organ involved (liver). With this information, what controls might be expected over the four pathways, A→D, A→M, D→M, and J→M?

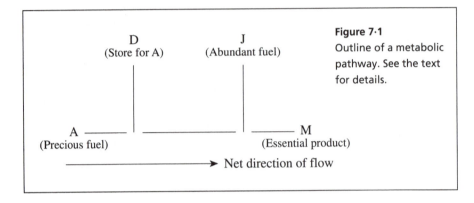

Figure 7·1
Outline of a metabolic pathway. See the text for details.

Because the function of three of the four pathways is to make M, the concentration of M ([M]) will probably control its own formation from A, D, and J. But how? Does M simply accumulate until the stock is full? Will more M then lead to synthesis of A, D, or J? How does J get used preferentially to A or D? Do supplies of A simply fill up stores of D until they are full? To address questions such as these, an understanding of the energetics of enzyme-catalyzed reactions is useful.

ENZYME-CATALYZED REACTIONS

A typical enzyme-catalyzed reaction interconverts two successive compounds in a pathway. This interconversion often involves cofactors that add or take away chemical entities such as phosphate or hydrogen; other enzyme-catalyzed reactions simply rearrange, split, or combine intermediates of the pathway, with or without cofactors.

Controls over metabolic pathways are exerted by modification of the activities of individual enzymes. The question is, which enzymes? It may be helpful to think of pathways by analogy with the flow of water through rivers and canals. Are pathways like networks of canals—all on one level so that water can flow backwards and forwards in response to addition or withdrawal at various tributaries? Or is a better analogy a network of rivers, each flowing in one direction with rapids, waterfalls, and slow, flat sections? Chapter 1 indicates that the latter is much more likely and also identifies the need for mechanisms that cause flow against some of the one-way parts (e.g., using energy to "pump the water back uphill").

EQUILIBRIUM AND NONEQUILIBRIUM REACTIONS

> *Controls are exerted mainly at reactions catalyzed by enzymes that are not at equilibrium.*

The analogy of the river introduces the concept of equilibrium and nonequilibrium into pathways and their control. Metabolic pathways contain reactions or sequences that are reversible physiologically (the flat sections) and other reactions that are irreversible (the rapids and waterfalls) unless extra energy (a pump) is used to push the flow upstream. Widening or deepening a flat part of a river will not significantly affect the rate of flow through it; such controls are effective only at the waterfalls or rapids. Thus, significant control is effective only in the irreversible parts of the pathway.

All enzymes are capable of catalyzing the interaction between their substrates and products in either direction. Following the analogy of the river, even Niagara Falls could be reversed if the level of Lake Ontario were to rise enough. Similarly, within each pathway, only some enzymes keep their substrates and products close to chemical equilibrium (called near-equilibrium reactions); at other reactions, the concentrations of products never rise high enough to approach chemical equilibrium (called nonequilibrium reactions). This distinction is important because controls are exerted at the nonequilibrium reactions, which are not reversible in vivo. Because each pathway contains at least one nonequilibrium reaction, pathways generally function in only one direction.

Bringing this information back into the metabolic pathways in Figure 7·1 allows the chart to be redrawn (Figure 7·2). The links between A, D, J, and M become directional, and the formation and breakdown of D are separated into two distinct pathways, both irreversible. Figure 7·2 also illustrates the probable con-

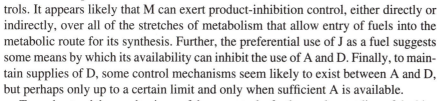

trols. It appears likely that M can exert product-inhibition control, either directly or indirectly, over all of the stretches of metabolism that allow entry of fuels into the metabolic route for its synthesis. Further, the preferential use of J as a fuel suggests some means by which its availability can inhibit the use of A and D. Finally, to maintain supplies of D, some control mechanisms seem likely to exist between A and D, but perhaps only up to a certain limit and only when sufficient A is available.

To understand the mechanisms of these controls, further understanding of the biochemistry of the pathway is needed. Figure 7·3 illustrates a more detailed metabolic pathway, containing all of the reactions of the pathway (but no cofactors). At which enzymes are controls likely to be exerted? From the above analogy with the river, the controls are likely to be exerted at the nonequilibrium reactions.

Where in a pathway are these nonequilibrium reactions likely to be found? Pathways occur inside cells and often interconvert fuels from the blood or in stores. The concentrations of intermediates in the cells are often minute compared with the amounts of fuels in stores. Because it would seem undesir-

Figure 7·2
Possible controls over the outlined metabolic pathway. Note that M is likely to control its own formation from all fuels, A is likely to promote its own storage as D, and D is likely to limit its own formation from A. Metabolism of J can lead to high levels of M and thereby prevent the conversion of A or D to M. Accumulation of A favors the synthesis of D, up to a given limit.

Figure 7·3
Detailed metabolic pathway and identification of equilibrium and nonequilibrium reactions. For details concerning the sites of equilibrium and nonequilibrium reactions, see the text. The nonequilibrium reactions are indicated by one-way arrows.

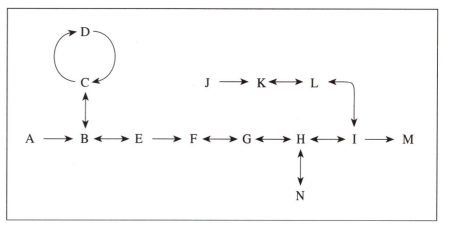

able for the concentration of the first intermediate of a pathway to depend strongly on the concentrations of the circulating fuels or stores, it seems likely that the first reaction is subject to controls and is thus nonequilibrium. Hence, reactions AB, CD, DC, JK, EF, and IM are likely to be nonequilibrium. Further likely sites for control are near branch or junction points so that each physiological function can be individually controlled. However, BC, for example, must be near-equilibrium since it clearly operates in both directions to make or degrade D.

The displacement from equilibrium is determined by measuring the concentration of all reagents in each reaction, including cofactors, and inserting them in the following equation:

$$\frac{[P_1][P_2]}{[S_1][S_2]} = X$$

The value of X can then be compared with the equilibrium constant for the reaction (determined in vitro). For flow to occur from substrates to products, X must be less than K_{eq}. If X is close to K_{eq}, the reaction is close to equilibrium; if X is much less than P (a ratio smaller than 10^{-2}), the reaction would appear to be nonequilibrium. Though educated guesses can be made, only measurement can clearly identify which reactions should be classed as nonequilibrium. Information on the equilibrium and nonequilibrium reactions of metabolic pathways, along with the physiological information on the priorities of the fuels, allows endless speculation over control mechanisms. To resolve this speculation, we must learn more about the properties of the nonequilibrium reactions.

To take the next step in analyzing the control mechanisms, we shall consider the properties of enzymes and then return to our hypothetical pathway.

PART B
PROPERTIES OF ENZYMES

> *Enzymes are catalysts. In general, the rate of enzyme-cat-*
> *alyzed reactions increases as the [S] rises, up to a maximum*
> *velocity.*
>
> *Key enzymes are regulated acutely by inhibitors or activa-*
> *tors and by phosphorylations and dephosphorylations.*
>
> *Long-term control occurs when new enzyme molecules are*
> *synthesized.*

Almost all of the reactions of energy metabolism are catalyzed by enzymes; very few are spontaneous (e.g., glycation of proteins during hyperglycemia and formation of acetone from acetoacetate; see Chapter 4).

KINETICS

The activity (v) of a typical simple enzyme (E) is determined by the concentration of substrate ([S]) up to the point at which the [S] is high enough to produce its maximum velocity (V_{max}) (Figure 7·4); the value of the [S] at which the enzyme is half saturated with S, and thus where $v = \frac{1}{2} V_{max}$, is called the K_m. The Michaelis-Menten equation, linking v, V_{max}, K_m, and the [S], is in the margin.

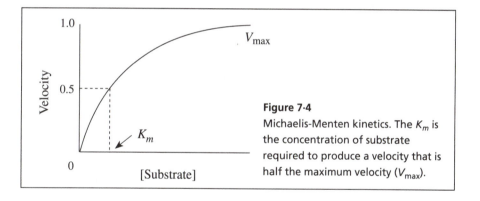

Figure 7·4
Michaelis-Menten kinetics. The K_m is the concentration of substrate required to produce a velocity that is half the maximum velocity (V_{max}).

Some enzymes, including a number of rate-controlling enzymes, show sigmoid kinetics (Figure 7·5). The response of v to the [S] is consistent with a substrate-activation mechanism; it provides the simplest physiological type of control over the rate of a reaction. As long as v is less than V_{max}, changes in the [S] will affect v; if the [S] is high enough to saturate E, changes in the [S] will have no effect on v because the enzyme will be at its V_{max}.

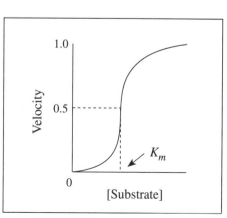

Derivation of the Michaelis-Menten Equation

The Michaelis-Menten equation describes the initial rate of an enzyme-catalyzed reaction. It assumes a reaction mechanism. (E = enzyme, S = substrate, P = product, and ES = substrate bound to the enzyme.)

$$E + S \leftrightarrow ES \rightarrow E + P$$
$$\text{(1)} \qquad \text{(2)}$$

Reaction 1 is assumed to be rapid enough to be at equilibrium. Reaction 2 is relatively slow and cannot be at equilibrium because P has not been formed (initial rate). Hence, the rate of formation of P depends on the [ES], according to $v = k$ [ES] (k is a rate constant).

Equilibrium in reaction 1 can be expressed as

$$K = \frac{[E][S]}{[ES]}$$

where the concentration of the total enzyme $[E_t] = [E] + [ES]$.

When S is saturating the enzyme, $[ES] = [E_t]$ and $v = V_{max}$. Substituting to insert v and V_{max} gives

$$K = \frac{(V_{max} - v)[S]}{[v]}$$

Rearranging gives

$$v = \frac{V_{max}[S]}{K + [S]}$$

This K is called the Michaelis-Menten constant, K_m. It is the dissociation constant of the ES complex and is equal to the [S] when $v = \frac{1}{2} V_{max}$.

Figure 7·5
Sigmoid kinetics. Sigmoid kinetics often arise when an enzyme has multiple subunits that interact with each other so that binding of S to the first subunit activates binding of another molecule of S to the second subunit, and so on.

ACTIVATION AND INHIBITION

> *The activities of many enzymes can be changed by activation and inhibition, both of which can affect the V_{max} or the K_m of an enzyme.*

V_{MAX} CONTROLS

Agents that affect the V_{max} have the same effect as increasing or decreasing the amount of enzyme ([E]) (Figure 7·6); they simply alter the vertical axis of the curve relating v and the [S] and do not affect the K_m of the enzyme.

Figure 7·6

V_{max} controls. This figure shows the effects of V_{max} activators or inhibitors on the rate of an enzyme-catalyzed reaction at varying [S]. The V_{max} changes but the K_m remains constant. The rate of the reaction catalyzed by an enzyme at a certain [S] is affected by the V_{max} controls at all values of the [S].

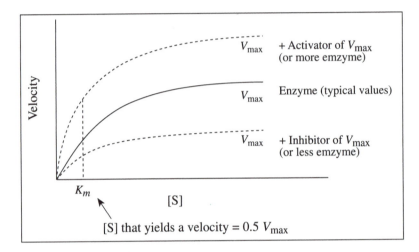

K_M CONTROLS

Agents that affect the K_m alter the horizontal axis of the curve relating v and the [S] but do not affect the V_{max} (Figure 7·7). For inhibitors, the mechanism has the effect of competing with S for the active site of the enzyme. Hence, this type of inhibition requires that the [S] rise if flux is to be maintained at this step.

Figure 7·7

K_m controls. This figure shows the effects of K_m activators or inhibitors on the rate of an enzyme-catalyzed reaction at varying [S]. Activators cause the K_m to decrease, thus increasing v at a certain [S] as long as $v < V_{max}$, and vice versa. K_m controls affect the rate of the reaction only if $v < V_{max}$.

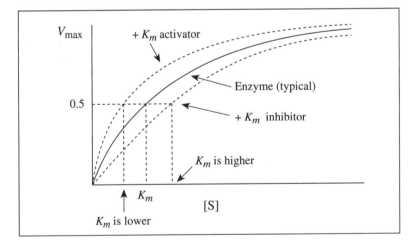

MECHANISMS FOR ACTIVATION AND INHIBITION

Three mechanisms that alter the activities of enzymes in a tissue can be identified: activation or inhibition through noncovalent binding of agents, covalent modification of the enzyme molecules, and synthesis or breakdown of enzyme proteins.

Activation or Inhibition by Noncovalent Binding

The V_{max} or K_m of an enzyme can be altered by noncovalent binding of inhibitors or activators to the enzyme molecule. These activators or inhibitors can be metabolites of the pathway itself (including the substrates or products of the enzyme), metabolites of other pathways, compounds formed purely for the purposes of control, inorganic ions, or even H^+. Because the binding of these activators or inhibitors is noncovalent, the extent of binding is governed by chemical equilibrium; the effects of the activators or inhibitors are therefore rapidly reversible when their concentrations fall. As a result, the factors that affect the concentrations of these activators or inhibitors exert the major control over these enzymes and do so extremely rapidly (fractions of a second).

Covalent Modification of Enzymes

The activities of enzyme molecules can be altered physiologically by chemical (covalent) modification. The most common means is by adding or removing a covalently bound phosphate, using protein kinases or phosphatases, respectively (Figure 7·8). These kinases and phosphatases must be under their own controls in order to achieve coherent physiological responses. The second-messenger mechanism for controlling the activities of protein kinases, described later in the chapter (see pages 183–84), allows hormones to signal their effects on intracellular metabolism.

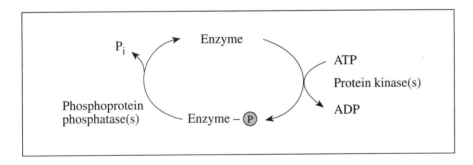

Figure 7·8
Chemical modification of enzymes. The activities of many controlling enzymes are usually modified by phosphorylation and dephosphorylation. Examples include pyruvate dehydrogenase, pyruvate kinase, glycogen synthase and phosphorylase, and hormone-sensitive lipase. The active forms of some enzymes are phosphorylated; other enzymes are inactivated by phosphorylation.

Synthesis and Breakdown of Enzyme Molecules

Like all body components, enzyme molecules are constantly being synthesized and destroyed; the usual half-life for enzyme molecules is a few hours to days. In theory, the concentration of a particular enzyme can be altered by changing the rate of its synthesis or breakdown. Controls over the rate of synthesis are well established and usually involve specific endocrine controls over the expression of the genetic information in the chromosomes.

The rates at which controls affect the concentrations of enzyme molecules are usually slow, often taking hours. Hence, these mechanisms mean little to the minute-by-minute control of energy metabolism but can be of major importance in the response of some pathways (e.g., the synthesis of fatty acids in the liver; Chapter 11, pages 227–29) to long-term changes in dietary status.

RETURN TO CONTROLS IN A PATHWAY

We can now return to the metabolic pathways outlined in Figures 7·1–7·3 and presented in Figure 7·9. Studies in vitro on the nonequilibrium reactions in Figure 7·3 allow biochemical explanations of the physiological characteristics of the interactions between A, D, J, and M.

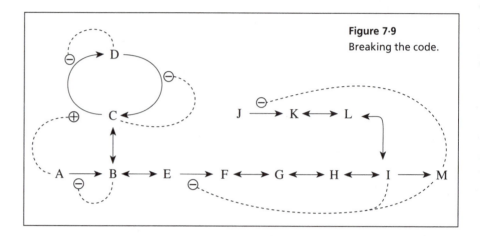

Figure 7·9
Breaking the code.

- The activity of the reaction EF (analogy with phosphofructokinase-1 in Chapter 8, pages 195–98) is the key to the control system. Because the reaction EF is inhibited by M (ATP), changes in the concentration of M or a related compound such as I (e.g., citrate) have immediate effects on the concentration of E.

- Since E is in equilibrium with B and C, both of which control the rates of their own formation, the activity of EF can control the rates of use of both A (glucose) and D (glycogen).

- If B (e.g., glucose 6-phosphate) controls AB (hexokinase) by changing the V_{max} of AB, this product-inhibition mechanism can exert absolute control over the rate of use of A as long as the [A] is much higher than the K_m.

- The effect of A in stimulating CD (e.g., glycogen synthase) and the effects of B in inhibiting AB allow an efficient mechanism for replenishing supplies of D when A and M are abundant. Because M does not inhibit AB, and because EF is the major control site, A is able to replenish D whatever the level of M may be.

CROSSROADS OF METABOLISM

The term "crossroad" indicates the critical point at which pathways intersect. The major crossroads are at pyruvate, at acetyl-CoA, and at glucose 6-phosphate (G6P), the junction point of glucose and glycogen metabolism (Figure 1·16, page 25). Since functions differ in consumer and maintainer organs, controls at these important points will also differ. Figure 7·10 illustrates the controls operating at these crossroads in the liver and in consumer tissues.

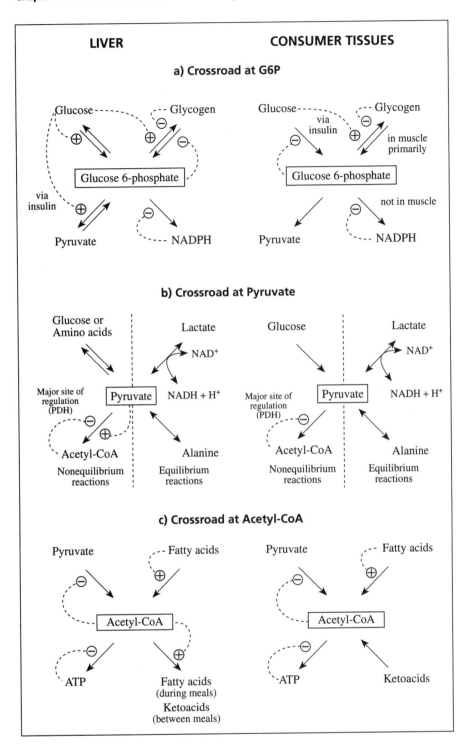

Figure 7·10

Controls at major metabolic cross-roads: a more detailed look. Energy metabolism centers around intermediates that are common to a number of pathways. The most important crossroads are at G6P, pyruvate, and acetyl-CoA. Events in the liver are on the left and events in the consumer tissues are on the right.

LIVER **CONSUMER TISSUES**

a) Crossroad at G6P

Glucose - - - - Glycogen   Glucose - - - - Glycogen
Glucose 6-phosphate   via insulin   in muscle primarily
Glucose 6-phosphate
via insulin
Pyruvate   NADPH   not in muscle
Pyruvate   NADPH

b) Crossroad at Pyruvate

Glucose or Amino acids   Lactate   Glucose   Lactate
NAD⁺   NAD⁺
Major site of regulation (PDH)   Pyruvate   NADH + H⁺   Major site of regulation (PDH)   Pyruvate   NADH + H⁺
Acetyl-CoA   Alanine   Acetyl-CoA   Alanine
Nonequilibrium reactions   Equilibrium reactions   Nonequilibrium reactions   Equilibrium reactions

c) Crossroad at Acetyl-CoA

Pyruvate   Fatty acids   Pyruvate   Fatty acids
Acetyl-CoA   Acetyl-CoA
ATP   Fatty acids (during meals) Ketoacids (between meals)   ATP   Ketoacids

PART C
MEMBRANES AND METABOLIC CONTROL

Case 7·1
Treatment of a Patient with Cholera
(Case discussed on page 184)

You travel as part of a medical team to a third-world country that has suffered from an earthquake. An epidemic of cholera has broken out. The standard therapy is oral salt plus glucose. Why does this therapy work?

FUNCTIONS OF MEMBRANES

> *Membranes are both barriers and routes for communication between compartments.*

MEMBRANES AS BARRIERS

Figure 7·11

The structure of membranes. The basic membrane structure consists of phospholipids arranged with their ionic (polar, hydrophilic) ends to the surfaces and their nonpolar (hydrophobic) ends in the middle. Specific protein molecules are attached to the surface of the phospholipid bilayer, are embedded in it, or traverse it. These protein molecules perform many functions (e.g., transport of materials across the membrane, binding of hormones and transmission of their signals across the membrane, catalysis of reactions, recognition roles in the immune response, and cell recognition).

Some membranes separate cells from the extracellular fluid (ECF); others separate intracellular organelles from the cytosol. The contents of intracellular organelles are very different from the surrounding cytosol, and the composition of the intracellular fluid (ICF) differs markedly from that of the ECF. For example, the concentration of proteins in the ICF is much higher, and the ICF contains enzymes, cofactors, and intermediary metabolites not found in the ECF. In contrast, the ECF contains unique proteins, such as albumin and the immunoglobulins. Moreover, 30-fold to 40-fold concentration gradients of Na^+ (higher in the ECF) and K^+ (higher in the ICF) occur across the cell membrane.

The barrier function of membranes is achieved by their lipid bilayer structure. Phospholipid molecules, which contain a nonpolar hydrophobic end and a polar hydrophilic end, are organized in membranes with the nonpolar parts toward the middle of the membrane and the polar parts on each surface facing the aqueous phase (Figure 7·11). The resulting bimolecular layer, approximately 8 nm thick, is impermeable to polar compounds, since they cannot cross the nonpolar membrane core. Most nonpolar materials cannot gain access to the membrane because they are not soluble in the aqueous medium on either side of the membrane. Only a few compounds (such as methanol, ethanol, or acetone) that are soluble in both polar and nonpolar solvents can readily diffuse across biological membranes. Important compounds such as urea and water cross most biological membranes via specific channels or carriers.

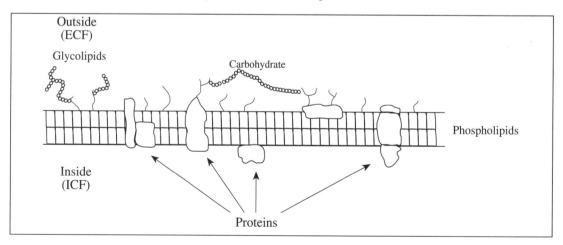

MEMBRANES AS A MEANS OF COMMUNICATION

Membranes also allow communication and transport of fuels between cells and the ECF and between organelles and the cytosol. Almost all enzyme-catalyzed reactions and energy reserves are found inside cells, but most cells derive their energy fuels from other cells. Inside cells, the ATP generation system, pyruvate dehydrogenase, and parts of the carbohydrate and fat systems are located in mitochondria.

This separation of functions between compartments requires specialized communication systems. Fuels, ions, and metabolic intermediates cross cellular or mitochondrial membranes in the various stages of energy metabolism. Plasma proteins, which are made in cells, must cross cellular membranes to be released to the ECF. Hormones in the blood, however, exert controls over intracellular contents without crossing the cell membranes. Cell membranes allow leakage of ions down concentration gradients for short periods of time, thereby allowing conduction of impulses.

Proteins on membrane surfaces and embedded in membranes allow communication across membranes. Some proteins span the membrane; others are found on one side only, though they may link up with other proteins on the other surface. The orientation of proteins in and on the membranes is very specific, so that each type of membrane is unique, with inner surfaces that are very different from the outer surfaces.

Because water can cross most phospholipid bilayers, but almost all the physiological solutes cannot, biological membranes are semipermeable; cells and organelles change their sizes in response to differences in osmotic pressure across these membranes (see Figure 4·8, page 116).

TRANSPORT ACROSS BIOLOGICAL MEMBRANES

> *Molecules cross biological membranes by:*
>
> 1. *simple diffusion (for a few uncharged nonpolar compounds);*
> 2. *facilitated diffusion (using carriers, pores, or channels specific to each molecule);*
> 3. *membrane fusion (for macromolecules or specialized compounds such as neurotransmitters; not discussed further).*

CARRIER-MEDIATED TRANSPORT

Specific molecules (up to molecular mass of approximately 300) can be carried across biological membranes by proteins embedded in the membrane (Figure 7·11). These transporting proteins are as specific in their action as any enzyme-catalyzed reaction. Transporters are generally of two types: facilitated-diffusion and energy-driven transporters.

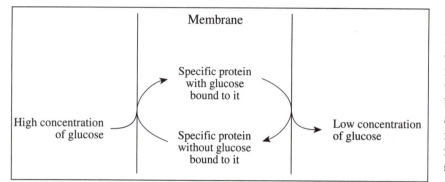

Figure 7·12
Facilitated-diffusion transport systems. This model is only one possibility; others involve a specific pore or channel. In any model, the carrier binds specifically to the compound that it transports.

Facilitated-Diffusion Transporters

Facilitated-diffusion transporters involve a carrier in the membrane (or a pore or channel) that can bind to and release the molecule to be transported on either side of the membrane, thus helping to move the molecule across the membrane. A model mechanism is shown in Figure 7·12. More complex facilitated-diffusion systems require the simultaneous transport of more than one different molecule or ion, either in the same direction or in the opposite direction. Examples include the cotransport of glucose and Na^+ in the intestine or the kidney tubule (Figure 7·13), and the ATP-ADP countertransport that moves ATP out of the mitochondrion where it is made to the cytosol where it is needed (see Chapter 10, pages 220–21).

Figure 7·13
Other facilitated transport systems. These transporters function only if both molecules are transported simultaneously.

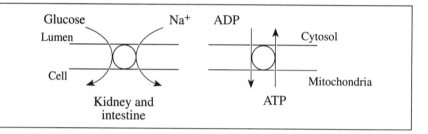

Energy-driven Transporters

Energy-driven transporters are fundamentally similar to facilitated-diffusion transport systems except that the energy of ATP hydrolysis is usually incorporated into the reaction, thus giving it the potential to pump molecules against a concentration gradient. Examples include the pumping of H^+ into the lumen of the distal portion of the kidney tubule (Figure 7·14) and the counterpumping of K^+ and Na^+ into and out of cells (respectively) by Na^+-K^+ ATPase (Figure 7·15).

Controls can be exerted over both facilitated-diffusion and energy-driven types of transporters. The entry of glucose into muscle is a simple facilitated-diffusion mechanism that is activated by insulin (which causes more transport units to be

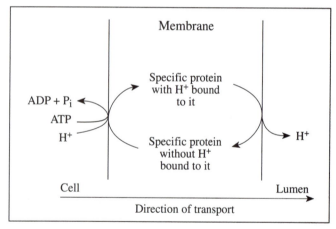

Figure 7·14
Energy-driven transporters. Energy from the hydrolysis of ATP drives the carrier system that pumps H^+ to the outside of the membrane when the kidneys excrete H^+.

Figure 7·15
Accumulation of glucose using a gradient of Na^+. Glucose is absorbed from the intestine and the kidney tubule against a concentration gradient. The driving force for the entry of glucose + Na^+ from the lumen is the very low [Na^+] in the cell, resulting from the activity of Na^+-K^+ ATPase.

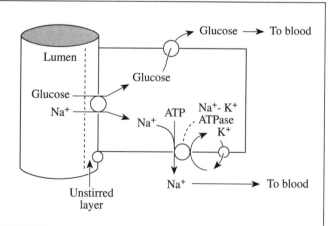

inserted into the cell membrane) or by the effects of anaerobic metabolism (existing glucose transport units become more active). Without these stimuli, the entry of glucose into muscle cells is slow and can limit the metabolism of glucose.

Question

(Discussion on page 271)

7·1 Glucose and Na^+ must be absorbed together, in equimolar amounts, through the cell membrane. The daily intake of glucose is approximately 1500 mmol; the total quantity of dietary Na^+ plus Na^+ secreted into the upper GI is about 600 mmol. How can this quantity of glucose be absorbed?

MEMBRANES AND HORMONE CONTROLS

> *Many hormones control intracellular functions without entering the cell. They react with specific receptors on the plasma membrane surface.*
>
> *The hormone signal is transmitted by second-messenger mechanisms.*

Many hormones are small proteins (insulin, glucagon, ACTH, growth hormone, etc.) that exert controls over intracellular functions; the concentrations of these hormones in blood are usually very low (nmol/l or less). They typically do not cross membranes to relay their messages; instead, they transmit messages into the cell from the outer surface using second-messenger mechanisms (Figure 7·16). Many second-messenger mechanisms have been identified; we shall only highlight an example that is common in energy metabolism.

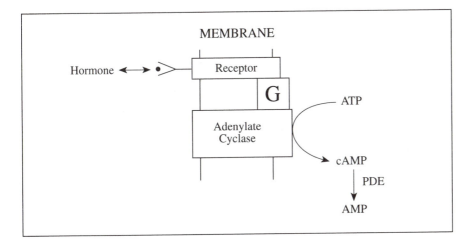

Figure 7·16
Hormonal stimulation of the activities of enzymes through cAMP. Hormones bind reversibly to specific receptors. The hormone-receptor complex, by binding GTP on a specific protein (the G-protein), transduces the message to activate adenylate cyclase. Cyclic AMP (cAMP) is formed and activates a specific protein, kinase A, which phosphorylates a controlling enzyme. (PDE = phosphodiesterase, an enzyme that inactivates cAMP by converting it to AMP.)

HORMONES AND RECEPTORS

Hormone receptors on the outer surface of the cell bind specifically to hormone molecules circulating in the blood. These receptors are proteins that recognize only one type of hormone, for which they have a very high affinity. The interaction between the hormone and the receptor is reversible, such that the concentration of the hormone-receptor complex falls when the concentration of the hormone in the blood decreases. The number of receptors on the surface of each cell can change as a result of the balance between synthesis and breakdown. Modification of the receptor protein (e.g., by phosphorylation and dephosphorylation) can change the affinity of the receptor for the hormone.

FORMATION OF SECOND MESSENGERS

Formation of the hormone-receptor complex on the outer surface of the cell stimulates an enzyme to produce a second messenger inside the cell. The original second messenger discovered was cyclic AMP (cAMP), formed from ATP (Figure 7·16); other second messengers have also been identified, the most important being inositol 1,4,5-*tris*phosphate, diacylglycerol, calcium, and cyclic GMP.

Because second messengers are made inside the cell by enzymes stimulated by the hormonal "first" messengers acting on the outside surface of the cell, the concentration of a second messenger can be many times higher than that of the hormone.

ACTIONS OF SECOND MESSENGERS

Second messengers such as cAMP usually stimulate enzymes that modify the activities of other enzymes, often through phosphorylation and dephosphorylation. Stimulation of cAMP-dependent protein kinase A by cAMP magnifies the effect of the hormone by orders of magnitude (e.g., the cascade of events in the hydrolysis of glycogen; Chapter 8, page 208).

When hormones such as adrenaline in some smooth muscles stimulate the formation of inositol 1,4,5-*tris*phosphate (IP$_3$), the endoplasmic reticulum releases Ca^{2+} into the cytosol, which normally has a very low concentration of Ca^{2+} (1 µmol/l); these Ca^{2+} can activate enzymes such as glycogen phosphorylase (Chapter 8, pages 208–9) and can also be taken up by mitochondria, where they activate enzymes of energy metabolism, such as pyruvate dehydrogenase (Chapter 9, pages 211–13) and 2-oxoglutarate dehydrogenase (Chapter 10, page 220).

Discussion of Case 7·1
Treatment of a Patient with Cholera
(Case presented on page 180)

Patients with cholera secrete much salt and water into the GI tract (cholera toxin stimulates this secretion); the loss of extracellular fluid can lead to death. Replacing the salt lost as a result of cholera cannot be achieved simply by administering oral salt solutions; the salt is not absorbed rapidly enough without glucose. The oral treatment of cholera is a sodium and glucose solution because both are transported together through the transport system (Figure 7·17).

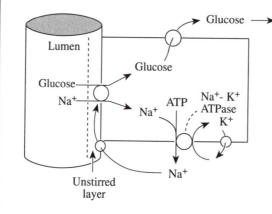

Figure 7·17

Mechanism for absorption of energy fuels. The unique features of this mechanism are an unstirred layer and a tight junction between cells that is permeable to sodium but not chloride. This configuration allows sodium to cycle in the microenvironment where glucose is absorbed; thus, one sodium ion carries many molecules of glucose.

PART D
COENZYMES, ENZYMES, AND CONTROLS

Case 7·2
Selective Hemolytic Anemia
(Case discussed on pages 187–88)

While filming a movie in an area endemic for malaria, an Italian actor (Gino Casetta) developed hemolytic anemia ("broken" red blood cells). The director thought it was caused by the drugs used to prevent malaria (these drugs influence the formation of NADPH), but the producer disagreed because two other actors, Jim Stone and Boris Mossowski, took the same medications without ill effects. Who was right, the director or the producer? Explain your answer. Why was the lesion restricted to red blood cells, with no pathological consequences seen in the liver or muscle?

COENZYMES

> Coenzymes are components of enzyme-catalyzed reactions. They help remove, add, or transport chemical groups. Examples are NAD^+, $NADP^+$, coenzyme A (CoA), and the adenine nucleotide (ATP).

The pathways described in Figure 1·16 (page 25) include only a few intermediates, omitting many other components of the reactions. Most enzyme-catalyzed reactions modify the intermediates of the pathway by adding or removing chemical groups. In many cases, these chemical groups are transferred to coenzymes that are free in solution; free coenzymes can transfer chemical groups to other compounds through reactions catalyzed by other enzymes. Many enzymes also contain bound coenzymes (called prosthetic groups) that are essential to the reaction. Many of these coenzymes and prosthetic groups contain vitamins of the B complex.

ATP, NAD^+, $NADP^+$, and CoA are the most important coenzymes in energy metabolism.

ATP AND TRANSFER OF PHOSPHATE AND ENERGY

ATP is mostly synthesized by the ATP generation system in mitochondria (Chapter 10), except when anaerobic glycolysis is used. Because most ATP is consumed outside mitochondria, ATP must be transported across the mitochondrial membrane. The ATP-ADP countertransporter in the inner mitochondrial membrane (see Figure 7·13, page 182) is therefore very active.

ATP is used in two general ways: as a source of phosphate or as a source of energy. Examples of its use as a source of phosphate are found in glycolysis (Chapter 8).

$$\text{Glucose} + \text{ATP} \longrightarrow \text{Glucose 6-phosphate} + \text{ADP}$$

Examples of the use of ATP as a source of energy include energy-driven transport mechanisms (Figures 7·14 and 7·15, page 182) and the first reaction of fatty acid oxidation (Chapter 11, page 231).

$$\text{Fatty acid} + \text{ATP} + \text{CoA} \longrightarrow \text{Fatty acyl-CoA} + \text{AMP} + \text{PP}$$

Because ATP is essential in energy metabolism, very sensitive control mechanisms are needed. The concentration of ATP in tissues remains quite stable despite very large changes in the rate of consumption of energy (Table 7·1). This stability suggests the presence of a "buffering" system for ATP.

In muscle, ATP participates in two equilibrium reactions, which are catalyzed by creatine phosphokinase and adenylate kinase. In resting muscle, most of the adenine nucleotides are in the form of ATP, and the concentration of CrP is high. In exercise, the concentration of CrP falls markedly and that of ATP falls slightly, indicating that CrP provides a buffering system for ATP.

$$\text{ATP + Creatine} \longleftrightarrow \text{ADP + Creatine phosphate (CrP)}$$

The adenylate-kinase equilibrium causes a large percentage change in the [AMP] with only a small change in the [ATP]. Activation of an enzyme (e.g., PFK1; see Chapter 8) by AMP therefore offers a sensitive mechanism that is consistent with inhibition by ATP.

$$\text{ATP + AMP} \longleftrightarrow \text{2 ADP}$$

Table 7·1

Changes in the [ATP] and the [AMP] in muscle

		Rest	**Exercise**	**% Change**
ATP	mmol/l	7.0	6.3	-10%
AMP	mmol/l	0.12	0.26	$+116\%$

The degree of change of the [AMP] is much greater than that of the [ATP]. A small percentage fall in the [ATP] causes a large percentage rise in the [AMP], since the two compounds are linked through equilibrium by adenylate kinase.

NAD$^+$ AND NADP$^+$ AND TRANSFER OF H

Oxidation (addition of oxygen or removal of hydrogen) and reduction (addition of hydrogen or removal of oxygen) are very common in intermediary metabolism. Some of the enzymes catalyzing these reactions use oxygen directly and use coenzymes usually containing riboflavin; many of these enzymes exchange H with the soluble cofactors, NAD$^+$ or NADP$^+$. A typical reaction of this type is:

$$\text{A + NAD(P)H + H}^+ \longleftrightarrow \text{AH}_2 \text{ + NAD(P)}^+$$

Examples can be found throughout energy metabolism (see Chapters 8–12).

NAD$^+$ and NADP$^+$ have generally different functions. NADPH is the usual source of H for anabolic pathways such as synthesis of fatty acids. NADPH is formed primarily by three specific pathways or enzymes: the pentose phosphate pathway, malic enzyme, and NADP$^+$-linked isocitrate dehydrogenase (the latter is found only in mitochondria). Not all organs have these pathways. For example, red blood cells have only the pentose phosphate pathway and need NADPH to prevent damage by oxidants (see Case 7·2 for an example).

NAD$^+$ and NADH are more generally used in metabolism, accepting H from the oxidation of carbohydrates, amino acids, and fats and also supplying H in glucogenesis. Most of the H used in the ATP generation system is transmitted to the electron transport pathway via NADH.

Some of the NADH used in the mitochondrial ATP regeneration system is formed in the cytosol; conversely, in glucogenesis and synthesis of fatty acids in the liver, NADH generated in the mitochondria must exist in order to transfer H across the mitochondrial membrane. Such mechanisms involve metabolic intermediates, since neither NADH nor NADPH can cross the mitochondrial membrane. Transport of H is therefore through carriage on metabolic intermediates, requiring transport of oxidized and reduced forms of metabolites (the detailed pathways are complicated).

NAD(H) and NADP(H) are involved in many pathways of intermediary metabolism. Many of the reactions involving NAD(H) are in equilibrium and hence do not exert control. However, other NAD-linked dehydrogenases, such as PDH, are very tightly regulated.

TRANSFER OF ACYL GROUPS

Coenzyme A (CoA) is used extensively in the metabolism of acids of the fat system (see Chapter 1, pages 25–27, and Chapter 11). The activation of fatty acids or ketoacids for metabolism requires the consumption of ATP; formation of acetyl-CoA from pyruvate (through PDH, see Chapter 9) and of acetoacetyl-CoA from acetoacetate (by transfer of CoA from succinyl-CoA in the tricarboxylic acid cycle; see Chapter 10, page 217) spares the consumption of extra ATP.

Compounds bound to CoA often must cross the mitochondrial membrane. In the synthesis of fatty acids in the liver, acetyl-CoA produced in the mitochondria is used in the synthesis of fatty acids in the cytosol. Since CoA cannot cross mitochondrial membranes, special transport mechanisms exist. The transfer of acetyl groups out of the mitochondria for synthesis of fatty acids is via citrate (Chapter 11) formed using the first reaction of the tricarboxylic acid cycle. Citrate leaves the mitochondria on a carrier and is reconverted to acetyl-CoA in the cytosol by ATP-citrate lyase. Similarly, in the oxidation of fatty acids in the liver and muscle, fatty acyl groups are first bound to CoA in the cytosol, but the oxidation of these fatty acyl groups occurs in mitochondria. The transfer of fatty acids into mitochondria uses a dedicated-transport mechanism involving carnitine. Fatty acids are transferred from their CoA compounds to form carnitine derivatives, which cross the membrane and regenerate fatty acyl-CoA inside the mitochondria.

✒ Discussion of Case 7·2
Selective Hemolytic Anemia
(Case presented on page 185)

The director was right. The antimalarial drug affected Gino much more than Jim or Boris.

Red blood cells use NADPH to defend themselves against damage from toxic oxygen radicals. Since adult red blood cells do not have mitochondria, they cannot regenerate the reduced form of $NADP^+$ (NADPH) by the two pathways that involve mitochondrial metabolism ($NADP^+$-linked isocitrate dehydrogenase and the malic enzyme pathway, part of which is in mitochondria). Hence, $NADP^+$ must be reduced using the pentose phosphate pathway.

Glucose 6-phosphate dehydrogenase (G6PDH), the first enzyme of the pentose phosphate pathway, is normally present in great abundance. Flux through G6PDH depends on the ratio of $NADP^+$ to NADPH (more $NADP^+$ equals more flux). Although most healthy individuals are not limited in their need to regenerate NADPH by the amount of G6PDH, black males are especially prone to an inherited deficiency of G6PDH, as are Mediterraneans.

One hypothesis for such a high incidence of a deficiency of G6PDH is that it helps in some way to protect against malaria caused by *Plasmodium falciparum*; perhaps reduced glutathione is required for optimal parasite growth. Therefore, the individual of Mediterranean descent (Gino Casetta) could have had a much lower activity of G6PDH. A drug that decreases the activity of G6PDH will reduce flux to a clinically evident amount in a person who has a low level of G6PDH.

The lesion is restricted to red blood cells because they depend on the pentose phosphate pathway to regenerate NADPH. The liver has two other ways to make NADPH

(NADP$^+$-linked isocitrate dehydrogenase and the malic enzyme pathway). Muscle cells do not have (or need) a pentose phosphate pathway, so they are not likely to be affected by a deficiency of G6PDH.

CHAPTER 8
THE CARBOHYDRATE SYSTEM

OVERVIEW

Main forms in the circulation:	*Glucose and lactate*
Main storage form:	*Glycogen*
Key intracellular intermediates:	*Glucose 6-phosphate and pyruvate*
Main function:	*Primary source of fuel for the brain*

Carbohydrates, the major dietary source of energy, dominate energy metabolism. During meals, they are usually the primary source of energy. Between meals and without meals, provision of glucose for the brain is the major concern of energy metabolism, but fat-derived fuels provide most of the energy for the body.

The carbohydrate system contains three major components:
1. glucose, the energy fuel that usually has the highest concentration in the blood;
2. lactate, the major precursor of glucose in the circulation;
3. glycogen, the storage form of carbohydrates in the body.

Glycolysis, glucogenesis, and synthesis and breakdown of glycogen are the most important pathways of the carbohydrate system in energy metabolism. They not only interconvert glucose, glycogen, lactic acid, and pyruvic acid (Figure 8·1) but also bring into metabolism other common dietary carbohydrates—fructose, galactose, and glycerol. These pathways are central to three metabolic processes in energy metabolism: the regeneration of ATP, the storage and release of glucose to maintain concentrations of glucose in blood, and the storage of excess dietary glucose as fats (triacylglycerols).

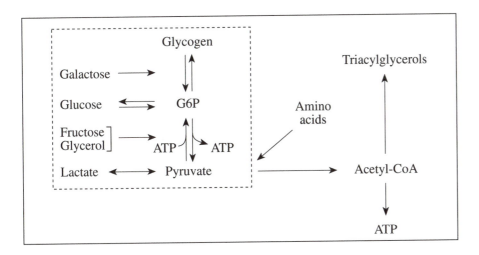

Figure 8·1
The carbohydrate system. The carbohydrate system is enclosed in the dashed box. Glycolysis converts glucose to pyruvate or lactate. Glucogenesis makes glucose or glycogen from precursors, mainly amino acids, lactate, and glycerol. Synthesis and breakdown of glycogen link with glycolysis and glucogenesis. The key branch point in these processes is glucose 6-phosphate (G6P).

PART A
GLYCOLYSIS

Glycolysis, which converts glucose to lactate, is the only way to regenerate ATP out-side the ATP generation system and is thus the only source of ATP in tissues that are unable to use O_2 (red blood cells) or are rendered anoxic (e.g., rapidly exercising mus-cle). Glycolysis (forming pyruvate) is also essential in aerobic regeneration of ATP (in the brain and muscle) and in conversion of glucose to triacylglycerols (in the liver).

Case 8·1   Harry Had a Heart Attack   page 199

PART B
GLUCOGENESIS

Glucogenesis converts pyruvate, lactate, or other precursors (such as glycerol) to glu-cose or glucose 6-phosphate. Glucogenesis occurs in the liver and kidneys; it is the sole source of glucose for the brain in prolonged fasting and is also used to convert dietary proteins to glucose or glycogen when the intake of proteins supplies carbon in excess of the body's immediate needs.

PART C
SYNTHESIS AND BREAKDOWN OF GLYCOGEN

Synthesis and breakdown of glycogen are of major importance to energy metabolism in the liver and muscle. Stores of glycogen in the liver provide glucose for the brain between meals in the normal dietary cycle and for up to 20 hours after the most recent meal. Stores of glycogen in muscle provide fuel for regeneration of ATP in muscle during exercise.

Abbreviation	Enzyme
F1,6Pase	Fructose 1,6-*bis*phosphatase
GK	Glucokinase
G6Pase	Glucose 6-phosphatase
HK	Hexokinase
LDH	Lactate dehydrogenase
PC	Pyruvate carboxylase
PDH	Pyruvate dehydrogenase
PEPCK	Phosphoenolpyruvate carboxykinase
PFK1	Phosphofructokinase-1
PFK2	Phosphofructokinase-2
PK	Pyruvate kinase

Abbreviation	Metabolic intermediate
$F1,6P_2$	Fructose 1,6-*bis*phosphate
$F2,6P_2$	Fructose 2,6-*bis*phosphate
F6P	Fructose 6-phosphate
G6P	Glucose 6-phosphate
PEP	Phosphoenolpyruvate

PART A
GLYCOLYSIS

OVERVIEW OF PHYSIOLOGICAL FUNCTIONS AND CONTROLS

GLYCOLYSIS AND REGENERATION OF ATP

Function: *Regeneration of ATP.*
Control: *Product inhibition by ATP and related compounds—a very effective control.*

Glycolysis converts glucose to lactate or pyruvate in all tissues in order to regenerate ATP (Figure 8·2 and Table 8·1). As expected from the purpose of the pathway (formation of ATP), product inhibition (negative feedback) through ATP and related compounds (ADP, AMP, citrate, etc.) is the major means of control. Also, as expected from the need to preserve glucose whenever possible, the controls over glycolysis for regeneration of ATP are very effective.

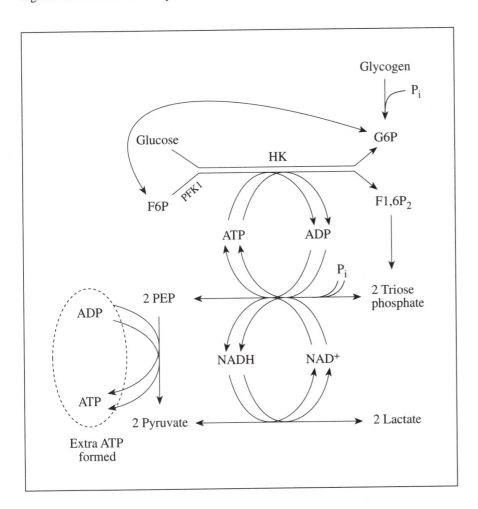

Figure 8·2
Regeneration of ATP in glycolysis. ATP is consumed in two kinase reactions (HK and PFK1) and is resynthesized before phosphoenolpyruvate (PEP) is formed. Two more molecules of ATP are formed in the pyruvate kinase reaction (one per pyruvate formed), yielding a net regeneration of ATP. Inorganic phosphate, added at the triose level and during the breakdown of glycogen, ends up in ATP.

To illustrate one control system in some detail, we shall describe glycolysis and its regulation because the links between function and control are relatively straightforward.

191

Table 8·1

Ways of looking at glycolysis

---

1. The Overall Pathway

A 6-carbon molecule (hexose) is converted into two 3-carbon molecules (pyruvate or lactate).

All intermediates carry phosphate groups to retain them inside cells.

Before the 6-carbon molecule is processed, it must have a phosphate group attached to either end (one for each 3-carbon molecule).

---

2. The Energy Story

All additions of phosphate at the hexose level require ATP; hence, two phosphates from two molecules of ATP are used per molecule of glucose.

Inorganic phosphate is added at the triose level.

Two molecules of ATP are generated for each triose phosphate formed. Hence, glycolysis generates a net yield of two molecules of ATP per molecule of glucose.

---

3. The Hydrogen Story

Each triose yields one NADH upon conversion to pyruvate; this NADH is reconsumed either when lactate is formed or when the reducing power of NADH is transported into the mitochondria for synthesis of ATP or for use in the synthesis of fatty acids.

Overall:

$2 \text{ NAD}^+ + \text{Hexose} \longrightarrow 2 \text{ (NADH} + \text{H}^+) + 2 \text{ (Pyruvate}^- + \text{H}^+)$

$2 \text{ (NADH} + \text{H}^+) + 2 \text{ (Pyruvate}^- + \text{H}^+) \longrightarrow 2 \text{ NAD}^+ + 2 \text{ (Lactate}^- + \text{H}^+)$

$2 \text{ (NADH} + \text{H}^+) + \text{O}_2 + 6 \text{ (ADP} + \text{P}_i) \longrightarrow 2 \text{ NAD}^+ + 6 \text{ ATP} + \text{H}_2\text{O}$

ANAEROBIC GLYCOLYSIS (GLUCOSE TO LACTATE)

Anaerobic glycolysis produces ATP faster than any other metabolic pathway.

The price to pay is 1 H⁺ per ATP if glucose is the substrate and 0.67 H⁺ per ATP if glycogen is the substrate.

Anaerobic glycolysis is the only source of ATP in red blood cells (which lack mitochondria and hence contain neither pyruvate dehydrogenase nor the ATP generation system). Organs such as muscle also rely on anaerobic glycolysis to regenerate their ATP when the oxidative ATP generation pathway is unable to meet the demands for ATP (see Event 3·1, page 88).

Anaerobic glycolysis uses 11 enzyme-catalyzed reactions to catalyze the sequence of reactions summarized in equation 1:

$$\text{Glucose} + 2 \text{ ADP}^{3-} + 2 \text{ P}_i^{3-} \longrightarrow 2 \text{ Lactate}^- + 2 \text{ ATP}^{4-} \qquad (1)$$

The ATP formed is used simultaneously in cellular work (equation 2):

$$2 \text{ ATP}^{4-} \longrightarrow 2 \text{ ADP}^{3-} + 2 \text{ P}_i^{2-} + 2 \text{ H}^+ \qquad (2)$$

The net reaction of anaerobic glycolysis when glucose is the substrate is therefore the formation of 2 H⁺ + 2 lactate anions per glucose consumed, or 1 H⁺ per ATP regenerated.

Anaerobic glycolysis in muscle can regenerate ATP much faster than the complete oxidation of glucose even though aerobic metabolism yields 5% of the amount of ATP per glucose (see Chapter 2, pages 44–48). Anaerobic glycolysis can therefore produce lactate anions and H⁺ extremely rapidly. The acid load can be absorbed to a certain extent by buffering or by metabolism of lactate plus H⁺ to neutral end products—glucose or CO_2. Reconversion to glucose (the Cori cycle) or to glycogen conserves the glucose carbon and, in effect, uses a glucose-lactate-glucose cycle to transfer energy (ATP) between organs (e.g., liver and red blood cells; see Figure 1·19, page 27).

Anaerobic glycolysis produces lactic acid much more rapidly than it can be removed by metabolic pathways.

AEROBIC GLYCOLYSIS (GLUCOSE TO PYRUVATE)

> *Aerobic glycolysis is the first step in pathways that remove carbon from the carbohydrate system.*

Aerobic glycolysis uses glucose to regenerate ATP and to synthesize triacylglycerols. The pathway is identical to that for anaerobic glycolysis (Figure 8·3) except that the pyruvate is oxidized to produce many molecules of ATP (Table 8·2). The net reaction is:

$$\text{Glucose} + 2\,\text{ADP}^{3-} + 2\,\text{P}_i^{2-} + 2\,\text{NAD}^+ \longrightarrow$$
$$2\,\text{Pyruvate}^- + 2\,\text{ATP}^{4-} + 2\,\text{NADH} + 2\,\text{H}^+$$

Because the pyruvate is used in PDH, aerobic glycolysis is a part of the route that destroys precursors of glucose. Activity of PDH must therefore be strictly controlled.

PDH is discussed in Chapter 9, pages 211–13.

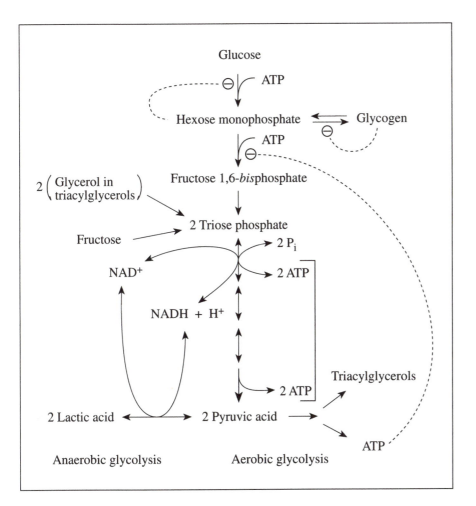

Figure 8·3
Outline of the pathway of glycolysis. Glucose is combined with two phosphate groups, using two molecules of ATP. The hexose *bis*phosphate is then split into two 3-carbon molecules, which are metabolized to generate two molecules of ATP per triose phosphate. Structures and enzymes are provided on page 282.

The NADH + H⁺ formed in aerobic glycolysis are reconverted to NAD⁺, usually through incorporation of H into H_2O in the regeneration of ATP (Chapter 10). Since pyruvate and NADH + H⁺ are used as quickly as they are formed by aerobic glycolysis, no significant acid load results from aerobic glycolysis.

In consumer tissues such as muscle or the brain, the purpose of aerobic glycolysis is to regenerate ATP. The controls over the pathway are therefore identical to those for anaerobic glycolysis (discussed below). In the liver and adipose tissue, where glucose is primarily converted to triacylglycerols, extra controls, depending mainly on insulin, are in place to overcome the inhibition of the pathway caused by product inhibition via ATP and related compounds.

Table 8·2

ATP regenerated in aerobic glycolysis

- *Glycolysis generates two molecules of ATP per molecule of glucose.*
- *Stoichiometry: Glucose + $6 O_2$ $\longrightarrow$ $6 CO_2$ + $6 H_2O$*
- Six molecules of ATP are produced per molecule of O_2, or 36 ATP/glucose.

Other Fuels

The pathway of glycolysis also links into other metabolic fuels or stores (Figure 8·3). For example, synthesis and breakdown of glycogen link into the hexosemonophosphate stage of glycolysis. Another fuel, the glycerol released from triacylglycerols, enters the pathway of glycolysis at the triose phosphate level, as does dietary fructose.

Questions

(Discussions on page 271)

8·1 Some patients develop severe liver, kidney, and intestinal problems very soon after eating sucrose. Why? You may want to think of this problem as follows:

1. What does sucrose contain?

2. Which component of sucrose is likely to cause the problem?

3. Why do symptoms develop rapidly?

4. Why are the problems restricted to some organs?

5. Are these patients likely to have high dental bills?

8·2 The appearance of sucrose in the urine after consumption of a drink containing sucrose provides a rapid and easy screening test for stomach ulcers. Why does it work?

BIOCHEMICAL ASPECTS OF CONTROL OVER GLYCOLYSIS

CONTROL OVER GLYCOLYSIS FOR REGENERATION OF ATP IN THE BRAIN, MUSCLE, AND KIDNEYS

> *Control over glycolysis is exerted at two nonequilibrium kinases: hexokinase (HK) and phosphofructokinase-1 (PFK1).*

The conversion of glucose to fructose 1,6-*bis*phosphate ($F1,6P_2$), a hexose *bis*phosphate, is catalyzed by two kinases, hexokinase (HK) and phosphofructokinase-1 (PFK1). Both kinases use ATP to add a phosphate to an end of the hexose molecule, and both are normally maintained far from thermodynamic equilibrium. Between these two kinases is an equilibrium reaction that interconverts glucose 6-phosphate (G6P) and fructose 6-phosphate (F6P) (Figure 8·4). The outline is very similar to the example shown in Figure 7·9, page 178.

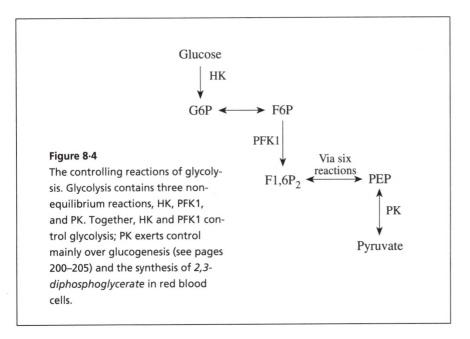

Figure 8·4
The controlling reactions of glycolysis. Glycolysis contains three non-equilibrium reactions, HK, PFK1, and PK. Together, HK and PFK1 control glycolysis; PK exerts control mainly over glucogenesis (see pages 200–205) and the synthesis of *2,3-diphosphoglycerate* in red blood cells.

2,3-diphosphoglycerate (2,3-DPG): A chemical that is made in very high concentrations in red blood cells and is related to the glycolytic intermediate 1,3-diphosphoglycerate. By binding to hemoglobin, 2,3-DPG increases the affinity of hemoglobin for O_2. High levels of 2,3-DPG can make O_2 less available to tissues because bound hemoglobin does not readily release O_2 at the tissue level.

The major control over glycolysis in all tissues (except the liver) is exerted at the first two nonequilibrium reactions of the pathway: HK and PFK1. These kinases function together as a single, very flexible control system. In brief, the effects of ATP are exerted at PFK1; the resulting effects on the concentration of hexosemonophosphates are passed on to HK.

Control of PFK1

> *Control of PFK1 is by the [ATP] and related compounds.*

Both ATP (an inhibitor) and AMP (an activator) control PFK1 (Figure 8·5). As indicated in Chapter 7, these synergistic effects of AMP and ATP greatly enhance the sensitivity of control, since, at normal physiological concentrations, the percent change in the [AMP] is far greater than that of the [ATP] (Table 7·1, page 186).

Two other factors may exert control over PFK1.

1. Citrate inhibits PFK1 in vitro and thus may offer a control mechanism related to the supply of fatty acids or ketoacids (alternative fuels); control by citrate may be important in muscle.

2. A high [H$^+$] is a strong inhibitor of PFK1 and may protect muscles from an excess accumulation of H$^+$ during anaerobic glycolysis.

Figure 8·5

Control of PFK1. (a) ATP is both a substrate and a K_m inhibitor of PFK1; the physiological [ATP] is well into the inhibitory range. (b) While citrate augments inhibition by ATP, AMP lessens this inhibition. PFK1 is inhibited by an increase in the [H$^+$] in the physiological range; inhibition by H$^+$ affects the V_{max}, not the K_m.

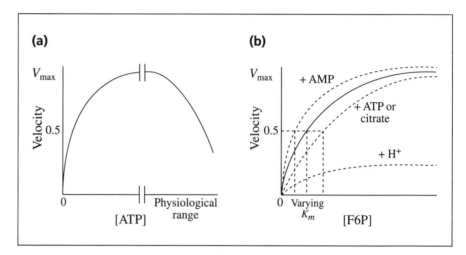

Figure 8·5(b) shows that the effects of ATP, AMP, and citrate on PFK1 are K_m-type controls with respect to the substrate F6P. Hence, with a constant rate of supply of F6P, inhibition of PFK1 will cause the [F6P] to rise to reestablish the rate of flow. Since F6P and G6P are in equilibrium with each other, any change in the [F6P] is matched by a proportional change in the [G6P].

> Inhibition of PFK1 leads to a rise in the [G6P].

Control of HK

> *Control of HK is through product inhibition by G6P.*

Figure 8·6 shows that HK is inhibited by its product, G6P, through a V_{max} type of mechanism. Because the K_m of HK for glucose (0.05 mmol/l) is much lower than the normal concentration of glucose in cells, HK usually is saturated with glucose, its substrate; thus, the [G6P] is able to exert an absolute control over the rate of entry of glucose into glycolysis in consumer organs (Figure 8·7).

Figure 8·6

Kinetics of hexokinase and glucokinase. (a) Hexokinase, found in the brain, muscle, and kidneys, has a high affinity for glucose; inhibition by G6P lowers its V_{max}. (b) Glucokinase, found in the liver, has a low affinity for glucose and is not inhibited by G6P.

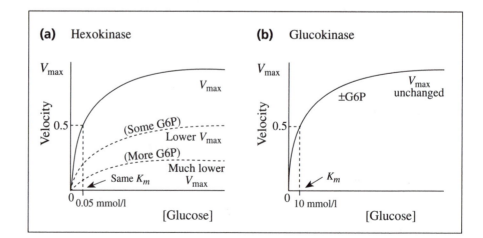

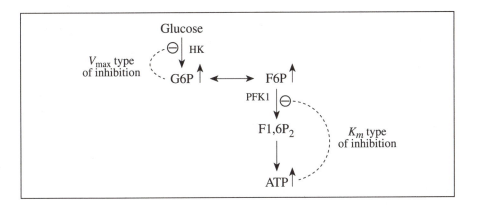

Figure 8·7

The HK-PFK1 control system. As ATP and related compounds (the products of glycolysis) build up, PFK1 is inhibited in a competitive fashion; this inhibition leads to a rise in the [F6P]. Since F6P and G6P are linked by a near-equilibrium reaction, the [G6P] also rises. G6P inhibits HK, lowering its V_{max} so that a rise in the [G6P] exerts absolute control over the rate of glycolysis.

INSULIN AND CONTROL OF THE ENTRY OF GLUCOSE INTO TISSUES

The first metabolic effect of insulin to be identified—its ability to stimulate the entry of glucose into muscle cells—is still the most frequently cited mechanism by which insulin lowers the concentration of glucose in blood. Since the HK-PFK1 system described above is rigidly controlled through product inhibition by ATP and related compounds, the antipolytic action of insulin—not the acceleration of glucose transport—is the predominant way that insulin controls glycolysis. Under normal physiological conditions, insulin will not have an important direct effect on the rate at which glucose is oxidized in muscle. Of note, other factors, such as hypoxia, may lead to a marked acceleration of the transport of glucose into muscle cells, as well as an increased rate of glycolysis whether or not insulin is present.

CONTROL OVER GLYCOLYSIS FOR SYNTHESIS OF TRIACYLGLYCEROLS IN THE LIVER

> *When levels of glucose are elevated, insulin causes the liver to override product inhibition of glycolysis by ATP.*

Glycolysis provides pyruvate, which, in the liver, can be used for the synthesis of tri-acylglycerols. During the synthesis of triacylglycerols, the [ATP] is sufficiently high to restrict flow through PFK1. Hence, the liver must "ignore" the usual product inhibition of PFK1 by ATP and related compounds, thereby ensuring that flux through glycolysis removes excess glucose. These override controls involve both major control sites. HK is replaced by glucokinase (GK), an enzyme found almost exclusively in hepatocytes (and β cells of the pancreas), and a further insulin-sensitive control is added for PFK1 (see the discussion of fructose 2,6-*bis*phosphate below).

PROPERTIES OF GLUCOKINASE

> *Flux through GK depends on the concentration of glucose in the normal physiological range.*
> *Flux through GK is not inhibited by G6P.*

In hepatocytes, GK is the major (perhaps the only) enzyme catalyzing the conversion of glucose to G6P. GK has a low affinity for glucose (K_m = 5–10 mmol/l) and, as shown in Figure 8·6(b), is not inhibited by G6P. GK therefore responds to the concentration of glucose in the physiological range and is not limited by the rate of use of G6P.

Because insulin stimulates the synthesis of new enzyme molecules of GK, a second, slower level of control over GK arises and is consistent with the substrate activation mechanism for the uptake of glucose by the liver.

CONTROL OF PFK1 IN THE LIVER

> *Fructose 2,6-bisphosphate controls PFK1 in the liver.*

A high concentration of glucose in blood causes the level of insulin to increase and the level of glucagon to decrease. This hormonal setting lowers the [cAMP], which, in turn, activates phosphofructokinase-2 (PFK2), an enzyme found primarily in the liver that produces fructose 2,6-*bis*phosphate ($F2,6P_2$) from F6P (Figure 8·8). $F2,6P_2$ activates PFK1.

When the concentration of insulin falls and the concentration of glucagon rises, PFK2, which is also activated by F6P (its substrate), becomes a phosphatase and destroys $F2,6P_2$, thus removing the stimulation of PFK1.

Figure 8·8

Substrate activation of glycolysis in the liver. A meal causes the concentration of glucose in blood to increase. The result—a high concentration of insulin and a low concentration of glucagon—promotes synthesis of GK. The high concentration of glucose also directly increases flux through GK. Hence, the [G6P] and the [F6P] both increase. The high level of insulin and low level of glucagon inhibit adenylate cyclase, thus decreasing the [cAMP]. This decrease stimulates PFK2 (as does the increased [F6P]), forming more $F2,6P_2$. The increased [$F2,6P_2$] activates PFK1, thus allowing more glucose through to pyruvate. The control by $F2,6P_2$ overcomes inhibition by ATP.

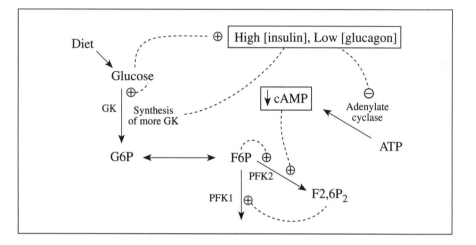

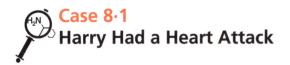

Case 8·1
Harry Had a Heart Attack

Harry had crushing chest pain for six hours. At the hospital, his doctor suspected that he was having a heart attack. Suddenly, Harry's heart stopped beating. The doctor called for help and started to work. How long would Harry live if nothing was done? What could save him? Explain in biochemical terms.

Relevant data:

- Normal consumption of oxygen is 12 mmol/min, and three molecules of ATP are formed per atom of oxygen.

- Death occurs after the accumulation of 1000 mmol of H^+.

Discussion of Case 8·1
Harry Had a Heart Attack

PATHWAYS FOR REGENERATION OF ATP

After his heart and brain have used their very tiny stores of ATP (and creatine phosphate), Harry must make all his ATP from anaerobic glycolysis. No fuels from the fat system can be used because their metabolism requires oxygen.

QUANTITY OF ATP NEEDED PER MINUTE

Harry must make as much ATP by anaerobic glycolysis as he did with oxidative metabolism before his heart stopped (72 mmol of ATP per minute).

Quantitative Analysis

- Each atom of oxygen permits three molecules of ATP to be regenerated, and there are two atoms of oxygen per molecule of O_2. Therefore, each molecule of O_2 yields six molecules of ATP.

- Harry consumes 12 mmol of O_2 per minute, equivalent to 72 mmol of ATP per minute.

QUANTITY OF H^+ PRODUCED

- In anaerobic glycolysis with glucose as the substrate, 1 H^+ is made per molecule of ATP regenerated. Therefore, in producing 72 mmol of ATP per minute, anaerobic glycolysis must produce 72 mmol of H^+ per minute.

- Since an adult can buffer 1000 mmol of H^+, producing 72 mmol of H^+ per minute will exhaust the body's buffering capacity in 14 minutes.

- Death usually occurs well before 14 minutes because the brain has a higher requirement for ATP and a lower buffering capacity than organs such as muscle. Also, since the brain contains little glycogen, it will quickly run out of glucose and will be unable to regenerate enough ATP.

THERAPY

Quickly supply his brain with oxygen. Use cardiac massage, support breathing, and hope to "buy time" to reverse his cardiac problem (probably an arrhythmia).

PART B
GLUCOGENESIS

OVERVIEW OF PHYSIOLOGICAL FUNCTIONS AND CONTROLS

> *Glucogenesis is the pathway that synthesizes glucose or G6P from pyruvate. Glucogenesis is not simply the reverse of glycolysis.*
>
> *Control of glucogenesis is largely by the supply of precursors to the liver and to the kidney cortex.*
>
> *Hepatocytes contain important controls for glucogenesis.*

Glucogenesis converts precursors of glucose (lactate, intermediates of the tricarboxylic acid cycle, glycerol, some amino acids) to glucose. Glucogenesis shares most of its intermediates and many reactions (the equilibrium reactions) with glycolysis but has unique nonequilibrium reactions that reverse the effects of the three nonequilibrium reactions of glycolysis (Figure 8·9). The full pathway of glucogenesis is found in the liver and kidneys; muscle may contain all but the last reaction of glucogenesis, and thus can only form G6P (which is converted to glycogen) from pyruvate. It is important to consider which major substrate (lactate or amino acids) is used in this pathway since the former primarily removes lactate (plus H^+) but the latter makes new glucose (using amino acids). This difference is discussed in more detail below.

Figure 8·9

Control sites in glucogenesis and glycolysis. Different enzymes are used in glucogenesis to reverse flow through HK (or GK), PFK1, and pyruvate kinase (PK). The enzymes are glucose 6-phosphatase (G6Pase), fructose 1,6-*bis*phosphatase (F1,6Pase), and pyruvate carboxylase (PC) plus phosphoenolpyruvate carboxykinase (PEPCK). Whereas glycolysis forms two molecules of ATP per glucose, glucogenesis uses six molecules of ATP per glucose formed.

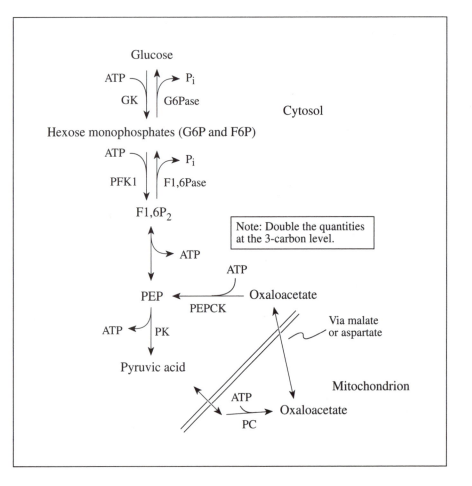

GLUCONEOGENESIS AND GLUCOPALEOGENESIS

> *Gluconeogenesis: synthesis of new glucose largely from amino acids.*
>
> *Glucopaleogenesis: resynthesis of glucose from lactate derived from glucose (the Cori cycle).*

In the liver and kidneys, glucogenesis is very important in ensuring that precursors of glucose are conserved as glucose in all dietary conditions except when dietary glucose is in excess. Precursors of glucose are predominantly lactic acid, glycerol, and amino acids (from dietary proteins or from body stores—mainly muscle). Because amino acids are not made from glucose, glucogenesis from amino acids is referred to as "gluconeogenesis" (Figure 8·10). Because lactic acid in the body is generated largely from glucose, we refer to glucogenesis from lactate as "glucopaleogenesis."

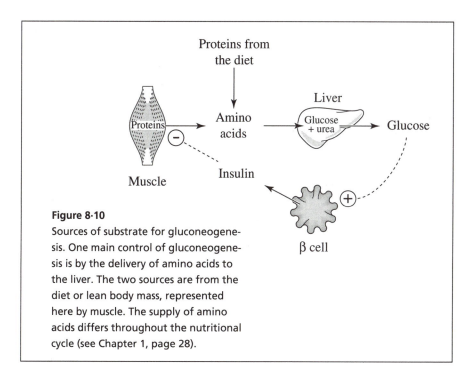

Figure 8·10
Sources of substrate for gluconeogenesis. One main control of gluconeogenesis is by the delivery of amino acids to the liver. The two sources are from the diet or lean body mass, represented here by muscle. The supply of amino acids differs throughout the nutritional cycle (see Chapter 1, page 28).

Question

(Discussion on page 272)

8·3 Is the conversion of glycerol derived from adipose tissue triacylglycerols better described as gluconeogenesis or glucopaleogenesis?

CONTROL OF GLUCONEOGENESIS AND GLUCOPALEOGENESIS

The activities of gluconeogenesis and glucopaleogenesis vary with dietary state and are influenced by both the level of insulin (and counter-insulin hormones) and the supply of precursors. We shall consider each pathway separately (Table 8·3).

Table 8·3

Comparison of gluconeogenesis and glucopaleogenesis

	Gluconeogenesis	Glucopaleogenesis
Substrate	• Amino acids	• Lactate (from glucose)
Mitochondrial phase	• Pyruvate to oxaloacetate	• Pyruvate to oxaloacetate
	• Obligatory link to synthesis of urea	• Not linked to the synthesis of urea
Cytosolic phase	• Synthesis of PEP: - Half via urea cycle - Half via same path as glucopaleogenesis	• All PEP formed as shown in Figure 8·11
	• Same last two steps: - F1,6P$_2$ to F6P - G6P to glucose	• Same last two steps: - F1,6P$_2$ to F6P - G6P to glucose
Source of cytosolic NADH	• Malate dehydrogenase	• Lactate dehydrogenase

Glucopaleogenesis

As depicted in Figure 1·19, page 27, the hepatic portion of the Cori cycle converts lactate derived from anaerobic glycolysis in other organs back to glucose. The maximum rate of this pathway is usually determined by the rate of delivery of lactate to the liver, although the upper limit is set by the delivery of oxygen to the liver and by the activity of enzymes. This pathway does not convert lactic acid to glucose at a rapid rate.

The major control of glucopaleogenesis is exerted at the pyruvate crossroad (Figure 8·11) primarily by the activity of PDH. If PDH is inhibited (from a low net concentration of insulin and high availability of fatty acids), pyruvate will flow through pyruvate carboxylase (PC) and phosphoenolpyruvate carboxykinase (PEPCK), driven by high levels of pyruvate and amplified by intracellular signals. Alternatively, during meals, high levels of glucose and insulin promote flux through PDH and thus lipogenesis.

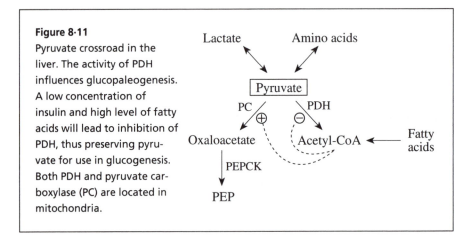

Figure 8·11
Pyruvate crossroad in the liver. The activity of PDH influences glucopaleogenesis. A low concentration of insulin and high level of fatty acids will lead to inhibition of PDH, thus preserving pyruvate for use in glucogenesis. Both PDH and pyruvate carboxylase (PC) are located in mitochondria.

Gluconeogenesis

> *During meals, gluconeogenesis helps in the disposal of dietary proteins.*
>
> *Between meals and without meals, gluconeogenesis provides glucose for the brain.*

Because gluconeogenesis from amino acids has two separate functions, it has two separate controls. Gluconeogenesis is closely linked to the synthesis of urea

(Figure 8·12). During meals, when the diet supplies more protein than is needed for immediate regeneration of ATP, the amino acids not immediately used for synthesis of new proteins must be removed. The liver contains the necessary enzymes. If glucose is also in excess, it is converted to triacylglycerols, as are glucogenic amino acids. If glucose is not in large excess, gluconeogenesis from amino acids occurs, controlled primarily by the rate at which the diet supplies amino acids to the liver.

Between meals and without meals, the brain must be supplied with the glucose it needs to regenerate ATP. As the concentration of glucose in the circulation falls, the concentration of insulin also declines. The lower level of insulin sends two sets of signals—first, to lean body mass (mainly muscle), causing a net breakdown of proteins and a higher delivery of amino acids into the circulation; second, to the liver, promoting glucogenesis over glycolysis. At this point, control is exerted indirectly by product inhibition (a low concentration of glucose means faster gluconeogenesis).

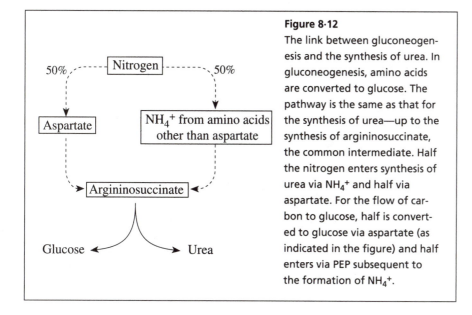

Figure 8·12
The link between gluconeogenesis and the synthesis of urea. In gluconeogenesis, amino acids are converted to glucose. The pathway is the same as that for the synthesis of urea—up to the synthesis of argininosuccinate, the common intermediate. Half the nitrogen enters synthesis of urea via NH_4^+ and half via aspartate. For the flow of carbon to glucose, half is converted to glucose via aspartate (as indicated in the figure) and half enters via PEP subsequent to the formation of NH_4^+.

OTHER FEATURES OF GLUCONEOGENESIS

Gluconeogenesis in the kidneys aids in the elimination of H^+. Glutamine, an amino acid, is converted to NH_4^+, HCO_3^-, and glucose. The NH_4^+ are excreted in the urine, and the HCO_3^- return to the blood to restore concentrations in the BBS to normal. The rate of renal gluconeogenesis from glutamine is determined by the need for removal of H^+.

Gluconeogenesis in muscle regenerates stores of glycogen from the lactic acid that is formed during vigorous exercise.

Questions

(Discussions on page 272)

8·4 What are the relative proportions of glucose and urea synthesized during metabolism of proteins?

8·5 Urea and phosphoenolpyruvate (PEP) are formed in equimolar amounts in gluconeogenesis. How is this equality possible if each precursor of PEP has one atom of nitrogen from amino acids, but each molecule of urea is formed with two atoms of nitrogen?

BIOCHEMICAL ASPECTS OF CONTROL OVER GLUCOGENESIS

> *Extrahepatic control is via rate of delivery of substrates for glucogenesis.*
>
> *Intrahepatic control is largely at the pyruvate crossroad and depends on the activity of PDH.*
>
> *Less important (but necessary) regulation occurs at the F1,6P$_2$-F6P step and at the G6P crossroad.*

THE PATHWAY OF GLUCOGENESIS

Glucogenesis from pyruvate catalyzes the following net reaction:

$$2 \text{ (Pyruvate}^- + \text{H}^+) \; + \; 2 \text{ (NADH} + \text{H}^+) \; + \; 6 \text{ ATP}^{4-} \longrightarrow$$
$$\text{Glucose} \; + \; 2 \text{ NAD}^+ \; + \; 6 \text{ (ADP}^{3-} + \text{P}_i^{2-})$$

The anionic charges in this equation are not balanced; the imbalance disappears when the regeneration of ATP is taken into account.

Glycolysis and glucogenesis share many enzyme-catalyzed steps that are close to equilibrium and thus can be ignored when considering controls. However, several steps are unique to glucogenesis; these nonequilibrium reactions reverse the effects of the nonequilibrium reactions of glycolysis (Figure 8·9, page 200). Controls over glucogenesis and glycolysis are therefore closely linked to provide a balance between these two opposing pathways.

CONTROLS AT THE PYRUVATE CROSSROAD

The major determinant of the rate of glucogenesis is the balance of the opposing reactions between pyruvate and PEP; this balance is dictated by controls over PDH. The activity of the glucogenic enzyme PC depends in vitro on acetyl-CoA, which can have a major switching effect on the pyruvate crossroad because it inhibits PDH and activates PC. However, no significant controls are assumed to operate at the exit of oxaloacetate (as malate or as aspartate) from the mitochondria or at the formation of PEP from oxaloacetate.

CONTROLS AT PEP

Another site of control is the conversion of PEP to pyruvate. Control is primarily exerted on the glycolytic enzyme pyruvate kinase (PK). When active, PK ensures flow through glycolysis. PK is activated by "upstream" glycolytic intermediates (F1,6P$_2$) and inhibited by alanine, a glucogenic substrate. For glucogenesis to proceed, F1,6P$_2$ must be removed by controls at the PFK1-F1,6P$_2$ step (discussed below) and alanine must be provided. For effective control to be exerted, the influences of these pathway intermediates must act in concert with hormonal influences. The low level of glucose leads to low insulin and high glucagon levels. This hormonal setting, via a high level of cAMP in hepatocytes, leads to phosphorylation and inactivation of PK. This endocrine control system also affects PFK1 and PFK2, as well as PDH, and thus is common to all the control reactions of glycolysis. The same endocrine balance also decreases the rate of synthesis of regulatory enzymes.

In summary, the balance between PEP and pyruvate (Figure 8·13) is subject to controls from the endocrine balance, the availability of alternative fuels for regeneration of ATP (signaled by acetyl-CoA), and the availability of substrates for glycolysis and glucogenesis. The inhibitory effects of insulin (preventing the release of amino acids and fatty acids from stores) reinforce the overall effectiveness of the control mechanisms. The conversion of amino acids to glucose is favored when the concentration of insulin is low. In parallel, more fatty acids are available so that regeneration of ATP occurs via the fat system.

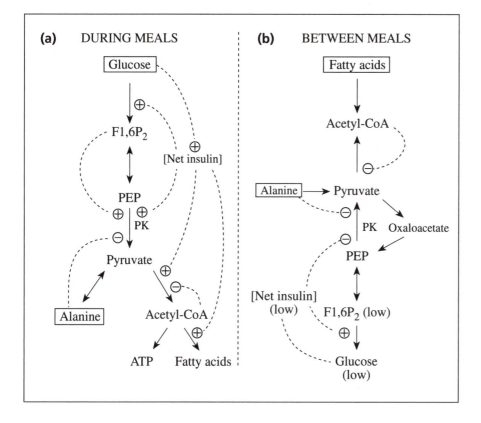

Figure 8·13

Controls over formation and use of PEP. Fuels provided in the diet or from stores are shown as rectangles. (a) During meals, the concentration of glucose rises. The result—a higher net concentration of insulin and more $F1,6P_2$—activates PK. The net effect is flux toward pyruvate (fatty acids are low at this time). (b) Between meals, levels of insulin are low, and the PEP formed from pyruvate or alanine is directed toward glucose. This movement occurs because high levels of alanine and low levels of $F1,6P_2$ inhibit PK, as does the increased level of cAMP (resulting from low concentrations of glucose and insulin). Also, the alternate fate for pyruvate is blocked (low levels of insulin and high levels of acetyl-CoA inhibit PDH). The net effect is movement up to $F1,6P_2$.

CONTROLS AT $F1,6P_2$

The next site of control in glucogenesis is at the interconversion of F6P and $F1,6P_2$ (the reactions between PEP and $F1,6P_2$ are near equilibrium). The controls over PFK1 depend most heavily on inhibition by ATP plus activation by AMP; in addition, in the liver, insulin, acting through the cAMP system and PFK2, can override product inhibition of PFK1 by ATP. The activity of fructose 1,6-*bis*phosphatase (F1,6Pase) is also controlled by AMP. Hence, glucogenesis is promoted when the [ATP] is high.

CONTROLS AT G6P

The final interconversion is between glucose and G6P. Because glucose 6-phosphatase (G6Pase) is located inside a compartment (endoplasmic reticulum), the ability of G6P to cross this membrane may be regulated. It is also possible that the controls over GK dominate the system.

Question

(Discussion on page 272)

8·6 What symptoms and signs would you expect if a patient had a defect in the glucogenic pathway? What might aggravate them?

PART C
SYNTHESIS AND BREAKDOWN OF GLYCOGEN

OVERVIEW OF PHYSIOLOGICAL FUNCTIONS AND CONTROLS

> Glycogen is the body's store of glucose.
> In the liver, glycogen supplies glucose for the brain.
> In muscle, glycogen is the substrate used to regenerate ATP during exercise.

PHYSIOLOGY

The physiological role of glycogen differs between the liver and muscle—the two organs in which glycogen is very important. Glycogen in the liver is a "buffer" of glucose. When glucose is presented to the body in large quantities, it can be stored as glycogen in the liver; when dietary supplies of glucose are insufficient, glycogen in an adult's liver (approximately 100 g) can provide the brain with glucose for up to 20 hours. Glycogen in muscle (approximately 20 g/kg of muscle) can provide the energy needed by muscle for a few minutes of strenuous anaerobic exercise (which is limited by accumulation of lactic acid before the glycogen is exhausted) or for one hour or more of high-level aerobic exercise (Figure 8·14).

Figure 8·14

Control over glycogenolysis (breakdown of glycogen) in the liver and muscle. The function (and therefore the control) of glycogenolysis differs between the liver and muscle. Whereas glycogenolysis in the liver yields glucose in the blood for use by various tissues, glycogenolysis in muscle leads to the regeneration of ATP in muscle. When glycogenolysis is stimulated in the liver, the pathway of glycolysis is inhibited to encourage release of glucose to the blood. In muscle, glycogenolysis and glycolysis must be activated together to provide the needed ATP.

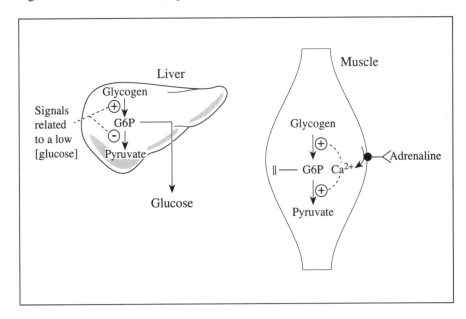

PATHWAYS

The pathways for synthesis of glycogen in the liver and muscle ensure that reserves of glycogen are well stocked when glucose is plentiful from the diet. Once these stores are full, the only other means of disposing of dietary glucose are oxidation (limited by turnover of ATP) and lipogenesis (usually a pathway with a slow flux).

CONTROLS

The controls over synthesis and breakdown of glycogen in the liver and muscle directly reflect the functions of the pathways in the two organs. In the liver, glycogen is deposited when the diet supplies an excess of glucose; glycogen is broken down when the brain must draw on stores of glycogen. When the liver breaks down glycogen to produce glucose, glycolysis must be inhibited (Figure 8·14). The controls over synthesis and breakdown of glycogen in the liver are therefore driven primarily by the concentration of glucose in blood (a high concentration of glucose causes a high level of insulin and a low level of glucagon, and vice versa).

In muscle, glycogen is also deposited when glucose is available from the diet and the muscles are resting (use of energy is low). The bulk of glucose removed during a meal is directed toward the synthesis of glycogen in muscle. This muscle glycogen (approximately 20 g per kg of muscle, or 450 g in an average adult male) is then used when the muscle demands energy. The controls over synthesis of glycogen in muscle are rather similar in principle to those in liver. A high level of glucose in the circulation leads to the release of insulin from β cells. The high level of insulin promotes the transport of glucose into myocytes and directs G6P towards glycogen.

The controls over breakdown of glycogen in muscle are very different from those in the liver. In muscle, glycolysis and breakdown of glycogen are activated simultaneously. Breakdown of glycogen in muscle is controlled through events related to muscular contraction. The stimulus for breakdown of glycogen in muscle and muscular contraction both involve Ca^{2+} as principal signals (the contraction-relaxation cycle in muscle consumes ATP).

BIOCHEMICAL ASPECTS OF THE METABOLISM OF GLYCOGEN

STRUCTURES AND PATHWAYS

Each molecule of glycogen consists of a primer point (a tyrosine residue in the protein glycogenin) with a single stalk, from which a branching structure grows. There are 6–9 molecules of glucose between each branch point. All the activity of adding or removing molecules of glucose occurs at the ends of the many branches, never at the single stalk.

The structure of glycogen is shown on page 280.

CONTROLS

Sensitive and effective controls are exerted over both synthesis and breakdown of glycogen. The enzymes subject to control are glycogen synthase and glycogen phosphorylase. Both enzymes are controlled through interconversion between active and inactive forms. As might be expected, factors that stimulate synthesis tend to inhibit breakdown, and vice versa. We shall briefly outline these very complex controls.

Glycogen Synthase

Glycogen synthase can be interconverted between two forms—one that is relatively inactive and can function only in the presence of G6P and another that is active even without G6P. Interconversion involves addition of phosphate (from ATP) or removal of inorganic phosphate (P_i). The phosphorylated form depends on G6P and is often referred to as the inactive form; it may be called "glycogen synthase D" (D for "dependent on G6P"). The D form is converted to the I (independent, or active) form by phosphoprotein phosphatase, which is inhibited by glycogen. The conversion of the I to the D form is by protein kinase, which is stimulated by cAMP (see Chapter 7, pages 183–84).

Figure 6 on page 283 illustrates the pathway of glycogen synthesis.

These controls over glycogen synthase allow further control through hormonal influences (acting through cAMP-dependent protein kinase; Figure 8·15), by substrate supply (acting through effects of the [G6P] on the D form) and by product inhibition (acting through the effects of glycogen on the phosphoprotein phosphatase).

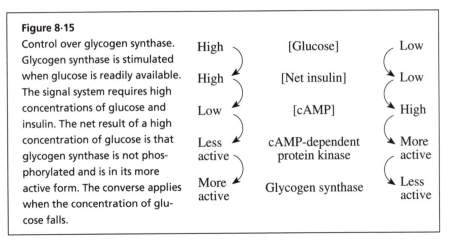

Figure 8·15
Control over glycogen synthase. Glycogen synthase is stimulated when glucose is readily available. The signal system requires high concentrations of glucose and insulin. The net result of a high concentration of glucose is that glycogen synthase is not phosphorylated and is in its more active form. The converse applies when the concentration of glucose falls.

Glycogen Phosphorylase

Glycogen phosphorylase can also be activated or inactivated by phosphorylation and dephosphorylation, respectively. Phosphorylase *b*, the inactive form, is activated to the *a* form by a specific enzyme, phosphorylase kinase, which itself exists in active (phosphorylated) or inactive (dephosphorylated) forms. The inactive phosphorylase kinase is activated by the hormone-regulated cAMP-dependent protein kinase mentioned above. Both active enzymes are inactivated by a family of phosphoprotein phosphatases. Thus, phosphorylase is subject to an extremely effective cascade of controls (Figure 8·16).

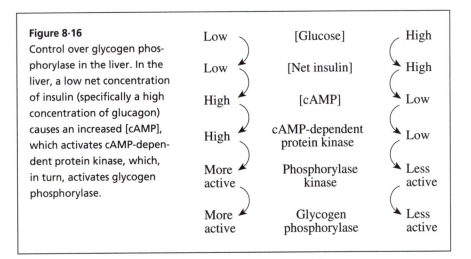

Figure 8·16
Control over glycogen phosphorylase in the liver. In the liver, a low net concentration of insulin (specifically a high concentration of glucagon) causes an increased [cAMP], which activates cAMP-dependent protein kinase, which, in turn, activates glycogen phosphorylase.

The amplified actions of hormones on the [cAMP] are further amplified through two successive steps—first, the activation of protein kinase; second, the activation of phosphorylase kinase. In addition, substrate supply contributes to the controls over phosphorylase. Muscle breaks down glycogen to regenerate ATP. A low [ATP] is signaled by an increased [AMP]; AMP activates phosphorylase *b*, the inactive form of phosphorylase. In the liver, a high concentration of glucose (> 7 mmol/l) decreases the activity of phosphorylase, leading to activation of glycogen synthase. These actions ensure reciprocal control of synthesis and breakdown of glycogen (Figure 8·17).

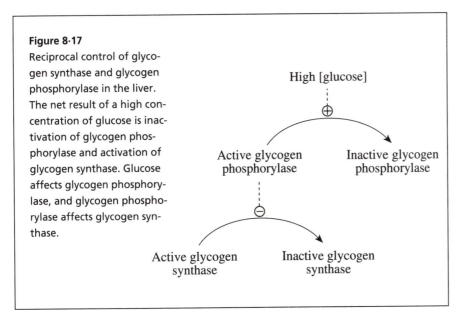

Figure 8·17
Reciprocal control of glycogen synthase and glycogen phosphorylase in the liver. The net result of a high concentration of glucose is inactivation of glycogen phosphorylase and activation of glycogen synthase. Glucose affects glycogen phosphorylase, and glycogen phosphorylase affects glycogen synthase.

High [glucose]

Active glycogen phosphorylase → Inactive glycogen phosphorylase

Active glycogen synthase → Inactive glycogen synthase

Questions

(Discussions on pages 273–74)

8·7 What symptoms and signs would you expect in the absence of glycogen phosphorylase in the liver?

8·8 What symptoms and signs would you expect in the absence of glycogen phosphorylase in muscle?

8·9 Glucose appears to be stored with 3 g of water per gram of glycogen. What are the molar proportions of glucose and water stored?

Can 1 g of glucose bind 3 g of water?

How might the water be bound to glycogen?

CHAPTER 9
THE PYRUVATE DEHYDROGENASE SYSTEM

OVERVIEW

Pyruvate dehydrogenase (PDH) converts pyruvate, a precursor of glucose, into acetyl groups (in the form of acetyl-CoA) that cannot be converted back to glucose. Because this irreversible reaction destroys precursors of glucose, it must be tightly controlled. PDH tends to be in its active form in all organs only when levels of glucose are high (during meals). It is in the active form in the brain except when ketoacids are abundant. The oxidation of fat-derived fuels produces acetyl-CoA and NADH—compounds that lead to inhibition of PDH in part by promoting its phosphorylation (a process that converts it to an inactive form). In the liver, this control over PDH can be overridden so that a surplus of glucose, acting via a high net concentration of insulin, leads to reactivation of PDH.

FUNCTIONS OF PDH

> *PDH destroys precursors of glucose.*
>
> *PDH catalyzes an irreversible reaction that is tightly controlled.*
>
> *Golden rule: burning fatty acids will conserve glucose by inhibiting PDH.*

The pyruvate dehydrogenase (PDH) system catalyzes a single reaction. PDH irreversibly converts pyruvate, a precursor of glucose, to acetyl-CoA, which cannot be converted to glucose.

$$\text{Pyruvate} + NAD^+ + CoA \longrightarrow \text{Acetyl-CoA} + NADH + H^+ + CO_2$$

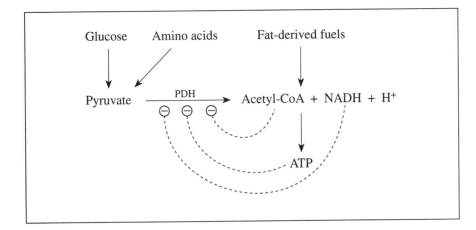

Figure 9·1

PDH and metabolic processes for regeneration of ATP. Acetyl-CoA and NADH, the substrates for the ATP generation system, can be made both from PDH and from oxidation of fatty acids. The need to conserve glucose (or its precursors) makes product inhibition of PDH by acetyl-CoA and NADH (and ATP) very likely.

When glucose is freely available (e.g., during a carbohydrate-rich meal), PDH must be active to allow all tissues to use glucose for regeneration of ATP (Figure 9·1) and to enable the liver to synthesize triacylglycerols (Figure 9·2). PDH should be active in an organ only when fat-derived fuels are not available or cannot be oxidized.

Between meals and without meals, PDH should be inactive, except in the brain, which always requires oxidation of some glucose for regeneration of ATP. Excess activity of PDH in the absence of dietary carbohydrates wastes valuable precursors of glucose, thereby shortening survival time in a prolonged fast. Because the body must conserve its supply of glucose, and PDH irreversibly destroys precursors of glucose, PDH has very precise controls, involving interconversion between active and inactive forms, as well as inhibition of the active form by its products.

Figure 9·2

PDH and the metabolic processes for conversion of glucose to triacylglycerols. In the liver, excess dietary glucose is converted to triacylglycerols. A high net concentration of insulin plays a pivotal role by stimulating glycolysis and by activating PDH.

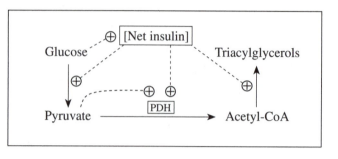

CONTROLS OVER PDH

CONTROLS COMMON TO ALL ORGANS

> *Products of PDH inhibit its activity.*

In all organs, PDH can be interconverted between inactive and active forms by phosphorylation and dephosphorylation (Figure 9·3). Phosphorylation of PDH by PDH kinase, an enzyme that is unique to the PDH complex, converts PDH to its inactive form; PDH kinase is stimulated by all the products of rapid oxidation of fatty acids, including a high [ATP] (through a low [ADP]), high ratios of NADH to NAD^+, and high ratios of acetyl-CoA to CoASH. PDH phosphatase, another control enzyme unique to PDH, activates PDH through dephosphorylation. PDH phosphatase can be stimulated by a higher concentration of intramitochondrial Ca^{2+}, resulting from a higher level of insulin in the circulation; in skeletal muscle, the stimulus for muscular contraction raises the concentration of Ca^{2+} in the cytosol and in mitochondria.

Figure 9·3

Regulation of PDH. PDH is converted from an active form to an inactive form by phosphorylation. Conversely, dephosphorylation activates PDH.

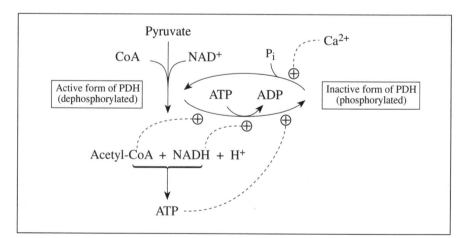

The active form of PDH is inhibited by its products, acetyl-CoA and NADH (Figure 9·4). These controls allow PDH to respond very precisely to the demand for acetyl-CoA and NADH in the regeneration of ATP (see Chapter 10), and they also permit the fat system to provide alternate sources of these intermediates (usually products of the oxidation of fatty acids or ketoacids), thereby preserving precursors of glucose. The relative importance of these controls is considered from a quantitative perspective in Question 9·1.

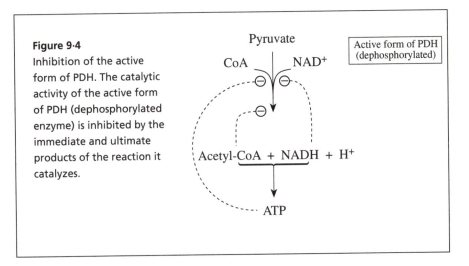

Figure 9·4
Inhibition of the active form of PDH. The catalytic activity of the active form of PDH (dephosphorylated enzyme) is inhibited by the immediate and ultimate products of the reaction it catalyzes.

CONTROLS IN THE LIVER

The liver must be able to override product inhibition of PDH.

In the liver, PDH may be used not only in the regeneration of ATP from glucose but also in the conversion of glucose to fatty acids for the synthesis of triacylglycerols. Because high levels of acetyl-CoA, NADH, and ATP normally inhibit the latter function, the liver must be able to override these product-inhibition controls. The major additional controls in the liver are the effect of pyruvate itself (which inhibits PDH kinase) and the effect of insulin (which signals high levels of glucose and leads to the activation of PDH phosphatase); both of these controls favor the conversion of PDH to its active form.

Question

(Discussion on page 274)

9·1 What signs and symptoms would lead you to suspect that a patient is suffering from a low activity of PDH?

What circumstances can cause this condition?

When is a low activity of PDH a) the greatest danger to the patient; b) the least danger to the patient?

How would you treat this patient?

CHAPTER 10
THE ATP GENERATION SYSTEM

OVERVIEW

> *The tricarboxylic acid cycle oxidizes acetyl-CoA to CO_2, producing NADH and $FADH_2$.*
>
> *The electron transport pathway oxidizes H atoms (on NADH and $FADH_2$) to H_2O, thereby regenerating ATP.*

The vast majority of regeneration of ATP in the body is carried out by the ATP generation system; a little is made by anaerobic glycolysis. The ATP generation system consists of two pathways: the tricarboxylic acid cycle (also called the citric acid cycle or the Krebs cycle, after its discoverer) and the electron transport pathway (also known as oxidative phosphorylation). The former pathway oxidizes acetyl groups (in acetyl-CoA) to yield the useful products NADH and $FADH_2$, and the latter pathway oxidizes hydrogen on NADH or $FADH_2$ to H_2O, with production of ATP. Both pathways are found only in mitochondria.

Because close to three molecules of ATP are produced per atom of oxygen consumed, the total amount of ATP that the body makes each day can be calculated. For the 2400 kcal diet, the 21.6 moles of O_2 consumed are roughly equivalent to 120 moles of ATP. Put another way, in one day, approximately 12 kg (27 lb) of phosphate are exchanged between ATP and ADP in a person burning 2400 kcal of fuels.

PART A
THE TRICARBOXYLIC ACID CYCLE

The tricarboxylic acid (TCA) cycle has two major components: the two-carbon acetyl group that is oxidized and the four-carbon catalytic carrier for the acetyl groups. The purpose of the TCA cycle is to make NADH. Control of this cycle is somewhat difficult to define because it is also a component of the pathways of gluconeogenesis, synthesis and degradation of amino acids, synthesis of fatty acids, and synthesis of porphyrin. The major control is activation of the key dehydrogenases (2-oxoglutarate dehydrogenase, also known as α-ketoglutarate, and NAD^+-linked isocitrate dehydrogenase) by Ca^{2+}, with negative feedback via NADH and selected molecules. The entry into the mitochondria of four-carbon intermediates of the TCA cycle is also tightly regulated, as is their departure from the mitochondria.

PART B
ELECTRON TRANSPORT PATHWAY

The electron transport pathway (ETP) consists of a chain of carriers that generate an electrochemical gradient for H^+ and link it to the generation of ATP. The purpose of the ETP is to provide the energy to regenerate ATP; the pathway generates ATP as fast as it can be hydrolyzed and then transports it out of mitochondria in exchange for ADP in the cytosol. Control of flux in this pathway is largely via the concentration of reduced cytochrome c and may be limited by the availability of oxygen.

Veterinary Case 10·1

How Can the Hummingbird "Hum"? page 221

PART A
THE TRICARBOXYLIC ACID CYCLE

PHYSIOLOGICAL FUNCTIONS AND CONTROLS

> *Four-carbon carriers are catalysts for the acetyl group that is oxidized.*
>
> *The principal products of the TCA cycle are NADH and FADH$_2$.*
>
> *Control of flux by activation of key steps (probably by Ca^{2+}) relates to the need for regeneration of NADH.*
>
> *The TCA cycle is also involved in many catabolic and anabolic pathways, each with its own specific controls.*

The complete TCA cycle oxidizes acetyl groups (as acetyl-CoA) to CO$_2$, releasing H atoms that are in the form of NADH and FADH$_2$ (Figure 10·1). The TCA cycle is the major source of the CO$_2$ excreted by living organisms.

Figure 10·1

The TCA cycle. The main function of the TCA cycle is to provide NADH and FADH$_2$ for the regeneration of ATP. The substrate that is oxidized is the two-carbon acetyl group of acetyl-CoA, derived from the oxidation of pyruvate (during meals) or fatty acids and ketoacids (between meals). The carriers of acetyl groups are four-carbon intermediates; they are neither destroyed nor created in the cycle.

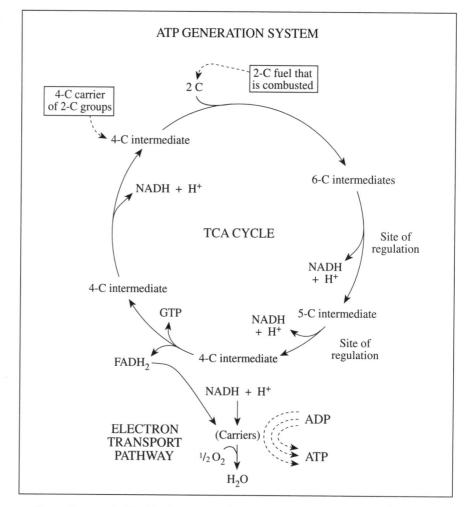

Controls over the TCA cycle seem to be exerted at NAD$^+$-linked isocitrate dehydrogenase and 2-oxoglutarate dehydrogenase. These enzymes are activated by a rise in intramitochondrial Ca^{2+}, and flux is limited by a rise in [NADH] and [ADP], thus reinforcing the links between the activities of the TCA cycle and the electron transport pathway.

The need for two successive controlling reactions for the TCA cycle may arise from the ability of 2-oxoglutarate to be interconverted readily with glutamate, an abundant amino acid; 2-oxoglutarate and glutamate play important roles in breakdown of amino acids and hence in gluconeogenesis from amino acids (see Chapter 12, page 240).

Other parts of the TCA cycle have functions besides provision of NADH for the electron transport chain. In the liver, synthesis of fatty acids uses citrate made by citrate synthase as a means for exporting acetyl groups (made from glucose) to the cytosol where they are needed. Synthesis of fatty acids is active when supplies of glucose, amino acids, and lactate are in excess, when NADH and ATP are freely available, and when 2-oxoglutarate dehydrogenase and NAD^+-linked isocitrate dehydrogenase are inhibited, thus conserving citrate for exit from the mitochondria. Between meals and without meals, citrate does not exit from the mitochondria, and acetyl-CoA and NADH are in plentiful supply from oxidation of fatty acids; intramitochondrial levels of acetyl-CoA rise, leading to ketogenesis (Chapter 11, pages 233–34).

Besides functioning in the breakdown of amino acids, glucogenesis, and the synthesis of fatty acids, parts of the TCA cycle also form components of other metabolic processes, such as synthesis of urea and of porphyrin (a component of hemoglobin). Hence, carbon that enters the TCA cycle can appear in many end products (Figure 10·2).

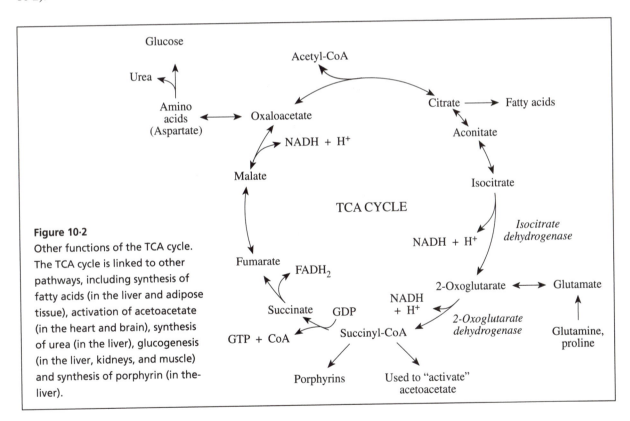

Figure 10·2
Other functions of the TCA cycle. The TCA cycle is linked to other pathways, including synthesis of fatty acids (in the liver and adipose tissue), activation of acetoacetate (in the heart and brain), synthesis of urea (in the liver), glucogenesis (in the liver, kidneys, and muscle) and synthesis of porphyrin (in the-liver).

Question

(Discussion on page 275)

10·1 Carbon in fatty acids cannot be made into glucose at an appreciable rate in humans. Nevertheless, radio-labeled carbon in fatty acids appears in glucose. How can these two statements be reconciled?

PART B
THE ELECTRON TRANSPORT PATHWAY

PHYSIOLOGICAL FUNCTIONS

> *The electron transport pathway (ETP) regenerates ATP by using the electrochemical gradient for H+ across the inner mitochondrial membrane.*
>
> *The major control over the ETP is via the concentration of reduced cytochrome c.*
>
> *The adenine nucleotide transporter exchanges cytosolic ADP for mitochondrial ATP.*
>
> *Creatine kinases may aid the diffusion of ADP in the cytosol of skeletal muscle.*

The ETP consists of a chain of electron carriers; electrons donated by NADH and FADH$_2$ move along this chain, eventually combining with oxygen atoms to yield H$_2$O (Figure 10·3). The formation of H$_2$O yields much energy. The ETP captures most of this energy by pumping H+ out of the mitochondria. The electrochemical gradient for H+ drives the synthesis of ATP from ADP + P$_i$ through a reaction that is conceptually the reverse of an ATP-driven transport enzyme (Figure 10·4). If H+ leak back into the mitochondria by another route, ATP is not synthesized and oxidative phosphorylation is said to be uncoupled.

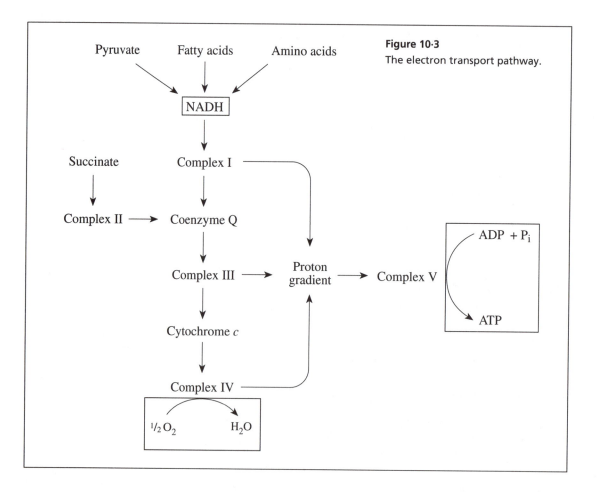

Figure 10·3
The electron transport pathway.

218

ATP is synthesized inside mitochondria by oxidative phosphorylation, but the majority of ATP is used outside the mitochondria. An exchange transport system is used to move the ATP across the mitochondrial membrane in exchange for ADP (see Figure 7·13, page 182).

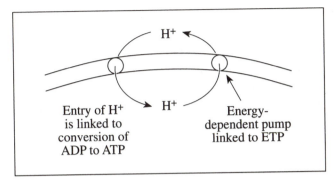

Figure 10·4
The synthesis of ATP. The parallel lines represent the inner mitochondrial membrane. In the chemiosmotic theory of ATP generation, H^+ generated during the oxidation of NADH are pumped out of mitochondria against a concentration and charge gradient. The H^+ then reenter mitochondria via a special transport unit down their electrochemical gradient; this action drives the synthesis of ATP.

Question

(Discussion on page 275)

10·2 Brown adipose tissue is an organ designed to generate heat during exposure to cold. Using your knowledge of the electron transport pathway, explain how this process works.

CONTROLS

CONTROL OVER THE ATP GENERATION SYSTEM

The ETP is the final common pathway in which NADH formed in the TCA cycle, in PDH, and in β oxidation of fatty acids is oxidized in the presence of oxygen to regenerate ATP (equations 1–3). The ETP traps close to 90% of the energy available in the regeneration of ATP. The pathway has three major segments, each of which produces ATP or NADH; the first two are near-equilibrium reactions and thus are not sites for regulation (equations 1 and 2). They produce the reduced form of cytochrome c, the substrate for the third step, the nonequilibrium step in the ETP (equation 3).

NADH + Oxidized cofactors + ADP + P_i $\longleftrightarrow$ Reduced cofactors+ NAD^++ATP (1)

Reduced cofactors + Oxidized cytochrome c + ADP + P_i $\longleftrightarrow$

   Reduced cytochrome c + Oxidized cofactors + ATP (2)

Reduced cytochrome c + ADP + P_i + $\frac{1}{2}O_2$ $\longleftrightarrow$ Oxidized cytochrome c + ATP (3)

CONTROL OVER THE ETP

It is now believed that increases in intracellular $[Ca^{2+}]$, which initiate muscular contraction, also stimulate the regeneration of ATP. With this mechanism, large changes in the concentrations of ADP and ATP are not needed to produce large changes in the rate of regeneration of ATP (Figure 10·5). The background for this statement follows:

- **Overall rise:** The magnitude of flux through the ETP must match the rate of use of ATP in steady state since this latter process regenerates the ADP needed to sustain flux through the ETP.

- **Control by the [ADP]:** The key question is whether or not the [ADP] controls flux directly. In some experiments conducted in mitochondria in vitro, this type of control does occur. Nevertheless, we believe that the [ADP] does not control events in vivo in organs such as the heart or skeletal muscle because the [ADP] is close to its K_m for oxidative phosphorylation. Thus, only a maximum of a twofold rise could occur by this mechanism (see Chapter 7, page 175)—not the 10-fold to 20-fold increases that are observed. Further, the [ADP] does not appear to change significantly in the transition between slower and faster rates of consumption of oxygen.

Figure 10·5

Link between ATP metabolism and muscle contraction. In response to nerve impulses, the sarcoplasmic reticulum releases Ca^{2+}, the signal that causes muscle to contract. Muscle contraction consumes ATP. The Ca^{2+} added to the cytosol are immediately taken up by mitochondria, causing rapid stimulation of the rate of regeneration of ATP. Hence, the same signal stimulates both use and regeneration of ATP, and large changes in the [ATP] and the [ADP] are not observed despite very large changes in the rate of turnover.

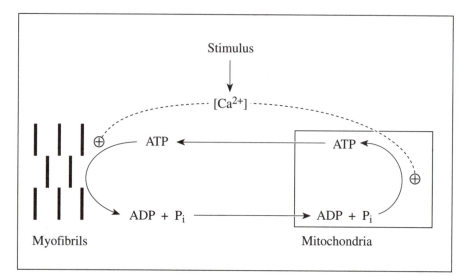

- **Control by the concentration of reduced cytochrome c:** This hypothesis is very attractive because the reduced form of cytochrome c is the substrate for a non-equilibrium reaction that is not nearly saturated with its substrate. What makes the concentration of reduced cytochrome c rise? Looking at equations 1 and 2, a rise in NADH and a fall in ADP or P_i in mitochondria could cause such an increase. The most likely regulator of the concentration of reduced cytochrome c in muscle is a rise in intracellular Ca^{2+}. As more Ca^{2+} enter mitochondria, three dehydrogenases (PDH, NAD^+-linked isocitrate dehydrogenase, and 2-oxoglutarate dehydrogenase) become active, leading to a rise in the ratio of NADH to NAD^+ in mitochondria.

The calculated V_{max} for oxidation of reduced cytochrome c is apparently 10-fold greater than the fastest rate of consumption of O_2 in vivo.

- **Change in the V_{max} of the ETP:** Some data suggest that this V_{max} rises, but the mechanisms are not clear. Perhaps the V_{max} is also regulated by the rise in intramitochondrial $[Ca^{2+}]$.

- **Limit by the P_{O2}:** The P_{O2} required for 50% flux through the ETP in vitro is close to 1 mm Hg. When conditions are set closer to those in vivo, this apparent K_m for O_2 is even lower. Hence, availability of O_2 will become a problem only if there is a defect in blood flow or if the demand for O_2 is so enormous that it exceeds the rate of delivery of O_2.

DIFFUSION OF ATP AND ADP

Skeletal muscle can rapidly increase its consumption of ATP (and O_2) by up to 20-fold from rest with only small changes in the [ATP] and the [ADP]. ADP is consumed in one area of the cell, the mitochondrion (located near blood vessels to shorten the distance for the diffusion of O_2), and is formed in another, the myofibril (located throughout the muscle cell). These two areas are too far apart for ADP to diffuse between them quickly enough (the [ADP] is very low). Two creatine kinase (CK) reactions "aid" this diffusion. CK is located both on the outer aspect of mitochondria and near myofibrils. The ratio of ATP to ADP is in equilibrium with the ratio of creatine phosphate (CrP) to creatine (Cr) (equation 4). If there is no hypoxia, the [H^+] in cells does not change appreciably despite large changes in turnover of ATP.

$$CrP^{2-} + ADP^{3-} + H^+ \longleftrightarrow Cr + ATP^{4-} \qquad (4)$$

Since the concentrations of CrP and Cr are very large, they can diffuse quickly and "transfer" ADP fast enough to permit an efficient system (Figure 10·6).

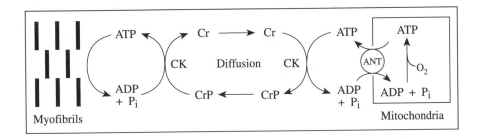

Figure 10·6
Use of CK to help ADP "diffuse" between myofibrils and mitochondria. For details, see the text. (ANT = Adenine nucleotide transporter.)

INTEGRATION OF CONTROLS WITH TRAINING

At the level of the mitochondrion in skeletal muscle, endurance training increases all of the components of the ATP regeneration system—the dehydrogenases of the TCA cycle (enzymes needed to form NADH) and the components of the ETP. In the cytosol of these cells, more myoglobin is present so that diffusion of O_2 can be faster. In the whole tissue, mitochondrial volume rises and mitochondria move closer to the capillaries so that there is a smaller distance over which O_2 must diffuse. In addition, the number of capillaries increases and they become more tortuous. Finally, the diameter of individual muscle fibers decreases, again to cut down the distance over which O_2 must diffuse.

Veterinary Case 10·1
How Can the Hummingbird "Hum"?

When a hummingbird hovers and flies rapidly, it consumes O_2 10–15 times faster than does the most elite human athlete. How can it deliver O_2 to muscle mitochondria so fast? How does the structure of its mitochondria differ, and how does it minimize the loss of energy?

Discussion of Veterinary Case 10·1
How Can the Hummingbird "Hum"?

COMPARISON OF DELIVERY OF O_2

The comparatively rapid delivery of O_2 to skeletal muscle involves four factors: the lungs, the cardiac output, the quantity of hemoglobin in blood to bind oxygen, and the efficient delivery of O_2 to muscle fibers. A hummingbird's lungs are extremely efficient because their very small air sacs (alveoli) increase the surface area for diffusion of O_2. The hummingbird's heart, relatively much larger than a human's, pumps more viscous blood at a fast rate. Blood circulation to muscle is different because there are many more (and more tortuous) capillaries per fiber and relatively small fiber bundles that cut down the distance for diffusion of O_2. The carriage of O_2 per liter of blood is increased because of a higher level of hemoglobin (see Veterinary Case 2·3, Neptune and the Seal, pages 75–76 and Veterinary Case 3·1, The Racehorse, page 95). In addition, the oxygen-hemoglobin dissociation curve (Figure 2·5, page 50) is shifted to the right so that O_2 can be down-loaded more readily at a higher Po_2 in tissues (to aid diffusion of O_2 to muscle mitochondria). Interestingly, the curve is not classically sigmoidal, but affinity is increased at a higher Po_2, thus increasing the efficiency of the loading of O_2 at the lungs.

COMPARISON OF MUSCLE

A hummingbird's muscle has more mitochondria that are closer to the capillaries so that the distance for diffusion of O_2 is reduced. Its muscle also has a much greater amount of myoglobin to aid in the diffusion of O_2. These adaptations are remarkable; the internal structure of each mitochondrion has two times the density of cristae compared with a human mitochondrion (the cristae contain some of the components of the TCA cycle and all of the components of the ETP).

CHANGES IN EFFICIENCY

A hummingbird's wings flap at an enormous rate. Without specific adaptations, great amounts of energy would be wasted at the start and stop of wing movement. To cut this potential waste of energy, the hummingbird has considerable elastic recoil at extremes of wing movement.

CHAPTER 11

THE FAT SYSTEM

PART A
OVERVIEW OF THE FAT SYSTEM

The fat system includes the pathways interconverting acetyl-CoA, fatty acids, triacylglycerols, and ketoacids. During meals, excess dietary fuels are deposited into energy reserves as triacylglycerols in adipose tissue. Between meals and without meals, triacylglycerols in adipose tissue become the major energy source for the body. The balance between deposition and use of triacylglycerols is controlled primarily by insulin.

PART B
SYNTHESIS OF FATTY ACIDS AND DEPOSITION OF TRIACYLGLYCEROLS

The principal fuels absorbed into the circulation from the diet are glucose and triacylglycerols (as chylomicrons). The liver converts glucose into acetyl-CoA, which it can then use to make fatty acids and triacylglycerols. These triacylglycerols are released to the circulation in particles called very-low-density lipoproteins (VLDL). The triacylglycerols in VLDL and in chylomicrons are deposited in adipose tissue by a pathway that releases fatty acids in the capillaries and then re-incorporates them into triacylglycerols in adipocytes. Insulin exerts the major control over the conversion of acetyl-CoA to fatty acids and triacylglycerols by activating acetyl-CoA carboxylase. The product of this reaction, malonyl-CoA, is a key regulator because it inhibits the entry of fatty acyl-CoA into mitochondria so that this long-chain acyl group will remain esterified and will not be oxidized.

Case 11·1   Why Is Rollo Obese?   page 229

PART C
BREAKDOWN OF TRIACYLGLYCEROLS AND OXIDATION OF FATTY ACIDS

Between meals and without meals, triacylglycerols in adipose tissue are hydrolyzed to fatty acids, which travel through the circulation bound to albumin and are oxidized in many tissues for regeneration of ATP. The pathway for oxidation of fatty acids in organs requires activation of the fatty acids to their CoA derivatives in the cytosol, transport into the mitochondria using the carnitine transport system, and oxidation to acetyl-CoA inside the mitochondria. When levels of fatty acids in the blood are high (because of rapid release from adipose tissue, permitted by a low net concentration of insulin), fatty acids are oxidized at a rapid rate. Conversely, when the concentration of fatty acids in blood decreases, the rate at which they are oxidized slows.

PART D
KETOACID FORMATION AND OXIDATION

Ketoacids are synthesized in the liver and provide a fat-derived, water-soluble fuel for the brain. Synthesis of ketoacids also protects the body from excess levels of fatty acids. The brain and kidneys oxidize ketoacids; their concentration in blood primarily determines their rate of oxidation, up to a limit set by the need of these organs for ATP.

Acetoacetate is made in mitochondria from acetyl-CoA; the rate of its synthesis is controlled by the rate of delivery of fatty acids and by the hormonal setting (stimulated by a low net concentration of insulin). β-hydroxybutyrate bears the same relationship to acetoacetate as lactate does to pyruvate.

PART E
THE ACETYL-COA AND FATTY ACYL-COA CROSSROADS IN THE LIVER

Acetyl-CoA, a key intermediate in the fat system, is used in:

1. the regeneration of ATP in all cells with mitochondria;
2. the synthesis of fatty acids in the liver (during meals);
3. the synthesis of ketoacids in the liver (without meals).

PART A
OVERVIEW OF THE FAT SYSTEM

The fat system must provide almost all of the energy used during fasting.

The fat system includes the pathways interconverting triacylglycerols, fatty acids, ketoacids, and acetyl-CoA (Figure 11·1). The key function of the fat system is the use of triacylglycerols in adipose tissue to provide most of the energy needed by the body when dietary glucose is not available. Because this source of energy is used between meals, it must be regenerated during meals. The fat system also provides ketoacids as a fuel for the brain during prolonged fasting.

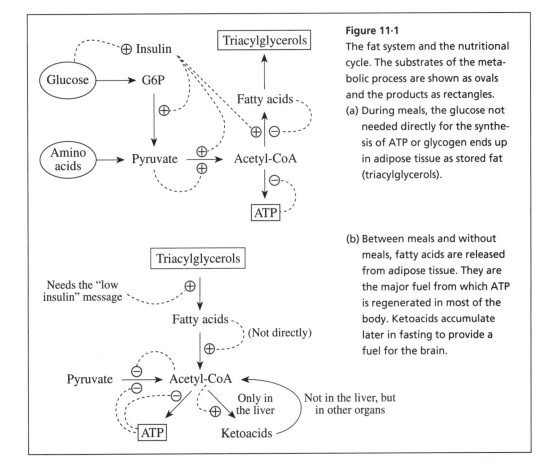

Figure 11·1
The fat system and the nutritional cycle. The substrates of the metabolic process are shown as ovals and the products as rectangles.
(a) During meals, the glucose not needed directly for the synthesis of ATP or glycogen ends up in adipose tissue as stored fat (triacylglycerols).

(b) Between meals and without meals, fatty acids are released from adipose tissue. They are the major fuel from which ATP is regenerated in most of the body. Ketoacids accumulate later in fasting to provide a fuel for the brain.

DURING MEALS

Energy not needed during a meal is converted to stores. When stores of glycogen are full, surplus glucose is converted to triacylglycerols.

Deposition of triacylglycerols from the blood into adipose tissue stores is the primary pathway of the fat system during meals. At least 25% of the calories in a normal western diet are supplied by triacylglycerols. In the intestinal tract, triacylglycerols are

225

hydrolyzed to fatty acids and monoacylglycerols. Resynthesis into triacylglycerols and packaging into chylomicrons occur in intestinal cells. Once released into the lymph, chylomicrons soon enter the blood. The triacylglycerols in chylomicrons can be taken up by some tissues (heart, skeletal muscle) for immediate oxidation, or they can be deposited into stores of triacylglycerols in adipose tissue; deposition into stores is favored when levels of insulin are increased.

Also during meals, dietary carbohydrate (glucose) may exceed the amounts needed for immediate regeneration of ATP or for replenishing stores of glycogen in the liver or muscle. Any excess can be converted into triacylglycerols in the liver, transported in blood as very-low-density lipoproteins (VLDL), and stored in adipose tissue (Figure 11·2). Deposition of triacylglycerols in adipose tissue (including their formation from glucose in the liver) is promoted by the relatively high levels of insulin caused by the intake of glucose.

Figure 11·2

Involvement of organs in esterification. Triacylglycerols in adipose tissue are formed from dietary carbohydrates (via glucose) and fatty acids. In the liver, glucose is converted to triacylglycerols, which are exported in VLDL particles. During digestion, dietary fat is split into fatty acids, which are converted to triacylglycerols in the intestine (packaged as chylomicrons). In the capillaries of adipose tissue, lipoprotein lipase (LPL) splits all triacylglycerols in both lipoprotein packages into fatty acids for uptake in adipocytes, where resynthesis of triacylglycerols will occur.

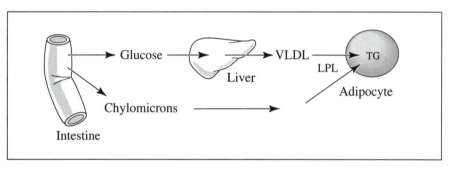

BETWEEN MEALS AND WITHOUT MEALS

Between meals, when dietary fuels have been absorbed and levels of glucose in blood have fallen, the body must rely on stored energy fuels to provide the needed ATP. The brain, which uses approximately 25% of the calories needed by the body at rest, consumes glucose released from glycogen in the liver. The energy for the rest of the body comes from fatty acids released from triacylglycerols in adipose tissue. Under primary control by insulin and its counter-hormones, hormone-sensitive lipase in adipose tissue breaks down triacylglycerols to fatty acids, which are released to the circulation and are oxidized in many tissues (except the brain) to regenerate ATP.

The low net concentration of insulin promotes both the release of fatty acids from adipose tissue triacylglycerols and the formation of ketoacids, which can be oxidized in the brain (mainly) and kidneys to regenerate ATP. Since stimulation of ketogenesis is delayed, a few days of complete fasting are required for the concentrations of ketoacids to build up to significant levels.

PART B
SYNTHESIS OF FATTY ACIDS AND DEPOSITION OF TRIACYLGLYCEROLS

SYNTHESIS OF FATTY ACIDS FROM ACETYL-COA

> *The conversion of glucose to fatty acids occurs in three stages: first, the generation of acetyl-CoA in mitochondria; second, the transfer of acetyl units into the cytosol; third, the polymerization of acetyl units into long-chain fatty acids.*

STAGE 1:
CONVERSION OF GLUCOSE TO ACETYL-COA IN THE LIVER

Synthesis of fatty acids from dietary glucose occurs mainly in the liver. The pathway proceeds via acetyl-CoA, which is formed from glucose via glycolysis and pyruvate dehydrogenase (PDH).

Control

The high concentration of glucose in the blood causes an increased flux through glucokinase because of the high K_m of this enzyme for glucose. The high level of fructose 2,6-*bis*phosphate (resulting from a high net concentration of insulin) overrides product inhibition at PFK1 via ATP (see Figure 8·8, page 198), thereby generating pyruvate at high rates. The high concentration of pyruvate and the high net concentration of insulin induce activation of PDH. The combination of low levels of fatty acids and activation of PDH leads to the formation of acetyl-CoA in hepatic mitochondria. Thus, synthesis of this acetyl-CoA is controlled by substrate activation coordinated by high levels of glucose and insulin.

STAGE 2:
TRANSPORT OF ACETYL GROUPS FROM MITOCHONDRIA TO CYTOSOL

Because synthesis of fatty acids occurs in the cytosol, acetyl-CoA must be moved out of the mitochondria where it is formed by PDH. The acetyl groups are exported from mitochondria in the form of citrate, which is made in mitochondria by citrate synthase, the first enzyme of the tricarboxylic acid (TCA) cycle. Citrate is transported across the mitochondrial membrane on a carrier and is broken down again to acetyl-CoA by ATP-citrate lyase (Figure 11·3).

Control

The major control exerted on this portion of the pathway is blockage of the alternate routes of degradation of acetyl-CoA. The most important option, flux through the TCA cycle, is blocked at isocitrate dehydrogenase and 2-oxoglutarate dehydrogenase (Chapter 10) because of high levels of NADH and low levels of ADP (which signal abundant ATP). Ketogenesis, an intramitochondrial pathway, does not occur since acetyl groups are readily transported out of the mitochondria via the citrate transporter; this transporter is more active when fatty acyl-CoA does not accumulate.

Figure 11·3

Formation of extramitochondrial acetyl-CoA. The acetyl-CoA that is formed in mitochondria from PDH is converted to citrate by citrate synthase. Citrate is transported out of the mitochondria and split back to acetyl-CoA by ATP-citrate lyase.

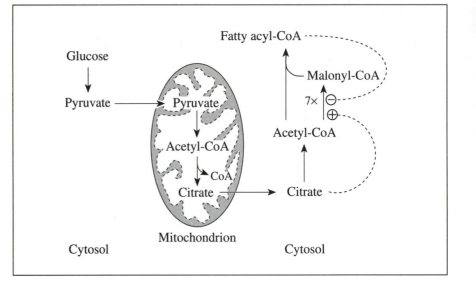

STAGE 3:
POLYMERIZATION OF ACETYL UNITS INTO FATTY ACIDS

Once outside the mitochondria, acetyl-CoA can then be incorporated into fatty acids. Seven out of eight acetyl-CoA molecules are first activated to malonyl-CoA by acetyl-CoA carboxylase (Figure 11·4). The two-carbon units derived from acetyl-CoA are then built into fatty acyl-CoA (usually 16 or 18 carbons). Fatty acyl-CoA, the product of fatty acid synthase, is then incorporated into triacylglycerols, using α-glycerophosphate (derived from glycolysis) to form the glycerol backbone (see Section Five, page 289).

Figure 11·4

Conversion of acetyl-CoA to fatty acids. In the synthesis of fatty acids, acetyl-CoA is first activated by acetyl-CoA carboxylase to malonyl-CoA, which is then used to add two-carbon units to the growing acyl chain. Acetyl-CoA carboxylase is a major control site for synthesis of fatty acids. (ACC = acetyl-CoA carboxylase.)

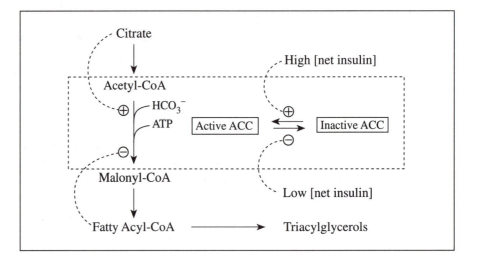

Control

Acetyl-CoA carboxylase, the first unique reaction of the pathway, is the major site of control over synthesis of fatty acids in the cytosol. Controls over acetyl-CoA carboxylase are parallel to those over PDH, though they differ in detail. Acetyl-CoA carboxylase exists in active and inactive forms. Insulin promotes the active form by causing acetyl-CoA carboxylase to be phosphorylated at a site that activates it. Other factors promoting the active form of acetyl-CoA carboxylase are citrate, a precursor (akin to the activation of PDH by pyruvate), and low levels of fatty acyl-CoA, the product of the pathway (similar to acetyl-CoA in PDH).

A second site of phosphorylation of acetyl-CoA carboxylase causes inhibition; this site is phosphorylated by cAMP-dependent protein kinase (Chapter 7, page 184), which is stimulated in the liver by glucagon through the cAMP second-messenger system. Because the levels of glucagon and insulin change in opposite directions, these two controls over acetyl-CoA carboxylase are mutually reinforcing.

Malonyl-CoA and Inhibition of Fatty Acid Oxidation. Acetyl-CoA carboxylase produces malonyl-CoA, another critical cytosolic signal. During meals, high concentrations of malonyl-CoA inhibit the entry of fatty acyl groups into mitochondria; hence, malonyl-CoA directs fatty acyl-CoA into the pathway for synthesis of triacylglycerols and away from that of oxidation.

> **Malonyl-CoA:**
> A compound in liver cytosol that increases in the presence of surplus glucose and signals that synthesis but not oxidation of fatty acids should occur.

STAGE 4:
DEPOSITION OF TRIACYLGLYCEROLS IN ADIPOSE TISSUE

Triacylglycerols in plasma, which come from the liver as VLDL or from the diet as chylomicrons, are deposited in adipose tissue. They are first broken down to fatty acids and glycerol (or monoglycerides) by lipoprotein lipase; the fatty acids enter the cell and are resynthesized into triacylglycerols by triglyceride synthase. The major control over deposition is the activation of lipoprotein lipase by insulin.

Case 11·1
Why is Rollo Obese?

Rollo is very obese. A liver biopsy revealed that the activity of fatty acid synthase was 10-fold greater than normal. Could increased activity of this multicatalytic protein be the cause of his obesity?

Discussion of Case 11·1
Why is Rollo Obese?

The level of glucose regulates the synthesis of fatty acids. A surplus of glucose supplies more malonyl-CoA, which increases flux through fatty acid synthase. Without a surplus of glucose, the substrate of fatty acid synthase will not increase in concentration. Accordingly, even a 10-fold excess of this enzyme should not "pull" carbon through the fatty acid synthase complex.

PART C
BREAKDOWN OF TRIACYLGLYCEROLS AND OXIDATION OF FATTY ACIDS

USE OF STORED TRIACYLGLYCEROLS FOR ENERGY

> *Stores of triacylglycerols in adipose tissue provide the vast majority of the energy required by the body between meals and without meals.*
>
> *Insulin inhibits the release of fuels from stores.*
>
> *Fatty acids and ketoacids inhibit oxidation of glucose and conserve precursors of glucose.*

Between meals and without meals, triacylglycerols in adipose tissue provide the majority of the energy used by the body (Figure 11·5). The liver, kidneys, and muscle use the fatty acids released from triacylglycerols as preferred fuels for regeneration of ATP. The liver also forms ketoacids from fatty acids; both the brain and kidneys can use ketoacids for energy. Fat-derived fuels inhibit the oxidation of glucose for regeneration of ATP and also prevent destruction of the precursors of carbohydrates by inhibiting PDH.

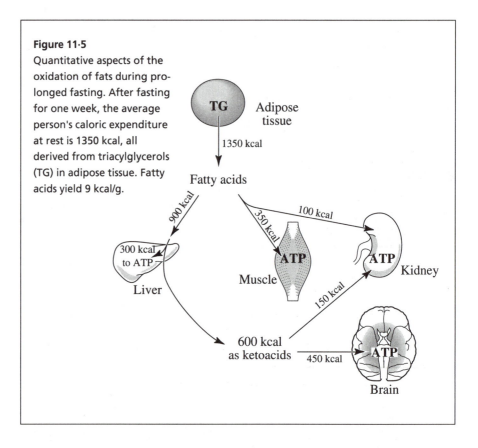

Figure 11·5

Quantitative aspects of the oxidation of fats during prolonged fasting. After fasting for one week, the average person's caloric expenditure at rest is 1350 kcal, all derived from triacylglycerols (TG) in adipose tissue. Fatty acids yield 9 kcal/g.

BIOCHEMICAL ASPECTS AND CONTROLS

BREAKDOWN OF TRIACYLGLYCEROLS

> *The major site of control in the release of fatty acids from adipose tissue is at the step catalyzed by hormone-sensitive lipase (HSL). A low level of glucose, via low insulin, promotes the release of fatty acids.*

Hydrolysis of stored triacylglycerols yields fatty acids and glycerol. Because fatty acids are strong detergents (they are the major constituents of soaps) and are relatively insoluble at physiological pH, they must be bound in all compartments. The fatty acids released from the breakdown of triacylglycerols in adipose tissue circulate in blood bound to albumin. They are carried across membranes by specific carriers, and they are transported through the cytosol by fatty acid-binding proteins (Figure 3·2, page 84).

Control

The activity of HSL, which regulates the release of fatty acids from adipose tissue, is mainly controlled by the effects of hormones, acting through changes in the [cAMP]. Synthesis of cAMP is stimulated in adipose tissue by adrenaline and ACTH (glucagon has minor effects) and is inhibited by insulin. The action of insulin in determining the [cAMP] in adipose tissue ensures that the release of fatty acids from stored triacylglycerols is closely attuned to the body's need for energy from this store (Figure 11·6).

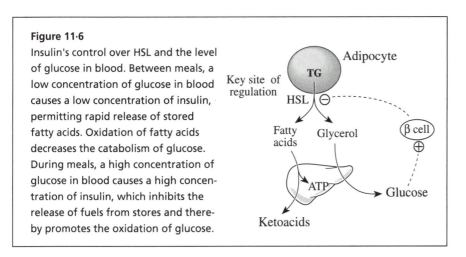

Figure 11·6
Insulin's control over HSL and the level of glucose in blood. Between meals, a low concentration of glucose in blood causes a low concentration of insulin, permitting rapid release of stored fatty acids. Oxidation of fatty acids decreases the catabolism of glucose. During meals, a high concentration of glucose in blood causes a high concentration of insulin, which inhibits the release of fuels from stores and thereby promotes the oxidation of glucose.

OXIDATION OF FATTY ACIDS

> *The rate at which fatty acids are oxidized depends on their concentration and, more importantly, on the signal for the entry of fatty acyl groups into mitochondria; this signal is a low level of malonyl-CoA.*
>
> *The rate of turnover of ATP sets a limit on the oxidation of fatty acids.*

In the cytosol, fatty acids are activated to their CoA derivatives, transported into mitochondria on a carnitine carrier, and oxidized in mitochondria to acetyl-CoA. In this process, called β oxidation, two carbon units are "cut off" in each spin of the cycle. The pathway is shown in Section Five, page 290.

All tissues that oxidize fatty acids use acetyl-CoA in the ATP generation system (see Chapter 10). In the liver, acetyl-CoA in excess of needs for ATP can be converted into ketoacids (see Part D below).

Control

The supply of fatty acids to cells exerts the first level of control over the oxidation of fatty acids. A high delivery of fatty acids is the result of a low level of insulin and thus the activation of HSL (Figure 11·6). A low net concentration of insulin occurs between meals and without meals when the blood glucose falls.

By causing a high intramitochondrial concentration of acetyl-CoA and NADH, oxidation of fatty acids can inhibit PDH and thus increase the share of acetyl-CoA derived from fatty acids (Figure 9·1, page 211). Since regeneration of ATP is the net result of β oxidation of fatty acids, the demand for ATP sets an upper limit to this pathway.

The fine-tuning of control over oxidation of fatty acids in the liver is via the concentration of malonyl-CoA in the cytosol. Malonyl-CoA, the inhibitor of the carnitine-linked fatty acyl group transfer system, regulates the entry of fatty acyl groups in hepatic mitochondria (Figure 5·2, page 128, and Section Five, page 289). When the concentration of glucose is low, the concentration of glucagon rises and the concentration of insulin falls, resulting in a higher concentration of cAMP in the cytosol of the liver. The increased level of cAMP causes phosphorylation and inhibition of acetyl-CoA carboxylase (Figure 11·4, page 228). The resulting low level of malonyl-CoA permits high activity of carnitine-palmitoyl transferase on the outer aspect of the inner mitochondrial membrane. Hence, the pathway for transport of fatty acyl groups into mitochondria becomes active. In contrast, when the concentration of glucose rises, the opposite hormonal changes occur and the level of malonyl-CoA rises, thereby preventing fatty acyl groups from gaining access to mitochondria for oxidation.

PART D
KETOACID FORMATION AND OXIDATION

Ketoacids are synthesized in the liver for use primarily in the brain during prolonged fasting.

Concentrations of ketoacids rise only after levels of insulin have remained low for at least two days.

PHYSIOLOGICAL FUNCTIONS OF KETOACIDS

Ketoacids provide a fat-derived source of energy for the brain. They are formed in the liver when the concentration of fatty acids in the circulation is high (Figure 11·7). They provide an alternative use for acetyl-CoA in liver mitochondria, thus allowing regeneration of intramitochondrial free CoA and hence faster oxidation of fatty acids. The rate of synthesis of ketoacids depends on the supply of intramitochondrial acetyl-CoA, which depends, in turn, on the rate at which fatty acids are delivered to the liver and whether the level of malonyl-CoA is low enough to stimulate their entry into mitochondria. The synthesis of ketoacids leads to the regeneration of ATP, which may help in setting the upper limit on ketogenesis.

Ketoacids are oxidized mainly in cells of the brain and kidneys. The last step of fatty acid oxidation splits acetoacetyl-CoA into two molecules of acetyl-CoA. For acetoacetate to be converted to acetoacetyl-CoA, a molecule of CoA must be donated from a special TCA cycle intermediate, succinyl-CoA (Figure 10·2, page 217).

The ability of the ketoacids to carry energy to other tissues is increased by formation of β-hydroxybutyrate, thus transferring extra H atoms to the other tissues. Quantitative aspects are indicated in Figure 11·5, page 230.

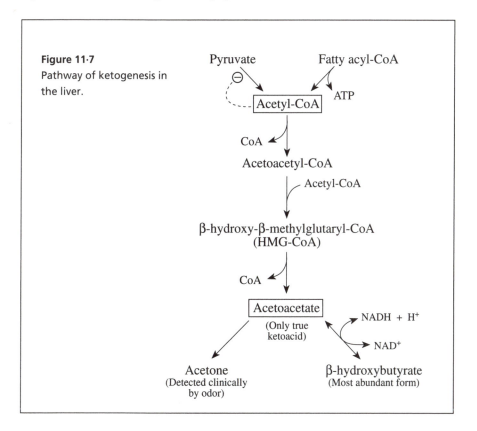

Figure 11·7
Pathway of ketogenesis in the liver.

BIOCHEMICAL ASPECTS AND CONTROLS

The stimulus for ketogenesis in the liver is a rapid supply of acetyl-CoA from oxidation of fatty acids; the controls over release of fatty acids from triacylglycerols and the oxidation of fatty acids (described above) therefore apply.

The ketogenic pathway converts two molecules of acetyl-CoA to acetoacetate. The major control over this pathway (at the acetyl-CoA crossroad) seems to be substrate supply; because three molecules of acetyl-CoA are used in the pathway, the potential influence of the [acetyl-CoA] or of the ratio of the [acetyl-CoA] to the [CoA] is likely to be very large. With a high rate of ketogenesis, the liver can regenerate most of its ATP from the NADH and FADH$_2$ provided by β oxidation of fatty acids, and the ATP generation system can operate largely without the tricarboxylic acid cycle.

The control over oxidation of ketoacids also appears to be simply substrate supply, up to a limit set by the need of the tissue for regeneration of ATP.

Question

(Discussion on page 275)

11·1   In a patient who lacks insulin, what quantity of ketoacids can the liver make per day? Assume that the supply of fatty acids is not rate-limiting and that the liver consumes 3000 mmol of O$_2$ per day.

PART E
THE ACETYL-COA AND FATTY ACYL-COA CROSSROADS

The management of the flow of acetyl-CoA inside liver mitochondria and of fatty acyl-CoA outside liver mitochondria must be very sophisticated (Figure 11·8).

INTRAMITOCHONDRIAL ACETYL-COA

The liver can derive intramitochondrial acetyl-CoA from both pyruvate and fatty acids. Because the oxidation of fatty acids prevents flux through PDH, acetyl-CoA will be formed from fatty acids rather than from pyruvate when levels of fatty acids are high. Conversely, low levels of intramitochondrial fatty acyl-CoA are required for a high flux through PDH.

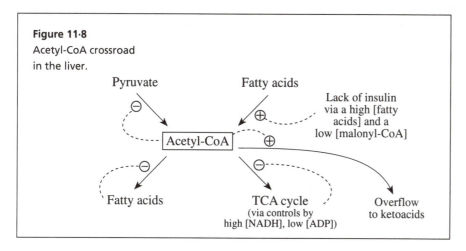

Figure 11·8
Acetyl-CoA crossroad in the liver.

The liver can use acetyl-CoA for regeneration of ATP, synthesis of fatty acids, and ketogenesis.

1. The use of acetyl-CoA for regeneration of ATP is controlled, as in all tissues, through product inhibition by ATP (really availability of ADP).

2. To synthesize fatty acids, the signals from a high availability of glucose are needed (a high net concentration of insulin). This pathway involves the exit of acetyl groups via citrate, as described in Figure 11·3, page 228.

3. The rate of formation of ketoacids is determined mainly by the [acetyl-CoA] in the mitochondria. It appears that the levels of acetyl-CoA that can be generated from PDH alone are not sufficient for significant ketogenesis; only rapid oxidation of fatty acyl-CoA with inhibition of both the TCA cycle and fatty acid synthesis can achieve the levels of acetyl-CoA required for rapid rates of ketogenesis.

EXTRAMITOCHONDRIAL FATTY ACYL-COA AND MALONYL-COA

Outside the mitochondria, the concentrations of fatty acyl-CoA and malonyl-CoA dominate the balance between synthesis and oxidation of fatty acids. A high concentration of fatty acyl-CoA, formed from high levels of fatty acids in blood, inhibits both the exit of citrate and the activity of acetyl-CoA carboxylase, thereby preventing the synthesis of malonyl-CoA. A low level of malonyl-CoA, the product of acetyl-CoA carboxylase, leads to the activation of the carnitine transfer system for entry of fatty acyl-CoA into mitochondria, thus augmenting the oxidation of fatty acids. Conversely, a high level of malonyl-CoA, which is a precursor for the synthesis of fatty acids, prevents the oxidation of fatty acyl-CoA (Figure 11·9).

Figure 11·9

Fatty acyl-CoA crossroad. Insulin promotes the synthesis of extramitochondrial acetyl-CoA and thus fatty acyl-CoA. The high concentration of malonyl-CoA resulting from a high net concentration of insulin prevents fatty acyl-CoA from entering mitochondria. Hence, synthesis of fatty acids is favored and oxidation of fatty acyl-CoA is inhibited.

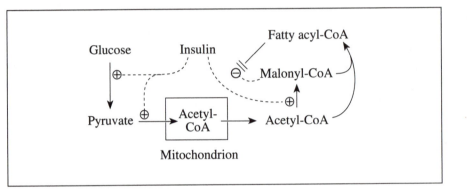

CHAPTER 12
THE PROTEIN SYSTEM

PART A
OVERVIEW OF PHYSIOLOGICAL IMPORTANCE AND CONTROLS

All proteins have specific functions and are only used as energy reserves because of the large amount available in the body. They become important in energy metabolism as precursors of glucose when dietary supplies of carbohydrate are low.

PART B
NITROGEN BALANCE

The nitrogen balance of the whole body indicates the balance between anabolism and catabolism of proteins. During the course of a normal dietary cycle, very dynamic changes in protein synthesis and degradation occur in individual organs (most importantly the liver, the site of initial metabolism of most amino acids). The nitrogen balance of the whole body, however, remains relatively constant.

PART C
PHYSIOLOGICAL USES OF AMINO ACIDS IN ENERGY METABOLISM

The major fate of the carbon in most amino acids is conversion to glucose (gluconeogenesis). The metabolism of glutamine, an amino acid, helps the kidneys eliminate H^+.

PART D
BIOCHEMICAL ASPECTS

Breakdown of amino acids occurs primarily in the liver. Half of their nitrogen is removed via conversion to NH_4^+ and the other half via conversion to aspartate; the products of this metabolism are urea and glucose. The syntheses of urea and glucose are linked because they share a common intermediate.

Veterinary Case 12·1

PART A
OVERVIEW OF PHYSIOLOGICAL IMPORTANCE AND CONTROLS

> *All proteins have specific functions.*
> *The use of body proteins as energy fuels removes functional components of the body.*

PROTEINS AND AMINO ACIDS

Unlike stored carbohydrates and fats, which function primarily as energy reserves, proteins in the body have specific functions. Proteins are significant as energy reserves because of the large amounts contained in the body (mainly in muscle) and because they are precursors of glucose.

Proteins consist of chains of amino acids; the sequences of the amino acids and the lengths of the chains are determined by genetic information encoded in DNA and are specific for each protein. As in all body components, synthesis and breakdown of proteins is constant; hence, amino acids are always entering and leaving energy metabolism. Because eight of the 20 amino acids cannot be made in the human body (Table 12·1) and must come from the diet, synthesis of proteins without dietary intake of these essential amino acids can only occur at the expense of other proteins in the body.

In terms of energy (ATP), proteins provide precursors for glucose, which is essential for regenerating ATP in the brain in the fasted state. Hence, there is always a net loss of proteins during fasting (see Case 1·9, pages 29-31); these proteins must be replaced when food again becomes available.

The balance between synthesis (anabolism) and breakdown (catabolism) of proteins is controlled by hormones. As long as amino acids are available, anabolism is stimulated by insulin (Figure 12·1), by some steroids (male or female), and by growth hormone. Catabolism is stimulated by counter-insulin hormones (e.g., glucocorticoids, thyroid hormone, glucagon, adrenaline) and also by many drugs.

Figure 12·1

Metabolism of amino acids within organs. When amino acids are available in the diet, the synthesis of proteins increases. A high net concentration of insulin triggers even more protein synthesis.

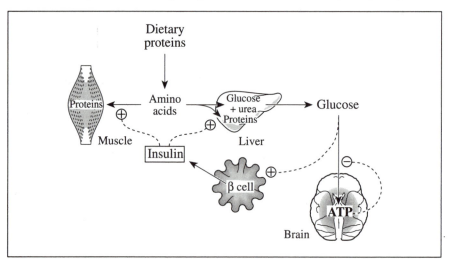

Data:
- Breakdown of 100 g of proteins yields 400 kcal.
- The liver utilizes a total of 300 kcal/day.

Because amino acids first encounter the liver as they enter the body via the portal vein, their initial metabolism occurs mainly in hepatocytes. Amino acids provide the liver with more kilocalories than it needs for its daily turnover of ATP. To temporarily hold the large supply of amino acids accompanying each meal, the body synthesizes new proteins (10% of which ultimately become albumin, i.e., 10 g/day), breaking them down later when they are needed. This synthesis of proteins allows the concentrations of amino acids in plasma to remain fairly constant, a necessary precaution because some amino acids function as neurotransmitters.

Table 12·1

Typical amino acid ratios in proteins

ALAnine	81	HIStidine	31	PROline	53
ARGinine	46	IsoLEucine* #	50	SERine	50
ASPartate + ASparagiNe	88	LEUcine* #	78	THReonine*	46
CYSteine	12	LYSine*	75	TRyPtophan*	23
GLUtamate + GLutamiNe	130	METhionine*	20	TYRosine	23
GLYcine	104	PHEnylalanine*	31	VALine* #	59

HYdroxyProline is made from proline during the synthesis of collagen.

The value beside each amino acid signifies the number of millimoles that is supplied per 1000 mmol of all amino acids in the proteins in a typical diet. The capital letters indicate the three-letter abbreviations (see also Figure 12·2).

* An essential amino acid—one that cannot be made by humans.

\# A ketogenic amino acid—one that cannot be made into glucose by humans. Some amino acids are part glucogenic and part ketogenic (e.g., TRP).

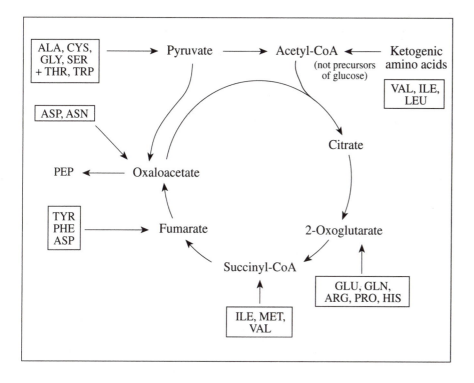

Figure 12·2

Sites of entry of some amino acids in gluconeogenesis. The metabolic fates for the common amino acids are indicated by the arrows from the enclosed rectangles. See Table 12·1 for abbreviations of these amino acids.

PART B
NITROGEN BALANCE

> *Nitrogen balance reflects the relative rates of anabolism and catabolism of proteins.*

The classical measure of the net status of synthesis or breakdown of proteins is the balance between the nitrogen consumed in the diet and that excreted in the urine; a positive nitrogen balance indicates a net anabolic state and is seen, for example, in growing children or in recovery from a catabolic state (see Case 1·1, pages 6–7). Positive nitrogen balance requires dietary consumption of proteins to provide the essential amino acids.

Even in negative nitrogen balance, with or without dietary proteins, synthesis of proteins occurs at relatively high rates because essential amino acids are recycled. In clinical medicine, measuring the rate of appearance of urea helps define the net rate of protein catabolism (see Case 4·6, page 119, for an example).

The above description considers the body as a single system; examining nitrogen balance at an organ level reveals a much more dynamic picture.

PART C
PHYSIOLOGICAL USES OF AMINO ACIDS IN ENERGY METABOLISM

> *During catabolism of amino acids, the carbon is primarily converted to glucose, the nitrogen to urea.*

GLUCONEOGENESIS

The amino acids released from breakdown of dietary or body proteins enter energy metabolism mainly as pyruvate (or precursors such as oxaloacetate, 2-oxoglutarate, and fumarate) or as acetyl-CoA (Figure 12·2); the former are referred to as glucogenic amino acids and the latter as ketogenic amino acids (Table 12·1). Most amino acids (80%) are catabolized in the liver while others (branched-chain and ketogenic amino acids) are catabolized in muscle and the kidneys. The dicarboxylic amino acids and their amides (GLU, GLN, ASP, and ASN) are the major fuels for the intestinal tract. Under optimal conditions, the body can convert 100 g of proteins to 60 g of glucose (some amino acids are ketogenic; only three-carbon units of amino acids can be used in gluconeogenesis since they must pass through pyruvate). The fates of amino acids derived from the diet or from breakdown of body proteins are therefore determined largely by the controls over pyruvate dehydrogenase and over the carbohydrate and fat systems (Chapters 9, 8 and 11, respectively). Insulin functions largely in these controls.

Amino acids derived from body proteins are essential sources of glucose for the brain, especially in the early stages of a prolonged fast (see Chapter 1, pages 29–31) after reserves of liver glycogen have been exhausted (18–24 hours after the last meal), but before ketoacids have become an alternate fuel for the brain.

EXCRETION OF H⁺ AND UREA

> *The metabolism of glutamine is of critical importance in excretion of H⁺ by the kidneys.*

A healthy North American adult in nitrogen balance consumes 75–100 g of proteins per day and synthesizes and breaks down 300–500 g of proteins per day. When 100 g of proteins are consumed, 16 g of nitrogen are added to the body. Almost all of this nitrogen is converted to urea and is excreted in the urine; most of the remainder is excreted as NH_4^+, allowing elimination of H^+ formed from intermediary metabolism. The amount of nitrogen excreted as NH_4^+ can increase markedly in chronic metabolic acidosis, resulting from the increased need to eliminate H^+ (see Cases 1·8, 1·9, and 2·8). In quantitative terms, normal adults excrete 400–500 mmol of urea (two atoms of nitrogen per molecule of urea) and 30–40 mmol of NH_4^+. In chronic metabolic acidosis, the excretion of NH_4^+ may rise to 200–300 mmol per day (Table 12·2).

Table 12·2

Daily urinary excretion of nitrogen wastes.

Compound		Normal	Chronic acid load
Urea	mmol (g N)	400–500 (12–15)	300–400 (9–12)
Ammonium	mmol (g N)	30–40 (0.4–0.6)	200–300 (2.8–4.2)
Creatinine	mmol (g)	10–14 (1.2–1.6)	10–14 (1.2–1.6)

Values represent the typical Western diet of an adult (consumption of 90 g of proteins per day). Usually less than 5% of the nitrogen excreted in the urine is in the form of NH_4^+. The rate of excretion of creatinine primarily depends on the muscle mass of the individual.

PART D
BIOCHEMICAL ASPECTS

Veterinary Case 12·1
Domesticated Cats Require Frequent Meals Containing Proteins

(Case discussed on page 246)

Tabby, a pet cat, must be fed a diet containing the amino acid arginine to prevent death from NH_4^+ toxicity. Why?

LOCATION IN ORGANS

> The liver is the site of final metabolism for most amino acids.
>
> In the breakdown of amino acids, half the nitrogen is converted to NH_4^+ and half to aspartate. Hence, deamination and transamination are key steps.
>
> The syntheses of urea and glucose from amino acids are linked because they share a common key intermediate.

Most amino acids are converted into common intermediates in the liver (Figure 12·3), the only organ that contains the many enzymes (6–7 per amino acid) needed for their metabolism. Metabolism of amino acids has several general characteristics.

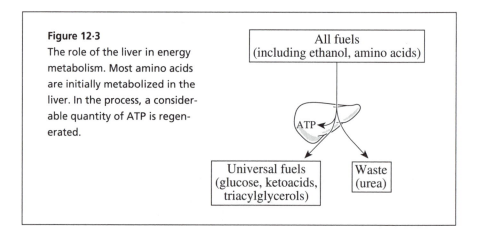

Figure 12·3

The role of the liver in energy metabolism. Most amino acids are initially metabolized in the liver. In the process, a considerable quantity of ATP is regenerated.

All fuels
(including ethanol, amino acids)

ATP

Universal fuels
(glucose, ketoacids,
triacylglycerols)

Waste
(urea)

1. The conversion of dietary proteins to glucose results in the net regeneration of ATP; therefore, the demand for regeneration of ATP in the liver can set an upper limit on gluconeogenesis.

2. As a result of constraints on regeneration of ATP in the liver, most amino acids of dietary origin are first converted into proteins and later released and oxidized; hence, gluconeogenesis (also ureagenesis since they are linked pathways) occurs steadily over the 24-hour daily cycle rather than in large surges shortly after meals.

3. Approximately 50% of the net regeneration of ATP in the liver is derived from hepatic gluconeogenesis. Most of the glucose formed is oxidized in other organs, such as the brain.

Other organs metabolize specific amino acids (Figure 12·4). Of note, the kidneys oxidize glutamine to make NH_4^+, which they excrete in the urine. This pathway is augmented in chronic acidosis. The intestinal tract is also a major site of production of NH_4^+ destined for the synthesis of urea. Because the intestinal tract is the primary site of metabolism of glutamine, glutamate, aspartate, and asparagine, it lessens the regeneration of ATP from amino acids in the liver.

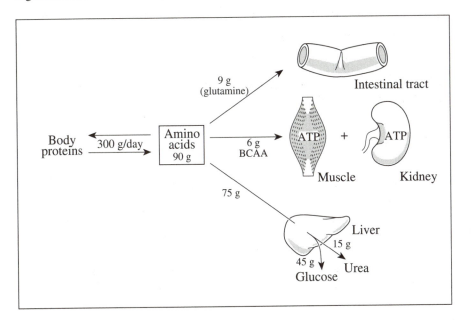

Figure 12·4
A quantitative look at the metabolism of amino acids. (BCAA = branched-chain amino acids.)

Muscle and the kidneys are major sites for the initial metabolism of branched-chain amino acids (isoleucine, leucine, valine). This function helps in diverting selected amino acids from oxidation in the liver. Their nitrogen returns to the liver, mainly in the form of alanine and glutamine, using carbon derived from metabolism of carbohydrates.

MAJOR REACTIONS AND PATHWAYS

Each of the 20 amino acids has a specific pathway for breakdown, and 12 of the 20 (the nonessential amino acids) can also be synthesized in the human body. The many pathways involved can be found in any standard text. Those for breakdown have some common features: transamination and deamination, metabolism of the carbon chain, and excretion of the nitrogen.

TRANSAMINATION AND DEAMINATION OF AMINO ACIDS

> *Transamination converts almost half the amino acids to glutamate.*
> *Glutamate is deaminated in mitochondria.*

Breakdown of close to half of the amino acids starts with removal of the amino groups (nitrogen) by transamination, leaving a carbon chain (for most amino acids) that then can be metabolized further—mostly to oxaloacetate or pyruvate. The amino group is converted to NH_4^+, a precursor for the urea cycle (Figure 12·5).

Transaminases

Transaminases usually transfer the amino group from amino acids to 2-oxoglutarate, an intermediate of the TCA cycle. The two major transaminases in the liver are aspartate aminotransferase and alanine aminotransferase, which catalyze the following reactions:

$$\text{Aspartate} + \text{2-Oxoglutarate} \longleftrightarrow \text{Oxaloacetate} + \text{Glutamate}$$

$$\text{Alanine} + \text{2-Oxoglutarate} \longleftrightarrow \text{Pyruvate} + \text{Glutamate}$$

The amino group can then be removed from glutamate by glutamate dehydrogenase:

$$\text{Glutamate} + NAD^+ \longleftrightarrow \text{2-Oxoglutarate} + NADH + H^+ + NH_4^+$$

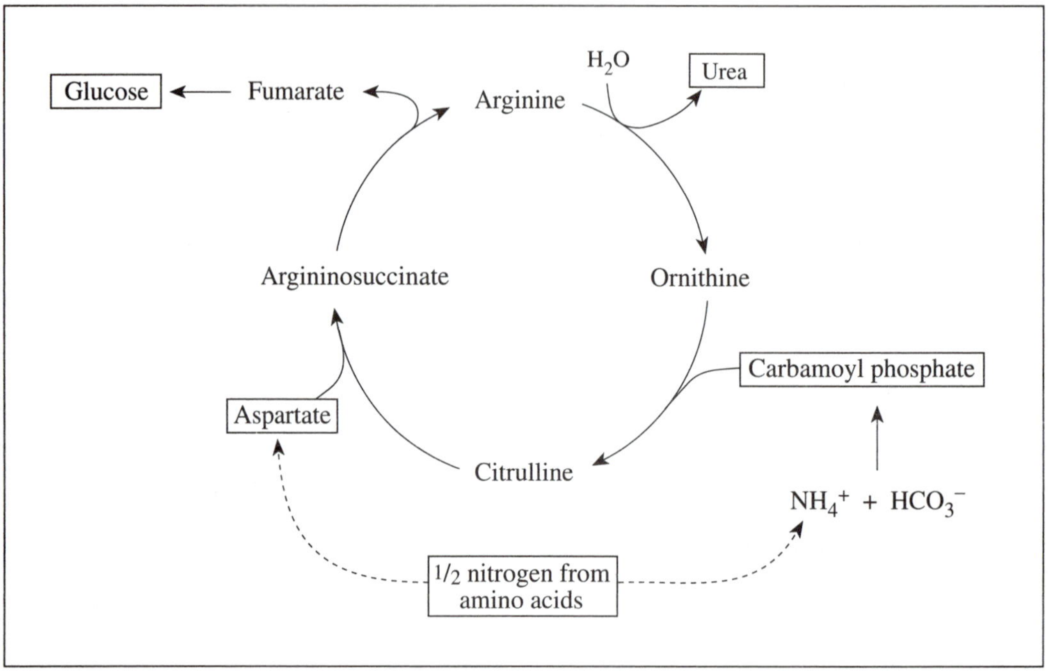

Figure 12·5
Urea cycle. Synthesis of urea occurs partly inside and partly outside the mitochondria. The pathways of gluconeogenesis and ureagenesis share a common near-final product, argininosuccinate, so their regulation is also shared.

METABOLISM OF THE CARBON CHAIN

After the amino groups are removed via transamination, half the carbon enters the gluconeogenic pathway as the products of transaminase reactions (i.e., as oxaloacetate and pyruvate).

The rest of the carbon enters the gluconeogenic pathway as fumarate, the ultimate product of aspartate metabolism in the urea cycle.

Oxaloacetate and pyruvate, the products of the major transaminases, are primarily converted to glucose via gluconeogenesis (Chapter 8). The ketogenic amino acids yield acetyl-CoA; hence, they cannot be converted to glucose because pyruvate dehydrogenase is functionally irreversible.

The branched-chain amino acids (isoleucine, leucine, and valine—the amino acids that cannot be made into glucose) are oxidized mainly in muscle. The nitrogen from these amino acids is released from muscle mainly as alanine and glutamine.

The breakdown of the sulfur-containing and the positively charged amino acids yields H^+, which constitute most of the H^+ that must be eliminated in conjunction with the excretion of NH_4^+.

FATE OF NITROGEN

> *Urea (made in the liver) is the main product in the catabolism of amino acids.*
> *The main acid-base product is NH_4^+ (made in the kidneys).*

In the urea cycle, half the nitrogen is derived from NH_4^+ and half from aspartate. Figure 12·5 shows the close linkage of the urea cycle to the glucogenic pathway. Flux through the early steps of activating NH_4^+ requires the expenditure of two molecules of ATP per NH_4^+ used (during the synthesis of carbamoyl phosphate). In addition, two more molecules of ATP are used during the synthesis of argininosuccinate. Even more energy (ATP expenditure) is required to transport compounds across the mitochondrial membrane during ureagenesis since part of the pathway is in the cytosol and part is in the mitochondria. Overall, close to 5 mmol of ATP are expended per mmol of urea synthesized.

$$2\ NH_4^+ + 2\ HCO_3^- + 4\text{-}5\ ATP \longrightarrow Urea + CO_2$$

Two other points merit mention with regard to urea synthesis. First, HCO_3^-, a substrate for the synthesis of urea, are synthesized later during the conversion of fumarate to glucose (Figure 12·5). This overall process, therefore, neither produces nor consumes HCO_3^-. Second, a special derivative of glutamate, N-acetylglutamate, is an activator of the synthesis of carbamoyl phosphate.

EXCRETION OF NITROGEN AS NH_4^+

The nitrogen excreted as NH_4^+ is carried to the kidneys as the neutral amino acid glutamine. The kidneys break down glutamine to NH_4^+ and HCO_3^-, converting the remainder of the carbon chain to glucose. The kidneys excrete the NH_4^+ (in exchange for Na^+ in the urine) and transfer the HCO_3^- to the plasma. The net result is elimination of H^+ (Figure 12·6). Please reexamine Veterinary Case 2·2, The Alligator and Excretion of Water, page 67.

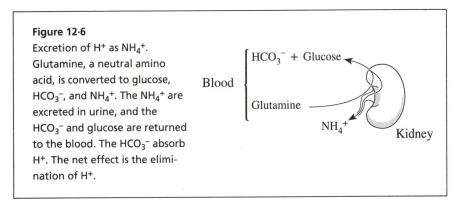

Figure 12·6
Excretion of H^+ as NH_4^+. Glutamine, a neutral amino acid, is converted to glucose, HCO_3^-, and NH_4^+. The NH_4^+ are excreted in urine, and the HCO_3^- and glucose are returned to the blood. The HCO_3^- absorb H^+. The net effect is the elimination of H^+.

Blood

HCO_3^- + Glucose

Glutamine

NH_4^+

Kidney

Discussion of Veterinary Case 12·1

Domesticated Cats Require Frequent Meals Containing Proteins

(Case presented on page 242)

Ornithine, citrulline, and arginine are the carriers of the urea cycle. If these compounds are not available, the urea cycle fails and NH_4^+ toxicity develops. A deficit of these compounds can occur if either input is low or if output is high.

INPUT

The input of citrulline or arginine provides the body with most of the carriers of the urea cycle. Citrulline is derived from two major sources, arginine from dietary or body proteins and de novo synthesis in the GI tract using glutamate, an abundant, nonessential amino acid (the product of transaminase reactions).

OUTPUT

Ornithine

Catabolism of ornithine by ornithine-α-aminotransferase depletes the pool of ornithine plus citrulline. The content of citrulline will therefore decline unless there is a continuous input of ornithine or citrulline.

Arginine

In some reactions, arginine is used but citrulline is not produced. The synthesis of creatine for skeletal muscle consumes arginine rapidly. Recall that creatinine, a product of the breakdown of creatine phosphate, is excreted in relatively large quantities each day (200 μmol/kg body weight).

THE CAT AND CITRULLINE

A kitten demands more arginine for synthesis of proteins than does an adult cat. A young domestic cat is also more susceptible to depletion of urea cycle carriers because it is not able to convert glutamate to citrulline in the intestinal tract at a sufficiently rapid rate. Thus, a kitten on an arginine-poor diet may suffer from convulsions within hours because this lack of citrulline results in NH_4^+ toxicity.

In normal physiology, citrulline does not enter (or leave) hepatocytes readily; it gains access to the general circulation from the intestinal tract. Citrulline is delivered to the kidneys and is reabsorbed in the proximal tubule, where it is converted to arginine and released into venous blood. This arginine enters the general circulation so that it is available for all organs for protein synthesis and in some tissues for the synthesis of nitric oxide, a vasodilator.

SECTION FOUR

Discussion of Questions

DISCUSSION OF QUESTIONS

Question 1·1

(Page 11)

The brain consumes about 10% more glucose than it needs to regenerate ATP, converting the excess to lactate. Why might this consumption be advantageous?

Consumption of a slight excess of glucose and its release as lactate might be seen as a safety valve to ensure sufficient fuel for the brain at all times, particularly when its need for ATP increases suddenly as a result of local stimulation. The lactate released from the brain circulates in the blood and, following a meal, is used as a fuel for other organs. The kidneys utilize lactate rather than glucose as their preferred fuel at this time because proximal tubule cells, though always exposed to a high supply of glucose (they reabsorb filtered glucose), are not well endowed with the enzymes necessary to oxidize glucose. This deficiency prevents them from oxidizing glucose when it would be disadvantageous to do so (between meals and during fasting). During fasting, the aim is to spare precursors of glucose (lactate) for the brain; hence, lactate is converted back to glucose in the liver (Figure 1·19, page 27).

Question 1·2

(Page 12)

Why should muscle use ketoacids sparingly?

The brain oxidizes ketoacids when a lack of glucose in the diet leads to a high concentration of ketoacids in the blood. During fasting, the brain needs approximately 50% of the ketoacids formed (at the maximum rate of formation). Since the maximum rate of ketoacid formation is limited by ATP-turnover constraints in hepatocytes, glucose for the brain will have to be formed from proteins in the body if muscle burns a large quantity of ketoacids. If proteins from muscle were to supply the brain with all the glucose it requires, about 1 kg of muscle tissue would be consumed per day (see the discussion of Case 1·4, page 14).

Question 1·3

(Page 14)

What nutritional therapy could have prevented Joe's death (Case 1·4)?

Joe died because he was oxidizing proteins instead of triacylglycerols for much of his energy needs. The goal of treatment should be to minimize this use of proteins (largely from muscle); it might include three components.

1. Provide all calories intravenously as glucose.

   One of the dangers of this treatment is that a large amount of water must be infused with the glucose, and the excretion of water is limited in trauma patients by high levels of antidiuretic hormone. A second danger is that if the use of glucose is not constant, hyperglycemia or hypoglycemia might occur.

   Ideally, Joe should receive enough glucose to meet as much of his caloric needs as possible. Joe is probably burning 50–60 kcal/hr, equivalent to 70–110 mmol (12–20 g) of glucose per hour; the rates could be higher if Joe has a fever. More glucose could be required if he is converting some of the glucose to triacylglycerols or excreting glucose in the urine, both of which could result from very high levels of glucose in blood; conversion to triacylglycerols will be stimulated further by exogenous insulin (see below).

Hypertonic solutions of glucose should be given intravenously to minimize the volume of water infused and hence the volume of urine produced. If treatment is successful, very little urea should be made from endogenous proteins, but, because at least 50 mosmol of solutes from salts or urea must be excreted along with each liter of urine, exogenous amino acids will also be needed.

2. Provide exogenous amino acids to promote net synthesis of proteins.

Amino acid mixtures plus insulin should be infused to minimize the net catabolism of body proteins. The insulin will promote conversion of glucose to glycogen and to triacylglycerols, meaning that more glucose (and water) must be infused; hence, care must be taken to ensure normal levels of glucose in blood. Also, by inhibiting lipolysis, insulin will lead to a smaller rate of oxidation of endogenous fatty acids, so again more glucose will be needed, in this case to regenerate ATP.

3. Provide exogenous fat-derived fuels.

Fat-derived fuels should also be given to minimize the need to oxidize glucose. Triacylglycerols can be administered (as emulsions) together with heparin (an activator of lipoprotein lipase) to promote release of their constituent fatty acids.

Using 1,3-butanediol provides a source of ketoacids and meets the hepatic demand for regenerating ATP. Infusion of an alcohol avoids the problem of having to add a cation and eliminates the need to add much additional water (alcohols cross cell membranes readily and thus do not cause internal water shifts resulting from changes in osmolality).

Question 1·4

(Page 21)

What changes will occur in the types of fuels oxidized in a patient who takes too much insulin? How long will the patient remain conscious?

The most important effect of insulin is to inhibit release of fatty acids from adipose tissue, resulting in limited availability of fatty acids for oxidation by the liver, kidneys, and muscle. Hence, glucose must be used for all of the energy requirements of the body.

A person at rest who uses only glucose to provide 2400 kcal/day requires 25 g of glucose per hour. The blood contains about 15 g of glucose; this source will therefore be exhausted in 40 minutes (an underestimate because the fatty acids do not disappear immediately on injection of excess insulin, and some glucose will still be released from liver glycogen and from glucogenesis). The brain, which depends on glucose to meet its high demands for ATP, will probably not start to suffer from effects of hypoglycemia until 60 minutes have elapsed (see Chapter 5 for a discussion of hypoglycemia).

Question 1·5

(Page 31)

What would you recommend that Henry (Case 1·9) add to his diet to minimize wasting of his muscles?

The loss of lean body weight occurs because amino acids are required as a source of glucose (Table 1·10, page 31). His brain's demand for glucose is only 20% of that when he was fed since, after five days of fasting, his blood contains 4–6 mmol/l of ketoacids. However, these ketoacids cause a load of H^+, which must be excreted in the urine as NH_4^+. Provision of these NH_4^+ demands breakdown of amino acids and hence consumption of lean body mass.

Henry might be advised to take an amino acid mixture to provide the essential amino acids that his body cannot make and to provide sufficient glutamine required for excretion of NH_4^+. However, he should avoid consuming excess amino acids because they are a source of calories and of glucose. Hence, they could stimulate release of insulin and decrease the concentration of ketoacids, which are essential to limiting the brain's use of glucose.

A second choice is to consume $NaHCO_3$ in amounts sufficient to absorb the H^+ that accumulate when ketoacid anions are excreted (see the figure below); this load of $NaHCO_3$ should equal the rate of excretion of NH_4^+ prior to its administration to achieve acid-base balance (see Case 2.7, page 66).

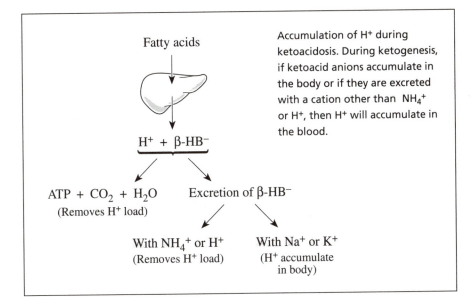

Fatty acids

$H^+ + \beta\text{-HB}^-$

$ATP + CO_2 + H_2O$
(Removes H^+ load)

Excretion of $\beta\text{-HB}^-$

With NH_4^+ or H^+
(Removes H^+ load)

With Na^+ or K^+
(H^+ accumulate in body)

Accumulation of H^+ during ketoacidosis. During ketogenesis, if ketoacid anions accumulate in the body or if they are excreted with a cation other than NH_4^+ or H^+, then H^+ will accumulate in the blood.

Question 1·6

(Page 31)

What will be the theoretical loss of carbohydrates, proteins, triacylglycerols, and weight if Henry (Case 1·9) exercises regularly, expending an extra 1500 kcal/day (assume it all comes from triacylglycerols)?

If all the extra 1500 kcal/day comes from the oxidation of fatty acids from triacylglycerols (9 kcal/g), the increased weight loss will be 167 g (1500 kcal/9 kcal/g), or 0.3 lb per day. The glycerol released by this extra breakdown of triacylglycerols will provide a further 17 g/day of glucose, more than sufficient (in theory) to cover the glucose that, without exercise, comes from proteins (Table 1·10, page 31). Whether or not this amount of proteins is actually spared will depend in part on the need to excrete NH_4^+ for acid-base balance (see the discussion of Question 1·5). Hence, the loss of weight may be less than that anticipated from the weight of the fatty acids that will be oxidized because the use of glycerol for gluconeogenesis will spare proteins and associated tissue water (heavy calories).

• To make 17 g of glucose, 30 g of proteins are needed (60% of the weight is converted to glucose).

• Muscle is 20% proteins by weight; therefore, 150 g of muscle are spared.

Question 1·7

(Page 33)

In what metabolic situations and in what tissues would active PDH be required a) all of the time; b) some of the time; c) never?

An organ requires active PDH when it is oxidizing glucose to regenerate ATP or when it is converting glucose to triacylglycerols. The brain always requires oxidation of some glucose and hence always needs some PDH in the active form. The liver and muscle use glucose only part of the time, and thus need PDH only when fatty acids are in short supply (i.e., in response to the actions of insulin). Only cells that never oxidize glucose to CO_2 (e.g., red blood cells) never need PDH.

Question 1·8

(Page 33)

If carbon from glucose must be conserved, especially in starvation, which is the most important point for control?

Carbon that passes through PDH can no longer be reconverted to glucose. Hence, PDH is the most important site for control over the absolute loss of carbon from the carbohydrate pool.

Question 1·9

(Page 33)

What types of controls would you expect over PDH in the liver?

When the supply of fuels from fatty acids is adequate for performance of biological work, PDH in the liver must limit the use of glucose for oxidation. You might therefore expect signals inhibiting PDH to be related to the availability of fatty acids for regeneration of ATP.

The liver must also be able to override product inhibition of PDH by ATP and related compounds. When supplies of glucose are in excess of the amount needed to regenerate ATP, PDH must be active to allow synthesis of fatty acids.

Question 1·10

(Page 33)

Look back over the previous cases. What were the roles of PDH in the various tissues in each case?

Have fun!

Question 1·11

(Page 35)

A super-marathon runner (50–100 miles) wants to eat enough fuel while running to compensate for energy used in a race (Chapter 3). Is this idea practical?

A super-marathon runner burns 8–12 kcal/min, which is equivalent to approximately 3 g (17 mmol) of glucose per minute. To match the consumption of energy, an equal amount must be absorbed from the diet. Though glucose can be eaten at this rate, the stomach will not release it rapidly enough to meet the body's metabolic needs during strenuous exercise. The stomach typically releases glucose at the constant, slow rates required by normal metabolism—about 2.5 kcal/min, or 3.5 mmol (0.6 g) of glucose per minute. In this way, an excessive input of glucose into the body is avoided during a normal demand for glucose. Hence, the super-marathoner's stomach will not allow maintenance of caloric balance during the race.

Question 1·12

(Page 36)

Could Sandy (Case 1·12) have had a high level of ketoacids in her blood before the panic attack?

Sandy did not respond to the fast in the same way that the other subjects did. Her levels of ketoacids in blood at 24 and 40 hours were fourfold higher than those of the others, and her concentration of glucose in blood at the 24-hour point was distinctly lower (Table 1·11). The high level of ketoacids indicates that the hormonal signals (low levels of insulin, high levels of glucagon) were driving ketogenesis by providing a high level of substrates (fatty acids) for this pathway and by signaling the liver to make more ketoacids (via a low level of malonyl-CoA, Chapter 11). Perhaps a higher level of adrenaline in Sandy also contributed to this set of signals by augmenting the rate of hydrolysis of triacylglycerols in adipose tissue.

REASONS FOR A LOWER LEVEL OF INSULIN

A lower level of glucose (to stimulate the release of insulin) and a higher α–adrenergic response (to inhibit the release of insulin from normal β cells) might have contributed to the lower level of insulin.

REASONS FOR SYMPTOMS OF THE PANIC ATTACK

As we pointed out in the discussion of Case 5·1 (page 126), it is not clear what signals the CNS about impending hypoglycemia, nor is it at all clear how intense the adrenergic response might be. Perhaps this response occurred earlier or to a greater degree in Sandy.

Question 1·13

(Page 36)

Why did Sandy's normal concentration of glucose in blood not cause a decrease in her level of ketoacids?

The α-adrenergic reaction caused by the panic attack (see above) prompted a rise in the concentration of glucose in plasma (adrenaline stimulates the hydrolysis of glycogen). Also, Sandy's store of glycogen in the liver, although low, was probably not completely exhausted. The higher levels of glucose in her blood might not have been able to elicit a release of insulin from the β cells of the pancreas because of inhibition of these cells by the α-adrenergic reaction.

Question 1·14

(Page 37)

Suppose Phil (Case 1·13) achieved rapid weight loss after taking diet pills. How would you determine the source of the weight he lost? If you obtained one of these pills, and had a research lab, animals, cells, mitochondria, etc., what tests would you perform to establish its biological effect(s)?

SOURCE OF WEIGHT LOSS

The loss of weight could have come from stored fats, carbohydrates, proteins, or from loss of intracellular fluid (ICF) or extracellular fluid (ECF).

Fats

To make this deduction, an assessment of caloric expenditure is necessary. Since Phil has no unusual physical activity, perhaps he is expending 2000–2400 kcal/day. Oxidation of fatty acids yields 9 kcal/g, so Phil can burn about 250 g of fatty acids. Since fatty acids are stored without water, his daily weight loss could only be 250 g, close to 0.5 lb per day. Hence, it is not possible for Phil to lose weight quickly by oxidizing body fat unless the drug increased biological work or uncoupled oxidative phosphorylation (i.e., increased consumption of oxygen markedly; see Chapter 10).

Carbohydrates and Proteins

Both of these compounds are retained in the body along with water—2 to 3 g of water per gram of fuel (Table 1·3, page 5). Since close to 4 kcal/g dry weight are produced with either fuel, the net yield will be close to 1 kcal/g wet weight. Accordingly, oxidizing these fuels can result in a weight loss of close to 2 kg/day with an energy expenditure of 2000–2400 kcal/day.

If proteins are oxidized, urea is the principal nitrogenous waste produced. The consumption of 100 g of proteins yields 400 kcal and approximately 500 mmol of urea. Thus, to know how much protein was oxidized, examine the rate of appearance of new urea in the body (rise in the concentration of urea × total body water) plus that excreted in the urine. If there is no loss of ECF or ICF volume and little loss of urea, then, by inference, the oxidation of carbohydrates (glycogen) is yielding most of the weight loss.

Loss of ICF and ECF Volumes

Water from the ICF is lost if the tonicity of body fluids rises (reflected by a rise in the $[Na^+]$ in plasma) or if particles are lost from the ICF (primarily by excretion of K^+ with phosphate; see pages 57–58 for a discussion of osmotic pressure). Loss of ECF volume depends on the degree of negative balance for Na^+ and is detected initially on clinical examination. Further details are beyond the scope of this text.

TESTING THE PILLS IN THE LABORATORY

If the weight loss is not due to excessive excretion of electrolytes and water in the urine, you could measure the effects of the pills on the consumption of oxygen and the fuels oxidized in tissues or subcellular particles.

Increased Consumption of Oxygen

In vitro studies of the effects of the drug on the consumption of oxygen in tissues, cells, or mitochondria could help explain its mechanism of action. The observation that the drug increases the consumption of oxygen could be explained in at least two ways:

1. The drug might increase the tissue's demand for ATP (for example, some drugs promote the leakage of ions across cell membranes down concentration gradients, thus forcing the pumps to expend ATP to recreate the gradients);

2. it might uncouple the synthesis of ATP from the consumption of oxygen.

Some "doctors" have prescribed uncoupling agents for weight loss; some of their patients have died, for obvious reasons.

Oxidation of Carbohydrates or Proteins vs Fatty Acids

A drug that inhibits the oxidation of fatty acids but not pyruvate could be identified by adding each substrate to cells or mitochondria and then measuring the consumption of oxygen. Alternatively, a drug that activates PDH (such as dichloroacetate) could have a similar effect (but not inhibit the oxidation of fatty acids in the absence of pyruvate or lactate).

Question 2·1

(Page 46)

The lab sends you the following results for a blood sample: $[H^+] = 60$ nmol/l, pH $= 7.22$, $Pco_2 = 50$ mm Hg, and $[HCO_3^-] = 32$ mmol/l. What do you conclude?

The results are not consistent with the Henderson equation,

$$[H^+] = \frac{23.9 \times Pco_2}{[HCO_3^-]}$$

so there must be an error. Ask for repeat values or be sure that all measurements were made on the same blood sample. Occasionally, arterial blood is used for pH and Pco_2 measurements, and venous blood is used for the measurement of HCO_3^-; the inconsistency in these numbers may be even too large for this type of error.

Question 2·2

(Page 52)

How would you decrease the quantity of CO_2 excreted by a patient suffering from chronic obstructive pulmonary disease (COPD)?

Because patients suffering from COPD have difficulty excreting CO_2 through their lungs, their need to excrete CO_2 must be decreased without lowering the rates at which they regenerate ATP. Oxidation of glucose gives more CO_2/ATP than oxidation of fatty acids (Table 2·6). Hence, the diet offered should be as low in carbohydrates as possible, with the bulk of energy coming from triacylglycerols.

Question 2·3

(Page 52)

Water and CO_2 are both produced in energy metabolism. What relative proportions are produced?

How might a clinician use this information in calculating water balance?

LINKING WATER AND CO_2 PRODUCTIONS

- Carbohydrates: The generic formula of carbohydrates is $C_n(H_2O)_n$. Oxidation of carbohydrates to CO_2 is shown in the equation below.

$$C_n(H_2O)_n + n\,O_2 \longrightarrow n\,CO_2 + n\,H_2O$$

Therefore, CO_2 and H_2O are formed with a 1:1 stoichiometry.

- Fatty acids: The generic formula of fatty acids is $CH_3 (CH_2)_n COOH$. Oxidation of fatty acids to CO_2 is shown in the equation below.

$$CH_3 (CH_2)_n COOH + (n+1) O_2 + 0.5 (n+2) O_2 \longrightarrow (n+2) CO_2 + (n+2) H_2O$$

  Therefore, CO_2 and H_2O are again formed with a 1:1 stoichiometry.

- Overall: Production of CO_2 and H_2O are linked with a 1:1 stoichiometry.

CALCULATION OF WATER BALANCE

The partial pressures of CO_2 and water vapor in alveolar air are very similar (40 and 47 mm Hg, respectively); hence, at a normal Pco_2 in arterial blood, these two compounds will be excreted with a near 1:1 stoichiometry. A clinician need not take into account loss of water from the lower respiratory tract and the production of water via metabolism, since these two values are equal.

Additional Point

Since the molecular weights of CH_2 and H_2O are virtually equal (16 vs 18), upon oxidation of fatty acids, the weight of water produced equals the weight of fatty acids oxidized.

We thank Dr. Man Oh, Department of Medicine, Downstate SUNY, Brooklyn, NY, for bringing these relationships to our attention.

Question 2·4

(Page 52)

What can be done to help the lungs excrete CO_2 if the rate of alveolar ventilation is low and fixed and you want to minimize the degree of acidemia?

The rate of excretion of CO_2 equals the product of the alveolar ventilation rate and the $[CO_2]$ in the expired air. If the alveolar ventilation rate cannot be increased, the only hope is to change the $[CO_2]$ in the expired air (proportional to the Pco_2 in blood). From the Henderson equation (described on page 253), increasing the $[HCO_3^-]$ will permit an increased Pco_2 with no change in $[H^+]$. Thus, an infusion of HCO_3^- will permit an increased rate of excretion of CO_2 without a change in $[H^+]$ or alveolar ventilation.

Question 2·5

(Page 56)

A normal person cannot buffer more than 1000 mmol of H^+. How can an elite athlete whose muscles, red blood cells, and brain produce 400 mmol of lactic acid per hour survive for more than four hours?

If the lactate anions accumulate in the body or are excreted without H^+, this acid load will be lethal. In the case of an elite athlete, lactic acid will not accumulate but will be converted back to glucose in the liver or oxidized to $CO_2 + H_2O$ in other organs, thus removing most of the load of H^+.

Question 2·6

(Page 56)

In hemodialysis, the solution used may contain an appreciable quantity of sodium acetate. Why?

Patients requiring hemodialysis have poor kidney function and hence cannot excrete H^+ produced during the normal metabolism of certain amino acids. Metabolic processes that consume anions such as acetate and produce neutral products such as $CO_2 + H_2O$ consume H^+. Hence, sodium acetate provides an energy fuel without a supply of H^+, and the Na^+ can be removed in the dialysis treatment.

Question 2·7

(Page 59)

How can therapy with insulin lead to a decrease in the concentration of D-lactate in Peggy (Case 2·4)?

A fall in the concentration of D-lactate implies either a decline in its rate of production or an increase in its rate of removal. Since it does not affect GI bacteria, insulin will not influence formation of D-lactate. Any effect must therefore be on utilization. D-Lactic acid is metabolized via flux through pyruvate dehydrogenase and the ATP generation system (see the figure below). By inhibiting lipolysis, insulin leads to a lower delivery of fatty acids (and thereby a lower oxidation of fatty acids); hence, the rate of oxidation of D-lactic acid and carbohydrates can increase. Therefore, if Peggy requires insulin to avoid life-threatening acidemia, she will have to receive enough glucose to avoid hypoglycemia (she might also require K^+ to avoid hypokalemia).

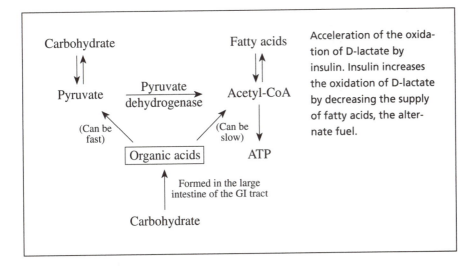

Question 2·8

(Page 60)

Why does ethanol create less of an acid load than methanol?

Ethanol is converted to acetic acid, which can be oxidized in the body to $CO_2 + H_2O$; methanol is converted to formic acid, which can only be oxidized very slowly and therefore must accumulate or be excreted. Fewer H^+ accumulate with the formation (and removal) of acetic acid because there is little net accumulation of this acid.

Question 2·9

(Page 60)

If both glucose and methanol are present at the same concentration in the blood (50 mmol/l), why will glucose but not methanol cause an osmotic diuresis?

GLUCOSE

Both glucose and methanol are filtered in the glomerulus. At normal levels of glucose in blood (less than 8 mmol/l), virtually all of the filtered load is reabsorbed; in contrast, during severe hyperglycemia (e.g., 50 mmol/l), the majority of the filtered load of glucose remains in the luminal fluid. Because salt and water are reabsorbed during severe hyperglycemia, the concentration of glucose rises markedly in the luminal fluid and a large concentration difference for glucose develops across the luminal membrane (i.e., some glucose remains in a very much smaller volume of luminal fluid). The glucose remaining in the lumen of the tubule acts osmotically to prevent reabsorption of some of the water containing electrolytes, thereby causing a diuresis.

METHANOL

No major concentration difference builds up for methanol because it is quite permeable across cell membranes. Hence, methanol will not be concentrated in the urine because it simply diffuses back into the blood as the water is reabsorbed from the tubule lumen, and no osmotic diuresis occurs.

Question 2·10

(Page 64)

How might clinicians use the rate of excretion of urea to indicate whether a patient is losing too much muscle mass (i.e., is in a very catabolic state)?

Urea is the major means of excreting nitrogen. If more nitrogen is excreted (as urea) than is consumed in the diet, the patient must be losing body proteins. Knowing this information early in the course of the illness provides an opportunity to change the nutritional therapy and reverse this catabolism before it is too late to do so (for a quantitative analysis, see the discussions of Cases 1·4 and 1·9, pages 14 and 29–31).

Question 2·11

(Page 65)

Would administration of $NaHCO_3$ be a good way to treat metabolic acidosis in a patient with renal failure?

Not really. Giving bicarbonate is an excellent way to remove H^+. The price to pay is the load of Na^+, which this patient cannot excrete. Accumulation of Na^+ can lead to the unwanted development of hypertension. Therefore, a technique such as hemodialysis to remove Na^+ is also required.

Question 2·12

(Page 65)

How long would it take an acid load to accumulate to lethal levels in patients who suffer from total anoxia, diabetic ketoacidosis, or complete renal failure?

How important are the kidneys in controlling the acidosis from anoxia or diabetes? Assume that a) death occurs after accumulation of 1000 mmol of H^+; b) the body normally consumes 12 mmol of O_2 per minute; c) ketoacids are produced at 1 mmol/min; d) the renal "excretion of H^+" is 70 mmol/day.

TOTAL ANOXIA

Since 6 mol of ATP are produced per mole of O_2 (the P:O is 3:1 and there are two atoms of oxygen per O_2), the rate of regeneration of ATP is 72 mmol/min when 12 mmol of O_2 are consumed per minute. If this ATP were to be produced anaerobically (equation below), 72 mmol of H^+ would be formed per minute, exhausting the capacity for buffering H^+ acutely in 14 minutes (ignore the valences of ATP, ADP, and P_i because hydrolysis of ATP initiated the glycolytic flux).

$$\text{Glucose} + 2\,(\text{ADP} + P_i) \longrightarrow 2\,\text{ATP} + 2\,\text{Lactate} + 2\,H^+$$

The brain would succumb much earlier (perhaps in 4 minutes) since it has a higher demand for ATP than other organs (except skeletal muscles during exercise), its buffering capacity is much lower than that of skeletal muscles, and insufficient glucose would be available to regenerate the ATP that it needs. The kidneys do not affect this condition because the rates of excretion of H^+ are very small (70 mmol/1440 min) compared with the rate of formation of H^+.

DIABETIC KETOACIDOSIS

Though the rate of ketogenesis can be close to 1 mmol/min, both the brain and the kidneys consume ketoacids, and the kidneys excrete some of them along with NH_4^+; net accumulation is therefore not more than 300 mmol/day (unless conditions such as coma or renal failure decrease the oxidation of ketoacids in the brain and kidneys). Thus, the physician has more than one day to control diabetic ketoacidosis once it begins to occur. Renal excretion of H^+ (as NH_4^+) is obviously important in slowing the net accumulation of H^+ in the blood.

COMPLETE RENAL FAILURE

It would take about two weeks for a patient to accumulate 1000 mmol of H^+ at a rate of 70 mmol/day. This patient would be likely to die from hyperkalemia long before the acidosis became lethal.

Question 2·13

(Page 72)

Some patients cannot synthesize urea because of an inborn defect in the urea cycle. How might they be helped to excrete waste nitrogen?

In normal metabolism, most of the body's nitrogen wastes are excreted as urea. The normal rate of excretion of urea is approximately 500 mmol/day (1000 mmol of nitrogen per day); NH_4^+, excreted at a maximal rate of 200 mmol/day, cannot substitute for urea because the rate is not high enough. A sufficient rate would require that the patient have chronic metabolic acidosis, which itself is harmful (retards growth, decreases bone mass, etc.). A new carrier for nitrogen is therefore required.

Case 2·8 offers a possibility. Metabolism of benzoate to hippurate for excretion traps nitrogen (derived from glycine); treatment with benzoate can therefore permit excretion of nitrogen. It is not possible, however, to meet the normal requirements for nitrogen excretion—1000 mmol/day (14 g)—via the excretion of hippurate. Hence, such patients must also be kept on a very low nitrogen-containing (protein) diet, perhaps one containing only essential amino acids or perhaps rich in their ketoacid derivatives; some amino acids are needed to allow for anabolism of proteins, especially during growth.

Question 3·1

(Page 90)

After finishing a race on two identical occasions, the concentration of lactate in an athlete's blood was 10 mmol/l. On the first occasion, the athlete lay down and rested immediately after the race; the concentration of lactate remained high for 30 minutes. On the second occasion, after the athlete "cooled-down" by jogging around the track for 30 minutes, the concentration of lactate was close to normal. How would you explain this difference in the concentration of lactate?

Lactate can be removed by conversion to glucose (in the liver and kidneys) or to glycogen (in the liver and muscle) or by oxidation to yield $CO_2 + H_2O$. Glucogenesis is a rather slow pathway; oxidation depends on the demand for ATP (as long as oxygen is available).

The liver consumes approximately 0.5 mmol of O_2 per minute. Since half a molecule of O_2 (yielding three molecules of ATP) is needed to convert one molecule of lactate to half a molecule of glucose (equation below), the maximal rate of conversion of lactate to glucose is 4 mmol/min (if no other process consumes O_2 in the liver); this rate will not be affected by light exercise. A similar story is true for the kidneys.

$$\text{Lactate}^- + H^+ + 3\ \text{ATP} \longrightarrow \text{½ Glucose} + 3\ (\text{ADP} + P_i)$$

At rest, a person consuming 12 mmol of O_2 per minute can burn at most 4 mmol of lactate per minute (three molecules of O_2 are needed to oxidize one molecule of lactate; equation below).

$$\text{Lactate}^- + H^+ + 3\ O_2 + 18\ (\text{ADP} + P_i) \longrightarrow 3\ CO_2 + 3\ H_2O + 18\ \text{ATP}$$

A sprint can generate as much as 400 mmol of lactate; thus, the athlete who rests and thereby relies on glucogenesis without much oxidation to remove lactate anions $+ H^+$ would take more than 100 minutes to dispose of the excess lactate. By jogging, the athlete increases the rate of oxidation fourfold or more, thus decreasing the time needed to oxidize lactate anions $+ H^+$.

Question 3·2

(Page 90)

We stated that about 60 mmol of ATP per kilogram of muscle was used during the 10-second sprint. How could the consumption of ATP have been measured?

Muscle biopsies before and just after a sprint provide figures for changes in glycogen, ATP, CrP, and lactate. Measuring levels of lactate in blood before and after will further augment the understanding of the amount of ATP regenerated. Measuring the fall in concentration of glycogen and CrP are most important in calculating the ATP yielded by anaerobic metabolism.

Question 3·3

(Page 90)

A runner must decide when to initiate the final sprint to the finish line. What are the trade-offs?

The simple answer is speed versus endurance. The sprint requires anaerobic regeneration of ATP, which can occur very rapidly but at the cost of a high production of H^+. Since 1 kg of skeletal muscle can buffer up to 40 mmol of H^+ (principally by non-BBS buffers) before function deteriorates significantly, a sprint producing 2 mmol of H^+ per kilogram of muscle per second can last only 20 seconds, after which, the metabolism in muscle is inhibited and the runner must slow down.

Question 3·4

(Page 90)

Each kilogram of muscle contains 20–25 g of glycogen, or 110–140 mmol of glucose. The molecular weight of glucose is 180. Explain the discrepancy.

To make glycogen, 1 mol of H_2O is removed per molecule of glucose. This removal reduces the molecular weight of glucose (normally 180) to 162 when it is stored in glycogen.

Question 3·5

(Page 91)

During vigorous exercise, a signal causes an increase in the rate of respiration without causing the Pco_2 and $[H^+]$ of arterial blood to shift out of the normal range. What might this signal be?

A direct nervous signal from muscle to the brain is possible. If carried to the brain through the blood, the signal must remain in the blood after passage through the lungs, since the brain is fed by arterial blood. Perhaps the most potent signal could be an elevated $[K^+]$ in plasma. A high level of K^+ will stimulate the respiratory rate provided that the Pco_2 remains close to or higher than the normal value of 40 mm Hg.

Question 3·6

(Page 91)

How does arterial blood differ in composition from venous blood during exercise?

First, venous blood has less oxygen. In addition, during aerobic exercise, mixed venous blood has a much higher Pco_2 and $[H^+]$ and a somewhat higher $[HCO_3^-]$ than arterial blood (see the table below). The Pco_2 can rise further when lactic acid is produced, for example, during the finishing kick of a race, when CO_2 is produced both by metabolism in the presence of oxygen and by buffering of H^+ by HCO_3^-.

Composition of arterial and mixed venous blood during steady-state strenuous exercise:

		Arterial Blood	Mixed Venous Blood
Pco_2	mm Hg	36	90
Po_2	mm Hg	100	10
$[H^+]$	nmol/l	40	80
pH		7.40	7.10
HCO_3^-	mmol/l	22	27
K^+	mmol/l	5.5	5.6
Lactate	mmol/l	3	3.1

Question 3·7

(Page 91)

How does the activity of PDH in resting muscle compare with that in vigorously exercising muscle?

During a 1500 m race, the muscles of a trained athlete can consume 150 mmol of O_2 per minute. At rest, the same athlete might consume 12 mmol of O_2 per minute, with consumption by muscle accounting for about one-third. Since glycogen provides most of the energy in a middle distance race, and a range of substrates can provide fuel for muscle at rest, the activity of PDH at rest can be as little as 1–2% of that in vigorous exercise.

Question 3·8

(Page 91)

What is the metabolic reason for a marathon runner "hitting the wall"?

During a race, the elite marathon runner uses essentially all the O_2 delivered to muscles. Carbohydrates provide the fuel needed for about two-thirds of the race before fatty acids become a major fuel. "Hitting the wall" may be related in part to the need to switch from aerobic to some anaerobic metabolism at a time when stores of glycogen in muscle have virtually run out. Anaerobic glycolysis causes an increased production of CO_2 that makes the $[H^+]$ in muscle rise so high that it compromises peak performance. Recall that in strenuous exercise, the capacity of buffers in muscle to absorb H^+ is already stretched to the limit. In addition, this high demand for ATP above that available from aerobic metabolism must be met from glucose in blood. "Hitting the wall" probably relates closely to the development of hypoglycemia, which has a negative effect on the marathon runner's delicate metabolic balances.

Question 3·9

(Page 94)

Will the pH in Eleanor's muscle differ from that of normal individuals during a sprint? If so, why?

In a normal person, muscle pH will fall during a sprint (Event 3·1) because anaerobic glycolysis from glycogen produces H^+. The breakdown of CrP will absorb some H^+. Since Eleanor cannot break down glycogen at a rapid rate, this production of H^+ does not occur; however, the CrP is still used to generate ATP. Hence, unlike the muscle of normal individuals, Eleanor's muscle may show a rise in pH (decrease in $[H^+]$) during a sprint.

Question 3·10

(Page 96)

If the horse in Veterinary Case 3·1 were to gallop for 24 hours, how much CO_2 would it produce?

The racehorse produced 200 mmol of CO_2 per second. Thus, in 24 hours, it could produce close to 18 000 000 mmol of CO_2. Since the molecular weight of CO_2 is 44, the weight of CO_2 produced in 24 hours would be close to one metric ton.

Question 3·11

(Page 96)

If the horse obtained extra red blood cells from its spleen (the original blood doper) what is the minimum weight of its spleen at rest, and what signal do you think caused its spleen to contract?

The amount of red blood cells in blood doubles with strenuous exercise; the normal weight of red blood cells in blood is 40 kg (40% of a blood volume of 100 liters). Hence, the spleen must weigh at least 40 kg at rest. Adrenaline would be a good candidate as the signal for the spleen to contract.

Question 3·12

(Page 96)

In what ways might the production of lactic acid be advantageous to the horse during the race? Are there also disadvantages?

ADVANTAGES OF LACTIC ACID

The key to success in aerobic exercise is delivering more O_2 to tissues. A high $[H^+]$ might increase delivery of O_2 to muscle cells by dilating blood vessels (a local effect) or by causing the affinity of hemoglobin for O_2 to decline. A high $[H^+]$ shifts the hemoglobin-O_2 dissociation curve to the right so that O_2 is released at a higher Po_2 (Figure 2·5, page 50). A higher Po_2 aids diffusion of O_2 into mitochondria.

DISADVANTAGE OF LACTIC ACID

The high $[H^+]$ in muscle cells causes H^+ to bind to proteins, making their charges more positive and thereby possibly inhibiting their functions.

Question 3·13

(Page 96)

What adaptations occurred in the horse's muscle to permit it to extract so much O_2 during the race?

The demand of muscle for large amounts of O_2 during exercise causes increases in temperature, Pco_2, and $[H^+]$—all changes that encourage release of O_2 from hemoglobin in muscle capillaries. Training increases the number and tortuosity of the capillaries and also increases the myoglobin content of muscle, thus accelerating the diffusion of O_2 through the tissue. Training also increases the number of mitochondria in muscle cells, especially near the blood vessels.

Question 4·1

(Page 103)

Why is ketoacidosis a rare event in patients with Type II diabetes mellitus?

Although the answer is unknown, the following discussion illustrates ways to think of the problem.

LOW PRODUCTION OF KETOACIDS

Patients with noninsulin-dependent diabetes mellitus (NIDDM) can have a lower rate of production of ketoacids in the liver. Among the hypotheses to consider are:

1. lower delivery of fatty acids to the liver (not supported by data);
2. lower formation of fatty acyl-CoA or reduction of its entry into hepatic mitochondria (again, not supported by data);
3. a higher rate of regeneration of ATP in hepatic mitochondria from other pathways; gluconeogenesis, for example, may lead to a decrease in the maximal rate of ketogenesis because it generates ATP in hepatocytes.

HIGH UTILIZATION OF KETOACIDS

Ketoacids might be overutilized in patients with NIDDM. Oxidation in the brain and kidneys is not as likely since oxidation in these organs is less than the rate of production in the liver. In contrast, if ketoacids were oxidized in muscle, an organ that does not usually oxidize them, the degree of ketoacidosis would be smaller.

Question 4·2

(Page 105)

Under what circumstances might there be excessive breakdown of proteins in the body?

Excessive breakdown of proteins might be due to decreased synthesis of proteins or might result from enhanced oxidation (or excretion) of amino acids.

DECREASED SYNTHESIS OF PROTEINS

Synthesis of proteins depends on the supply of amino acids and on signals to promote synthesis. Low levels of anabolic hormones (e.g., insulin) or high levels of hormones with a catabolic signal (e.g., corticosteroids) will tip the balance towards increased net breakdown of proteins.

INCREASED OXIDATION OF AMINO ACIDS

Either a low availability of other fuels to oxidize (e.g., fatty acids, ketoacids) or specific activation of enzymes favoring oxidation of specific amino acids (e.g., branched-chain ketoacid dehydrogenase) could lead to a limitation of amino acids available for synthesis of proteins. If one essential amino acid is not available, protein synthesis cannot occur, and there will be net breakdown of proteins with increased production of urea (see the discussion of Case 2·7, page 66, for a consideration of glutamine in chronic fasting).

Question 4·3

(Page 105)

The urine of a hyperglycemic patient contains 300 mmol of glucose per liter. How much muscle must this patient be breaking down to permit the excretion of one liter of urine if all the glucose is derived from proteins in muscle?

- The molecular weight of glucose is 180.

- One kilogram of muscle contains approximately 200 g of proteins, which can yield 120 g (667 mmol) of glucose.

- Therefore, to excrete 300 mmol of glucose, the body must break down approximately 450 g (1 lb) of muscle.

Question 4·4

(Page 105)

Hyperglycemic patients usually excrete a predictable amount of glucose in their urine; why might the amount excreted be less than expected?

The rate of excretion is the difference between the amount of glucose in the glomerular filtrate and the amount reabsorbed in the nephron. A low glomerular filtration rate might arise from contraction of ECF volume (a result of chronic hyperglycemia) or from intrinsic renal disease. It is not clear if there are any conditions with excessive reabsorption of glucose other than a low ECF volume.

Question 4·5

(Page 108)

Assume that a compound X inhibits its own formation and stimulates its own use (see page 16 and also Chapter 7, pages 171–174) and that the capacities for its formation (V_1) and use (V_2) are also subject to external controls.

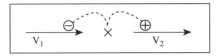

What influences on V_1 and/or V_2 will cause the [X] to rise or fall? What conditions would lead to an uncontrolled increase in the [X]? What conditions are required for the [X] to change to new steady states?

The effects of X on its own formation and use are fully consistent with efficient homeostasis; the [X] will rapidly settle down to a dynamic steady state, which will change as the external controls on V_1 and V_2 are exerted. The [X] will increase:

1. if V_1 is stimulated;

2. if V_2 is inhibited by external controls;

3. if both 1 and 2 occur;

4. if V_1 is stimulated more than V_2, and vice versa.

An uncontrolled increase in [X] will occur if the effect of product inhibition on V_1 is removed and V_2 reaches its V_{max}.

Question 4·6

(Page 108)

Nitrogen makes up one-sixth of the weight of protein, and 100 g of protein can yield 60 g of glucose. How many millimoles of glucose can be formed per millimole of urea or ammonium excreted in urine?

As outlined in the response to Question 4·3, 100 g of protein is equivalent to 333 mmol of glucose, and since one-sixth of the weight of protein is nitrogen (16 g in this case), approximately 1200 mmol of NH_4^+, or about 600 mmol of urea, will be formed (the molecular weight of nitrogen is 14, and urea contains two nitrogen atoms). Therefore, 1 mmol of urea in urine is roughly equivalent to the formation of 0.5 mmol of glucose; 1 mmol of NH_4^+ equals 0.25 mmol of glucose.

Question 4·7

(Page 110)

What major pathways besides ketogenesis lead to the regeneration or utilization of ATP in the liver? How might changes in the rates of these pathways affect the rate of ketogenesis?

Ketogenesis from fatty acids yields ATP because the oxidation of fatty acids yields one NADH and one $FADH_2$ per acetyl group released. The synthesis of glucose from amino acids yields ATP, as does the synthesis of fatty acids from glucose. The synthesis of glucose from lactate or pyruvate, however, consumes ATP.

Since the production of ketoacids yields ATP, the maximal rate of this pathway in the liver will be limited by the rate at which the liver uses ATP. Most of the liver's consumption of ATP is due to the demands for maintenance of its intracellular ionic environment; the demands of metabolism are not large by comparison. Nonetheless, any other regeneration of ATP (e.g., glucogenesis from amino acids) can inhibit ketogenesis; a pathway that consumes ATP (e.g., glucogenesis from lactate) can allow faster rates of ketogenesis.

Question 4·8

(Page 113)

Your patient has diabetes mellitus and presents with moderately severe hyperglycemia. Two strategies have been proposed as adjuncts to insulin therapy to lower the concentration of glucose in the blood—inhibition of glucogenesis and inhibition of fatty acid oxidation. What are the dangers with these additional lines of therapy?

The problems with using medications to inhibit glucogenesis or oxidation of fatty acids arise when the insulin therapy has reduced the levels of glucose in blood to normal. Maintenance of normal metabolic homeostasis requires both of these pathways.

INHIBITION OF GLUCOGENESIS

If glucogenesis is inhibited, two problems might arise: low levels of glucose for the brain and an accumulation of lactic acid, especially after a brief bout of exercise. To avoid hypoglycemia and damage to the brain, glucose will have to be administered several hours after meals when the actions of insulin become evident.

INHIBITION OF FATTY ACID OXIDATION

If oxidation of fatty acids is inhibited, the whole body will depend on glucose for regeneration of ATP, glucogenesis will also be inhibited (since the liver needs oxidation of fatty acids to generate ATP and to signal the activation of glucogenesis), and exogenous glucose will certainly be needed.

If you follow this line of therapy, your patient will be more likely to develop problems of hypoglycemia with exercise and between meals. You may add to the troubles of your patient with excess pharmacological intervention.

Question 4·9

(Page 117)

An analysis of two samples of blood gives the following results:

	Glycated albumin	Glycated hemoglobin
Case 1	Normal	High
Case 2	High	High

What do these results reveal?

The different half-lives of albumin (1–2 weeks) and red blood cells (3–4 months) may indicate when, over the past few months, there were abnormally high levels of glucose in the blood. Case 1 indicates hyperglycemia more than one month ago, but not recently. Case 2 indicates recent and long-term hyperglycemia since both albumin and hemoglobin are affected.

Question 5·1

(Page 131)

How much glucose can be synthesized from one kilogram of muscle? (Assume muscle is 80% water and 20% proteins.)

Sixty percent of the weight of proteins can be converted to glucose. One kg of muscle contains 200 g of proteins and can therefore yield 120 g of glucose, about the amount used daily by the brain. For more information, see the discussions of Questions 4·3 and 4·6.

Question 5·2

(Page 131)

How much would the concentration of lactate in the blood rise if all the glucose in the body fluids were converted to lactate?

Glucose is distributed in about 50% of body water, including all the extracellular fluid (ECF) and the intracellular fluid (ICF) of organs that do not require insulin for the entry of glucose (all organs except muscle and adipose tissue). Lactate can enter all body fluids, but its concentration in the ICF is about one-half that in the ECF. If we assume, for ease of calculation, that lactate is distributed in twice the volume of glucose and that each molecule of glucose yields two molecules of lactate, the maximum rise in the concentration of lactate in blood will be the same as the fall in concentration of glucose in the blood.

Question 5·3

(Page 131)

How would a rapid infusion of 1 liter of D_5W affect the concentration of glucose in the blood?

Because D_5W is 5% glucose in water, 50 g (276 mmol) of glucose will be added and will distribute in 50% of body water. Assuming that the patient is a 50 kg woman of normal build (total water is 30 liters and glucose distributes in 15 of these liters), the increase in the level of glucose in blood will be approximately 18 mmol/l (324 mg/dl).

Question 5·4

(Page 131)

If the rate of filtration in the kidney (the glomerular filtration rate, GFR) is 180 liters per day in a normal adult, and a normal kidney can reabsorb 1800 mmol (324 g) of glucose per day, how high must the concentration of glucose be in the blood before glucose will appear in the urine?

If 1800 mmol of glucose are contained in 180 liters, the concentration in blood will be 10 mmol/l (180 mg/dl); this amount is the renal threshold for glucose, the highest concentration of glucose in blood that can be attained without glycosuria.

Question 5·5

(Page 131)

How long will it take to develop symptoms of hypoglycemia after an overdose of insulin?

Assume that the normal level of glucose in blood is 5 mmol/l and that symptoms of hypoglycemia start at 2 mmol of glucose per liter. Therefore, 3 mmol of glucose must be removed from each liter of fluid (excluding the intracellular fluid of muscle, where this concentration is very low). If a patient has 40 liters of body water, glucose will be distributed in half of this volume, or 20 liters. Hence, this person will have to burn 60 mmol of glucose (10.8 g) before developing clinical signs of hypoglycemia.

The excess of insulin closes off release of fat-derived fuels from triacylglycerols, after which all the body's energy must come from glucose. Release of glucose from liver glycogen and consumption of glycogen in muscle are also inhibited by insulin. If the body's normal metabolism uses 2400 kcal/day (100 kcal/hr, equal to the oxidation of 25 g of glucose), then 10.8 g of glucose will be used in approximately 30 minutes. (The time is actually longer because inhibition of the release of fatty acids is not complete, and some glycogen in the liver breaks down in response to hypoglycemia.)

Question 5·6

(Page 131)

Will glycogen in muscle be converted into a fuel for the brain during hypoglycemia?

Muscle can release to the blood only 10–15% of the glucose stored in its glycogen (representing the hydrolysis of the 1:6 linkages at branch points, page 279). Lactate can be released following the breakdown of glycogen in muscle. This release, activated by adrenaline, can be large during anaerobic metabolism. The liver can slowly convert this lactate to glucose for eventual use in the brain.

Question 5·7

(Page 131)

What essential differences between Cases 5·1, 5·2, and 5·3 affect the time taken for signs of hypoglycemia to develop?

In Case 5·1, Michael can use only his circulating fuels before hypoglycemia develops. In Case 5·2, the stores of glycogen in Jackie's liver are also available as glucose. In Case 5·3, Ed can use ethanol, ketoacids, glucose, and liver glycogen to meet his energy needs.

Question 5·8

(Page 136)

What are the essential metabolic differences between Cases 5·6 and 5·7 concerning the timing of hypoglycemia and the anticipated levels of fuels in the blood?

Because Alice (Case 5·6) cannot release glucose from her liver, she gets into trouble when her dietary glucose has been used; the hypoglycemia should allow lipolysis, but the ketoacids do not have time to accumulate to high enough levels to provide sufficient fuel for her brain. With the exception of her brain, her body should be able to use fatty acids for much of its energy requirements. Since she is an infant, her brain weight is approximately 7% of body weight, as opposed to about 2.5% for a young adult.

In contrast, Desmond (Case 5·7), whose brain is also about 6% of body weight, can release glucose from glycogen but cannot make glucose. Hence, the time taken for his symptoms of hypoglycemia to appear is much longer than that for Alice.

Question 5·9

(Page 138)

If a tumor converts half of the glucose that it consumes to lactic acid rather than carbon dioxide, what fraction of the tumor's ATP would be made anaerobically? Assume that there is no change in the rate of turnover of ATP in the tumor and that all the generation of ATP in the tumor is by aerobic or anaerobic metabolism of glucose.

Complete oxidation of glucose yields 36 molecules of ATP per glucose; anaerobic glycolysis produces two molecules of ATP per glucose. Therefore, if half the glucose used by the tumor is converted to lactate, $\frac{2}{38}$ (or $\frac{1}{19}$) of the energy used by the tumor will come from anaerobic glycolysis and the demand for glucose will be almost doubled.

Question 6·1

(Page 145)

How rapidly can the liver synthesize glucose from the lactate produced in muscle during a 10-second sprint?

After a 10-second sprint, muscle glycogen will have dropped by 14 mmol of glucose per kilogram of muscle, equivalent to 28 mmol of lactate per kilogram, or close to 300 mmol of lactate (from 10 kg of muscle).

The liver's consumption of oxygen (approximately 2 mmol/min, or one-sixth of the body's consumption at rest) indicates the maximal rate of removal of lactate. Since 2 mmol of lactate require 6 mmol of ATP (equivalent to 1 mmol of O_2) for conversion to glucose, the liver can consume at most 4 mmol of lactate per minute. Hence, the liver will take much more than one hour to synthesize glucose from the lactate produced during a 10-second sprint.

Question 6·2

(Page 145)

Assuming that the body uses 12 mmol of O_2 per minute and that the total amount of glucose in solution in the body is 100 mmol (5 mmol/l × 20 liters), how long can free glucose meet normal bodily demands for ATP through anaerobic glycolysis if all the glucose is used?

During anaerobic glycolysis, 100 mmol of glucose can regenerate 200 mmol of ATP (2 ATP/glucose). The O_2 consumption indicates that 72 mmol of ATP are regenerated per minute (6 ATP/O_2 × 12 mmol of O_2 per minute). Therefore, the body's free glucose can provide ATP for close to three minutes through anaerobic glycolysis.

What will be the final concentration of lactate?

Since lactate effectively distributes in twice the volume of body fluids as does glucose, the final concentration will be the same as the fall in the concentration of glucose (5 mmol/l).

What might happen to the pH of the blood?

The pH of the blood will drop as a result of the formation of 5 mmol/l of H^+, which will be buffered immediately, primarily by the bicarbonate buffer system in the ECF. Buffering will consume 5 mmol/l of HCO_3^-, which will fall in concentration to 20 mmol/l. Assuming that the lungs can clear the excess CO_2 and adjust the P_{CO_2} very rapidly, the $[H^+]$ could rise by close to 10% from its normal 40 mmol/l to approximately 45 mmol/l, resulting in a pH of 7.35.

When oxygen returns, what is the maximal theoretical rate at which the lactate can be oxidized to CO_2 and H_2O?

If the body were to regenerate 72 mmol of ATP per minute solely by the oxidation of lactate (18 mmol of ATP per lactate anion; equation below), only 4 mmol of lactate would be oxidized per minute.

$$\text{Lactate}^- + H^+ + 3\,O_2 + 18\,(ADP + P_i) \longrightarrow 3\,CO_2 + 3\,H_2O + 18\,ATP$$

Because formation by anaerobic glycolysis of all 72 mmol of the ATP needed per minute would have produced a net yield of 72 mmol of H^+ (and lactate anions), aerobic oxidation of all of this lactate would take 18 minutes at a rate of 4 mmol/min.

Question 6·3

(Page 147)

Would a patient with a level of hemoglobin in blood that is 25% of normal be expected to develop lactic acidosis either at rest or during exercise that triples the overall consumption of O_2? Assume that the normal concentration of hemoglobin is 2.25 mmol/l (140 g/l), the normal cardiac output at rest is 5 liters/min (which can rise to 15 liters/min when needed), and the normal consumption of oxygen at rest is 12 mmol/min. Given: 2 mmol of hemoglobin per liter can carry 8 mmol of O_2 per liter.

A patient with 25% of the normal hemoglobin level will carry close to 2 mmol of O_2 per liter. This patient's heart can therefore deliver 30 mmol of O_2 per minute at the cardiac output rate of 15 liters/min. Because this rate of delivery is well above the 12 mmol of O_2 normally needed per minute at rest, the patient can survive quite readily at rest. In contrast, when the O_2 consumption increases threefold during exercise, delivery of O_2 will still be 30 mmol/min. Since the demand for O_2 will be 36 mmol/min, ATP will be regenerated anaerobically and lactic acidosis will occur.

Question 6·4

(Page 158)

Lactate dehydrogenase occurs only in the cytosol of cells, whereas β-hydroxybutyrate dehydrogenase occurs only in the mitochondria. How can you calculate the ratio of NADH to NAD^+ in both cytosol and in mitochondria from measurements made on the venous blood leaving an organ such as the liver?

What assumption must be made to perform this calculation?

Both enzymes catalyze reactions that are dead ends of metabolism and thus can be assumed to be at chemical equilibrium, e.g.,

$$\text{Lactate} + NAD^+ \longleftrightarrow \text{Pyruvate} + NADH + H^+$$

$$\text{β-hydroxybutyrate} + NAD^+ \longleftrightarrow \text{Acetoacetate} + NADH + H^+$$

Hence, assuming the $[H^+]$ to be constant, the ratio of lactate to pyruvate = $K[NADH]/[NAD^+]$ in the cytosol (the $[H^+]$ is incorporated into the value for K, the equilibrium constant). Measuring the ratios of lactate to pyruvate and of β-hydroxybutyrate to acetoacetate in the venous blood leaving the organ and knowing the values of K will therefore allow calculation of the cytosolic and mitochondrial $[NADH]/[NAD^+]$, respectively.

Calculations such as these show that the $[NADH]/[NAD^+]$ in mitochondria is higher than in the cytosol, an intriguing observation since the hydrogen atoms carried on NADH are burned in mitochondria to yield ATP and H_2O. The technique depends on the assumption that the release of the monocarboxylate anions is influenced only by their concentrations in the relevant compartments, that the dehydrogenases are at equilibrium, and that the $[H^+]$ in the relevant compartments is constant.

Question 6·5

(Page 158)

The liver normally uses 3 mol of O_2 per day. Metabolism of palmitic acid (16 carbons) to ketoacids can be described by the following equations:

$$C_{16} \text{ Fatty acid} + 4 NAD^+ + 7 FAD \longrightarrow \text{Acetoacetate} +$$
$$3 \text{ β-hydroxybutyrate} + 4 NADH + 7 FADH_2$$

$$NADH + H^+ + \tfrac{1}{2} O_2 \longrightarrow H_2O + 3 \text{ ATP}$$

$$FADH_2 + \tfrac{1}{2} O_2 \longrightarrow H_2O + 2 \text{ ATP}$$

How fast can the liver make ketoacids (theoretically)?

Three mol of O_2 per day is equivalent to 18 mol of ATP per day; conversion of 1 mol of palmitate (C_{16}) to 4 mol of ketoacids (3 mol of β-hydroxybutyrate plus 1 mol of acetoacetate—the approximate ratio in blood) produces 26 mol of ATP from the NADH and $FADH_2$. If this conversion were to supply all the ATP required by the liver, 0.7 mol of C_{16} would be needed, producing 2.8 mol of ketoacids per day, or close to 2 mmol of ketoacids per minute.

Question 6·6

(Page 163)

How much ethanol can be metabolized by the liver per day if the product is a) CO_2; b) acetate; c) β-hydroxybutyrate? Assume that the liver consumes 3 mol of O_2 per day and that conversion of ethanol to CO_2, to β-hydroxybutyrate, and to acetate produces 16, 2.5, and 6 mmol of ATP per mmol of ethanol, respectively.

Consumption of 3 mol of O_2 per day means that the liver is using 18 mol of ATP per day. Since the liver cannot be pushed to produce excess ATP, the maximal rates of ethanol consumption by the liver are 1.1, 7.2, and 3 mol/day if the products are CO_2, β-hydroxybutyrate, and acetate, respectively (divide the rate at which the liver is using ATP by the ratios of ATP per ethanol).

$$\text{Ethanol} + 2\,NAD^+ + 2\,H^+ \longrightarrow \text{Acetate} + 2\,NADH\ (+\ 6\ ATP)$$

$$\text{Acetate} + 2\,ATP + CoA \longrightarrow \text{Acetyl-CoA} + AMP + 2\,P_i$$

$$\text{Acetyl-CoA} + \tfrac{1}{2}\,NADH \longrightarrow \tfrac{1}{2}\,\beta\text{-HB}$$

$$\text{Overall: Ethanol} \longrightarrow 2.5\,ATP + \tfrac{1}{2}\,\beta\text{-HB}$$

Question 6·7

(Page 167)

Can inhibiting the oxidation of fatty acids help decrease the severity of acidosis caused by overproduction of acids from the gastrointestinal tract?

Fatty acids are the preferred fuel for oxidation. Oxidation of anions to neutral end products (CO_2 and H_2O) removes H^+ and thus decreases acidosis. Inhibition of fatty acid oxidation may allow many more molecules of organic anions ($+\ H^+$) from the GI tract to be oxidized since they yield many fewer molecules of ATP per anion oxidized (see the figure below).

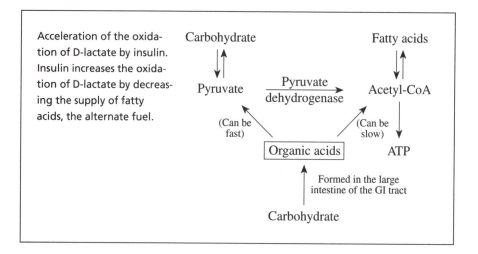

Question 7·1

(Page 183)

Glucose and Na⁺ must be absorbed together, in equimolar amounts, through the cell membrane. The daily intake of glucose is approximately 1500 mmol; the total quantity of dietary Na⁺ plus Na⁺ secreted into the upper GI is about 600 mmol. How can this quantity of glucose be absorbed?

Since one Na^+ is needed for each molecule of glucose, and there are insufficient Na^+ on the surface, there must be addition of Na^+ from the cells (or the fluid between them). In fact, Na^+ leak from between the cells back into the lumen to permit the very high rates of absorption of glucose (and amino acids).

Question 8·1

(Page 194)

Some patients develop severe liver, kidney, and intestinal problems very soon after eating sucrose. Why? You may want to think of this problem as follows:

1. What does sucrose contain?

2. Which component of sucrose is likely to cause the problem?

3. Why do symptoms develop rapidly?

4. Why are the problems restricted to some organs?

5. Are these patients likely to have high dental bills?

Sucrose is a disaccharide that is hydrolyzed to fructose and glucose during digestion. Since glucose is always present in the body, problems from ingestion of sucrose are likely to stem from fructose, which is metabolized rapidly in the liver, kidneys, and intestinal mucosa. Metabolism of fructose causes immediate problems in people with hereditary fructose intolerance (HFI), a disorder of metabolism. In HFI, fructose can be metabolized at its normal rapid rate to its first metabolite, using ATP in the process. The second enzyme, which allows fructose carbon to enter the glycolytic pathway, is missing (see Figure 5, page 283). Hence, fructose 1-phosphate accumulates, trapping phosphate and preventing regeneration of ATP. A patient with HFI will avoid candy and other sweet foods because their ingestion leads to marked discomfort. Consequently, such a person will rarely have tooth decay.

Question 8·2

(Page 194)

The appearance of sucrose in the urine after consumption of a drink containing sucrose provides a rapid and easy screening test for stomach ulcers. Why does it work?

Dietary sucrose is normally hydrolyzed in the small intestine to glucose and fructose, which can then be specifically absorbed into the blood and metabolized in tissues. Sucrose does not cross intact walls of the stomach or intestine; also, once in the blood, sucrose cannot be broken down in the body. Ulcers allow soluble stomach contents, including sucrose, to leak into the blood. Any sucrose entering the blood will appear in the urinary filtrate; because it cannot be reabsorbed from the nephron, filtered sucrose will appear in the urine.

The test for sucrose in the urine may be able to detect ulcers much earlier than endoscopy and hence may be especially valuable for patients who are particularly susceptible to gastric ulcers (e.g., from treatment with drugs such as nonsteroidal anti-inflammatory agents).

We thank Dr. J. B. Meddings and Dr. J. L. Wallace, University of Calgary, for preliminary information on the use of sucrose in diagnosis of stomach ulcers.

271

Question 8·3

(Page 201)

Is the conversion of glycerol derived from adipose tissue triacylglycerols better described as gluconeogenesis or glucopaleogenesis?

We consider it to be gluconeogenesis because it adds to the pool of glucose.

Question 8·4

(Page 203)

What are the relative proportions of glucose and urea synthesized during metabolism of proteins?

As outlined in the response to Question 4·3, 100 g of protein is equivalent to 333 mmol of glucose, and since one-sixth of the weight of protein is nitrogen (16 g in this case), close to 600 mmol of urea will be formed (the molecular weight of nitrogen is 14, and urea contains two nitrogen atoms). Therefore, 1 mmol of urea in urine is roughly equivalent to the formation of 0.5 mmol of glucose.

Question 8·5

(Page 203)

Urea and phosphoenolpyruvate (PEP) are formed in equimolar amounts in gluconeogenesis. How is this equality possible if each precursor of PEP has one atom of nitrogen from amino acids, but each molecule of urea is formed with two atoms of nitrogen?

The answer has a number of components:

1. Ketogenic amino acids, which cannot be converted to PEP, supply nitrogen but no precursors of PEP.

2. Some amino acids (e.g., glutamine, asparagine) contain more than one atom of nitrogen but yield only one molecule of PEP.

3. Some PEP can be made by a pathway not tied to synthesis of urea (glucopaleogenesis).

4. Some urea is converted back to $NH_4^+ + HCO_3^-$. Normally, close to 25% of the urea formed in the liver is broken down to $NH_4^+ + HCO_3^-$ by bacteria in the gastrointestinal tract, and these $NH_4^+ + HCO_3^-$ are reincorporated into urea in the liver

Question 8·6

(Page 205)

What symptoms and signs would you expect if a patient had a defect in the glucogenic pathway? What might aggravate them?

The glucogenic pathway converts pyruvate (lactate) to glucose. Hence, with a defect, lactic acid will accumulate and the supply of glucose will be low (hypoglycemia). Lactic acidosis will be greatest after exercise and during fasting (less oxidation of lactate). Hypoglycemia will be aggravated by fasting and exercise.

Question 8·7

(Page 209)

What symptoms and signs would you expect in the absence of glycogen phosphorylase in the liver?

If the enzyme that breaks down glycogen in the liver is missing, glycogen will accumulate (forming a large, distended liver). Since glycogen must supply glucose for the brain between meals, this defect will lead to signs and symptoms of hypoglycemia between meals.

Question 8·8

(Page 209)

What symptoms and signs would you expect in the absence of glycogen phosphorylase in muscle?

Glycogen is needed to regenerate ATP in muscle during vigorous exercise. Hence, a deficiency of glycogen phosphorylase in muscle should cause excessive fatigue and intolerance of such exercise (especially the sprint). Other signs would be a low rise in the concentration of lactate during exercise and a more alkaline pH in muscle during the first few seconds of exercise.

Question 8·9

(Page 209)

Glucose appears to be stored with 3 g of water per gram of glycogen. What are the molar proportions of glucose and water stored?

The molecular weight of water is 10-fold smaller than that of glucose (18 vs 180). Hence, there are 10-fold more molecules of water than of glucose when equal weights are present (really 9-fold since the molecular weight of glucose units in glycogen is 162). With 3 g of water per gram of glycogen, the molar ratio of water to glucose is another threefold larger (3×9, or 27:1).

Can 1 g of glucose bind 3 g of water?

Each glucose unit in glycogen has five oxygen atoms; direct binding of water to glucose would require more than five molecules of water to be bound to each oxygen atom—a very unlikely theory.

How might the water be bound to glycogen?

There are two major possibilities to consider.

1. Structured water:

    If water were "held" by ionic groups, "electrolyte-free water" could be bound in the glycogen particle. Proteins are the most likely source of these ionic groups. The glycogen particle contains many associated enzyme molecules (proteins); some of these enzymes catalyze the formation and breakdown of glycogen, and many others regulate its metabolism (via kinases, phosphatases, phosphatase inhibitors).

2. Solvent water:

    Intracellular water and ions could be held in some way within the glycogen particle, but such an arrangement is difficult to visualize.

POSSIBLE SIGNIFICANCE

During vigorous exercise, water is lost from the body (via sweat and the respiratory tract). Water also moves from the ECF into the ICF as a result of the osmotic force created by the increased concentration of lactate particles in the ICF of muscle (i.e., when lactic acid is produced and its protons are buffered on intracellular proteins). Water released from the glycogen particle when glycogen is consumed can minimize the decline in ECF volume resulting from the osmotic effects of lactate.

Quantities

- Water content:
  - Total glycogen in muscle = 450 g.
  - Three grams of water per gram of glycogen = 1350 g of water.

- Osmole gain:
  - The rise in intracellular lactate resulting from the breakdown of glycogen could be close to 10 mmol/l (10 mosmol/300 mosmol/kg H_2O, or 3.3%).

- Overall:
  - To avoid a shift of water (1 liter) into the ICF of muscle from the ECF, there must be a 3.3% rise in water in the ICF of muscle.

Question 9·1

(Page 213)

What signs and symptoms would lead you to suspect that a patient is suffering from a low activity of PDH?

PDH is needed to metabolize glucose but not fatty acids or ketoacids. Hence, symptoms of organ dysfunction appear when regeneration of ATP depends on the oxidation of glucose. Because of its great dependence on glucose, the brain is likely to be most sensitive to a low activity of PDH, unless ketoacids are available. Symptoms of brain dysfunction include confusion, coma, seizures, and adrenergic response. Other organs will also show symptoms of dysfunction (e.g., muscle fatigue) when their supplies of fatty acids and ketoacids are low.

What circumstances can cause this condition?

A low activity of PDH can occur with an inborn error of metabolism, or it may be secondary to a vitamin B_1 (thiamine) deficiency since this vitamin is a cofactor for PDH (see Case 6·5, Julian's Brain Was Pickled).

When is a low activity of PDH a) the greatest danger to the patient; b) the least danger to the patient?

The patient is in the greatest danger when fatty acids and ketoacids are not available (i.e., when levels of insulin are high after consumption of a carbohydrate-rich meal). The patient is in the least danger when the level of ketoacids is high (i.e., during starvation, when levels of insulin are low) or when alcohol (1,3-butanediol) is metabolized in the liver.

How would you treat this patient?

Provide vitamin B_1 in case a deficiency is the cause of a low activity of PDH (if so, the response will be very rapid). Try to provide ketoacids (in the form of 1,3-butanediol) as a fuel to avoid the need to infuse an anion with sodium.

Question 10·1

(Page 217)

Carbon in fatty acids cannot be made into glucose at an appreciable rate in humans. Nevertheless, radio-labeled carbon in fatty acids appears in glucose. How can these two statements be reconciled?

Radioactive (^{14}C-labeled) acetyl-CoA enters the TCA cycle as citrate. Citrate is not optically active since the two acetyl groups on either side of the central carbon are identical.

$$
\begin{array}{c}
\text{H} \quad \text{H} \quad \text{OH} \\
| \quad\quad | \quad\quad | \\
\text{HOOC}^* - \text{C}^* - \text{C} - \text{CH} - (\text{COOH}) \text{ Removed by OGDH} \\
| \quad\quad | \\
\text{H} \quad (\text{COOH}) \text{ Removed by ICDH}
\end{array}
$$

*Radioactive carbon.

However, because the citrate molecule can approach the active site of aconitase (the next TCA cycle enzyme) from only one side, the enzyme can distinguish the newly added acetyl group from that contributed by oxaloacetate; aconitase moves the OH away from the newly added acetyl group. Removal of CO_2 by isocitrate dehydrogenase (ICDH) and 2-oxoglutarate dehydrogenase (OGDH) leaves the carbons donated by acetyl-CoA on succinate and hence on oxaloacetate, both of which can be incorporated into glucose.

Question 10·2

(Page 219)

Brown adipose tissue is an organ designed to generate heat during exposure to cold. Using your knowledge of the electron transport pathway, explain how this process works.

The production of heat by metabolism is normally tied directly to the need of brown adipose tissue for regeneration of ATP and hence to the activity of the ATP generation system. Brown adipose tissue also has the capacity to uncouple the regeneration of ATP from the oxidation of fuels. By making a new channel for H^+ in the mitochondrial membrane, brown adipose tissue allows H^+ to enter mitochondria down their electrochemical gradient—a pathway that is not linked to the generation of ATP. Brown adipose tissue may then oxidize fuels in an uncontrolled manner, thereby generating heat independently of the need for regeneration of ATP.

Question 11·1

(Page 234)

In a patient who lacks insulin, what quantity of ketoacids can the liver make per day? Assume that the supply of fatty acids is not rate-limiting and that the liver consumes 3000 mmol of O_2 per day.

- The stoichiometry for ketogenesis is:

  Palmitate (typical C_{16} fatty acid) + 6 O_2 ⟶ 4 Ketoacids

- Oxygen consumption by the liver is 3000 mmol/day. Therefore, 500 mmol of palmitate can be oxidized in this fashion if 100% of the oxygen consumed is "dedicated" to ketogenesis (no other substrates can be oxidized, and there is no flux in the TCA cycle).

- Given the whole stoichiometry, only 2000 mmol of ketoacids can be formed in a day unless the rate of consumption of O_2 by the liver increases. Fewer ketoacids will be formed if other pathways generate ATP (e.g., gluconeogenesis; see Figure 12·3, page 242).

SECTION FIVE

Structures and Pathways

STRUCTURES AND PATHWAYS OF ENERGY METABOLISM

The purpose of this section is to summarize the structures of the chemical compounds involved in energy metabolism and the pathways through which they are interconverted. This extra detail might help some students appreciate the biology and physiology of energy metabolism.

GENERAL CHEMISTRY OF BIOLOGICAL MOLECULES

The major difficulty in presenting chemical structures is that they are always three dimensional, often flexible, and frequently very complex.

CHEMICAL BONDS

Biological molecules are built on a framework of carbon atoms (C), to which are added other atoms, most frequently hydrogen (H), oxygen (O), nitrogen (N), sulfur (S), and phosphorus (P). The atoms are joined together in compounds by chemical bonds. Table 1 shows the numbers of bonds (usually referred to as the valence) that can be formed by the various elements most often found in the molecules involved in energy metabolism.

Table 1
Numbers of chemical bonds usually made by elements in biological molecules

Element	Chemical symbol	Number of bonds
Carbon	C	4
Hydrogen	H	1
Oxygen	O	2
Nitrogen	N	3
Phosphorus	P	5

Most molecules of energy metabolism contain a number of simple groups (Table 2).

Table 2
Common chemical groups in biological molecules

Chemical Group	Chemical symbol	Placement
Hydroxyl	OH	Attached to a carbon atom
Aldehyde	CHO	At the end of a carbon chain
Ketone	C=O	In a carbon chain
Carboxylic acid	COOH	At the end of a carbon chain
Amine	NH_2	Attached to carbon atom
Sulfhydryl	SH	Attached to carbon atom
Phosphate	PO_4	Attached singly or in chains

THE STRUCTURES OF CARBON-CONTAINING COMPOUNDS

Carbon atoms, joined in chains, branched structures, or rings, form the core of biological molecules. The four chemical bonds linking carbon atoms to other atoms usually radiate from the carbon in the middle, toward the four corners of a tetrahedron. Hence, structures containing carbon are necessarily three dimensional. They also have the potential to be asymmetric. If each of the bonds from a carbon atom is linked to a different chemical structure, the compound can exist in two mirror-image forms, like opposite hands. These mirror images are called optical isomers because, in aqueous solution, they rotate plane-polarized light in opposite directions. Because the two isomers contain the same atoms and the same chemical bonds, they have the same routine test tube chemistry. However, because of the structural specificity of enzymes, optical isomers behave as different compounds in biological systems.

Because the carbon atom can enter into so many kinds of linkages, even a relatively simple chemical composition containing three carbon atoms, three oxygen atoms, and six hydrogen atoms ($C_3H_6O_3$) can exist in at least four very different compounds, some of which have pairs of optical isomers.

THE CARBOHYDRATE SYSTEM

STRUCTURES

Circulating Fuels

The major circulating fuels in the carbohydrate system are glucose and L-lactate; their structures are shown in Figure 1(a). Monosaccharides have the formula $C_n(H_2O)_n$, in which n is most frequently 5 or 6. Common hexoses, such as glucose, fructose, galactose, and mannose, all have the chemical composition $C_6(H_2O)_6$; they differ because of the orientations of the hydroxyl groups on the asymmetric carbon atoms. Ribose, a pentose (five carbons), is found in cofactors such as ATP, NAD^+, and $NADP^+$, and in nucleic acids.

L-Lactate is a three-carbon product of glucose metabolism; it is a carboxylic acid that has a hydroxyl group on carbon number two (C-2). Since this carbon is asymmetric, there are two optical isomers: the L form, which is the product of mammalian lactate dehydrogenase, and the D form, which is made by bacteria in the intestinal tract. When D-lactic acid is produced at rapid rates, metabolic acidosis can develop, as outlined in Case 2·4, page 53.

Key Intracellular Intermediates of the Carbohydrate System

We identified two intermediates as major crossroads of carbohydrate metabolism: glucose 6-phosphate and pyruvate (Figure 7·10, page 179); their structures are shown in Figure 1(b). Pyruvate, a three-carbon intermediate closely linked to L-lactate, has two fewer hydrogen atoms and thus has a keto-group on the middle carbon.

FIGURE 1

Structures in the carbohydrate system

(a) Circulating fuels in
the carbohydrate system.

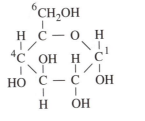

D-Glucose   L-Lactate

(b) Key intermediates in
the carbohydrate system.

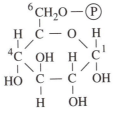

Glucose 6-phosphate   Pyruvate

Unique Bonds in the Carbohydrate System

The bonds linking glucose moieties in glycogen are illustrated in Figure 2. One
molecule of water is lost when bonds between glucose molecules are formed.

FIGURE 2

Major bonds in the carbohydrate system

The glucose molecules are linked
from C-1 of one glucose to C-4 of the
next glucose molecule in the main
chain; linkages are C-1 to C-6 at the
branch points.

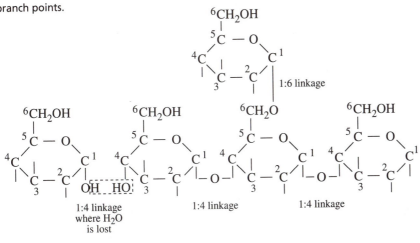

Storage Form of Carbohydrates

The storage form of glucose is glycogen, a polysaccharide that contains many molecules of glucose linked into a polymer. The key feature of glycogen is that it is branched (Figure 3). The linear portions of the structure consist of glucose molecules linked between C-1 and C-4; the branches are linkages between C-1 and C-6, as shown in Figure 2.

The molecule of glycogen begins on a primer, a tyrosine residue in a protein. With the continual formation of new growing points, it might be expected that a glycogen molecule could grow to a huge size. However, with increasing size, the molecule's outside edge becomes more and more crowded, and synthetic enzymes are prevented from gaining access to these growth sites. The observed maximal size is consistent with the frequency of branch points.

FIGURE 3

Glycogen

When an unbranched section has grown to approximately 14 residues, the branching enzyme transfers the terminal 6–8 residues to C-6 a few residues up the chain, thus forming another free C-4 growing point.

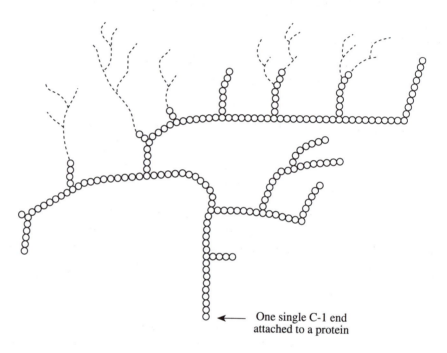

One single C-1 end attached to a protein

Other Compounds of Interest in the Carbohydrate System

Fructose: Fructose has the same chemical composition as glucose, but has a ketone group on C-2 instead of an aldehyde on C-1. This arrangement makes the compound chemically more reactive—a feature that has significance in hyperglycemic states when nonenzymatic glycation of proteins can cause long-term problems (see Chapter 4, page 117).

Disaccharides: Sucrose, the most common dietary simple carbohydrate, is a disaccharide formed from glucose and fructose. Lactose (in milk) contains glucose and galactose; mannose (in beer) contains two molecules of glucose.

Glycerol: Glycerol is a symmetrical three-carbon molecule with hydroxyl groups on each carbon. Glycerol forms the backbone of triacylglycerols (see Figure 9, page 286).

MAJOR PATHWAYS

The pathways of the carbohydrate system include glycolysis, glucogenesis, glycogen synthesis and breakdown, and the hexose monophosphate pathway.

Glycolysis and Glucogenesis

Glycolysis converts glucose to lactate or pyruvate; glucogenesis does the reverse. The two pathways share a number of enzymes that catalyze reactions that are always near their chemical equilibrium positions. The pathways differ in their nonequilibrium reactions; in addition, part of glucogenesis includes an excursion into a portion of the tricarboxylic acid cycle in mitochondria. The pathway of glycolysis is outlined in Figure 4; for simplicity, we do not show the detailed pathway of glucogenesis. Glucogenesis contains all the near-equilibrium reactions indicated by the reversible arrows in Figure 4. The pathway to "reverse" the flow past the nonequilibrium reactions of glycolysis is shown in Figure 8·9, page 200.

The intermediate glyceraldehyde (with a phosphate attached) and lactate (the end of the pathway) both have the same chemical composition ($C_3H_6O_3$), which is exactly half of the glucose molecule ($C_6H_{12}O_6$); hence, the second half of glycolysis extracts the energy equivalent of two ATP per lactate formed while merely rearranging the positions of the atoms attached to the carbon chain.

Although all tissues can carry out the pathway of glycolysis, only the liver and kidneys can perform the complete pathway of glucogenesis, releasing glucose into the blood. Muscle can convert lactate to glucose 6-phosphate and then to glycogen, but, because it lacks glucose 6-phosphatase, it cannot release glucose to the blood.

FIGURE 4

The pathway of glycolysis

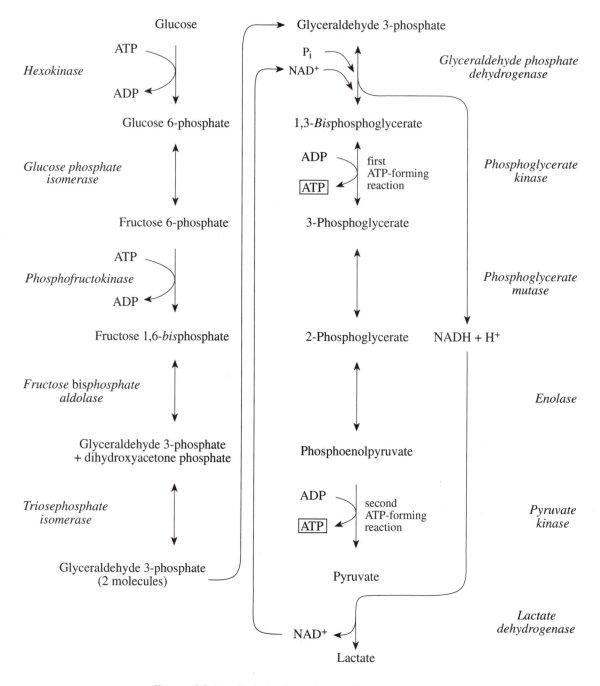

Entry of Other Carbohydrate System Fuels: Glycerol, released from triacylglycerols, enters glycolysis and glucogenesis at the triose phosphate level. Fructose also enters metabolism in the middle of the glycolytic pathway. Fructose is metabolized primarily in the liver, kidneys, and intestinal cells, through the pathway shown in Figure 5.

FIGURE 5

Entry of fructose into energy metabolism

A hereditary deficiency of fructose 1-phosphate aldolase causes hereditary fructose intolerance. In this condition, dietary fructose rapidly damages the tissues containing fructokinase because accumulation of fructose 1-phosphate traps phosphate and thus prevents regeneration of ATP (see Question 8·1, page 194).

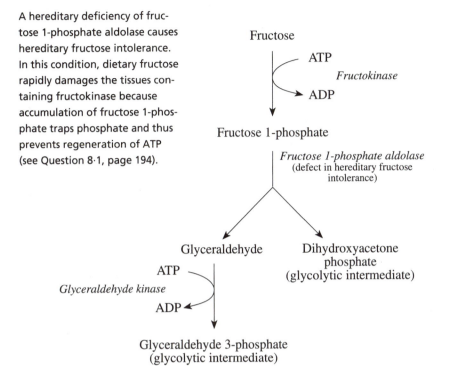

Fructose

ATP

Fructokinase

ADP

Fructose 1-phosphate

Fructose 1-phosphate aldolase
(defect in hereditary fructose intolerance)

Glyceraldehyde

Dihydroxyacetone phosphate
(glycolytic intermediate)

ATP

Glyceraldehyde kinase

ADP

Glyceraldehyde 3-phosphate
(glycolytic intermediate)

Synthesis and Breakdown of Glycogen

The pathways of synthesis and breakdown of glycogen are shown in Figure 6. The formation and breakdown of branch points involve further enzymes (not shown).

FIGURE 6

Synthesis and breakdown of glycogen

Uridine triphosphate (UTP) is similar to ATP; it is used to carry glucose to the growing glycogen chain. By attaching C-1 of its glucose to C-4 of the glucose on the end of the chain, it forms the unbranched chain portions of the glycogen molecule. For the breakdown of glycogen, inorganic phosphate (P_i) is added to the terminal glycosyl residue in glycogen by the enzyme glycogen phosphorylase.

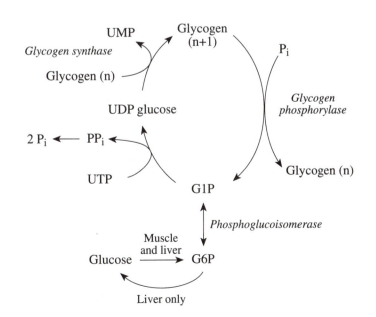

UMP

Glycogen (n+1)

P_i

Glycogen synthase

Glycogen (n)

Glycogen phosphorylase

UDP glucose

$2 P_i$ ← PP_i ←

Glycogen (n)

UTP

G1P

Phosphoglucoisomerase

Muscle and liver

Glucose → G6P

Liver only

THE PYRUVATE DEHYDROGENASE SYSTEM

Pyruvate dehydrogenase (PDH) is a multienzyme complex that oxidizes and decarboxylates pyruvate to an acetyl group that is held in a chemically active form by linkage to coenzyme A (Figure 7). The PDH complex contains five enzymes and three cofactors held together in a functional unit inside mitochondria. Three of the enzymes and the three cofactors participate in the reaction; the remaining two enzymes (PDH kinase and PDH phosphatase) carry out the phosphorylation and dephosphorylation, respectively, of PDH that are essential for its control (see page 212). The PDH reaction has two bound cofactors, thiamine pyrophosphate (derived from vitamin B_1) and lipoic acid, and also involves two nonbound cofactors, CoA and NAD^+.

FIGURE 7

Structures of acetyl-CoA, ATP, and citrate

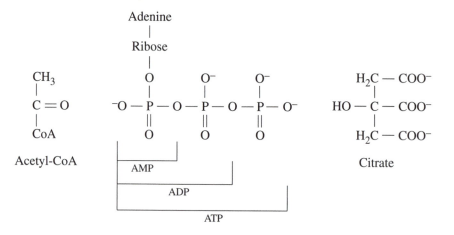

THE ATP GENERATION SYSTEM

The ATP generation system contains two sequential pathways, the tricarboxylic acid (TCA) cycle and the electron transport pathway.

TRICARBOXYLIC ACID CYCLE

The compounds of central importance to the TCA cycle (discussed on pages 216–17) are acetyl-CoA and citrate; ultimately, ATP will be formed via the electron transport pathway (Figure 7). Several other compounds are also important (Figure 8), including 2-oxoglutarate (which links to glutamate metabolism), fumarate (a byproduct of the synthesis of urea), and oxaloacetate plus malate (intermediates in glucogenesis).

FIGURE 8

Structures of 2-oxoglutarate, fumarate, malate, and oxaloacetate

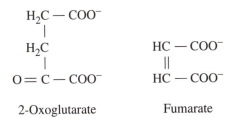

2-Oxoglutarate   Fumarate

Malate   Oxaloacetate

THE ELECTRON TRANSPORT PATHWAY

The electron transport pathway carries electrons along a chain of carriers to oxygen. The energy released from this pathway is used to create a proton gradient, which drives the synthesis of ATP. The overall process is also referred to as oxidative phosphorylation. The pathway is shown in Chapter 10 (Figure 10·3, page 218).

THE FAT SYSTEM

STRUCTURES

Almost all the fats are made from and broken down to acetyl groups (in the form of acetyl-CoA).

Circulating Fuels in the Fat System

The main circulating fuels of the fat system are fatty acids, ketoacids, and triacylglycerols (Figure 9). Acetoacetate and β-hydroxybutyrate are both called ketoacids though, chemically, only acetoacetate is a ketoacid.

Many compounds can be called fatty acids. Those relevant to energy metabolism typically contain even numbers of carbon atoms, (usually 16, 18, or 20) and 0–3 double bonds. However, many other fatty acids, some of which cannot be made in the body and must be supplied in the diet (essential fatty acids), are required for normal function.

FIGURE 9

Circulating fuels in the fat system

The major circulating fuels are fatty acids, triacylglycerols, and the so-called ketoacids.

$$H_3C - CH_2 - CH_2 - CH_2 - CH_2 - CH_2$$

Palmitate, a common fatty acid

$$| \\ CH_2 \\ | \\ CH_2 \\ | \\ CH_2 \\ | \\ CH_2 \\ | \\ CH_2 \\ | \\ CH_2 \\ | \\ CH_2 - CH_2 - COO^- + H^+$$

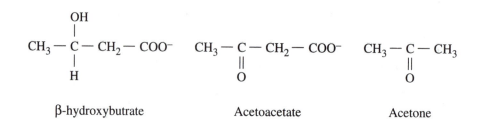

$$H_2C - O - \overset{O}{\overset{||}{C}} - (CH_2)_7 - CH = CH - (CH_2)_7 - CH_3 \quad \text{Oleic acid}$$

$$HC - O - \overset{O}{\overset{||}{C}} - (CH_2)_{14} - CH_3 \quad \text{Palmitic acid}$$

$$H_2C - O - \overset{O}{\overset{||}{C}} - (CH_2)_7 - CH = CH - CH_2 \quad \text{Linoleic acid}$$
$$CH = CH - (CH_2)_4 - CH_3$$

A triacylglycerol shown with three different
fatty acids for illustrative purposes (the glycerol
portion is contained in the box)

$$CH_3 - \overset{OH}{\underset{H}{\overset{|}{C}}} - CH_2 - COO^- \qquad CH_3 - \overset{}{\underset{O}{\overset{||}{C}}} - CH_2 - COO^- \qquad CH_3 - \overset{}{\underset{O}{\overset{||}{C}}} - CH_3$$

β-hydroxybutrate Acetoacetate Acetone

Storage Form of Fats

Triacylglycerols, the major storage form of the fat system, also circulate in the blood in the form of the lipoproteins VLDL and chylomicrons. Triacylglycerols contain three fatty acids, each linked to glycerol by an ester bond (Figure 10). Because there are so many different fatty acids, an enormous variety of triacylglycerol molecules exists.

FIGURE 10

Ester bond

Water is removed from a hydroxyl
and carboxyl group.

$$\text{CH}_3-(\text{CH}_2)_{14}-\overset{\overset{\displaystyle O}{||}}{C}-O-\text{CH}_2$$

Ester bond

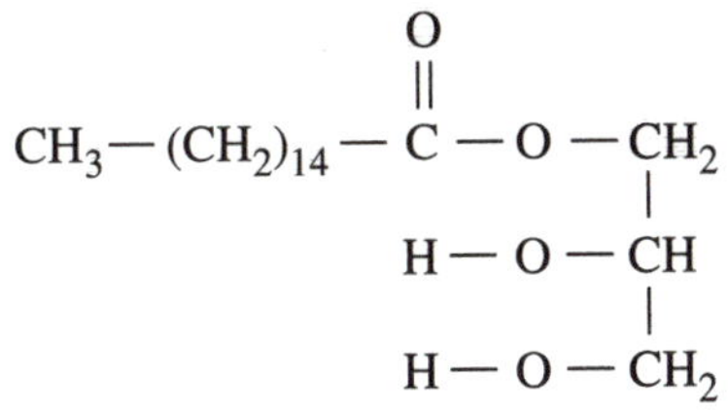

$$\text{H}-\text{O}-\text{CH}$$
$$\text{H}-\text{O}-\text{CH}_2$$

Key Intermediates of the Fat System

Acetyl-CoA (Figure 7), fatty acids (Figure 9), and malonyl-CoA (Figure 11) are the key intermediates of the fat system.

FIGURE 11

Structure of malonyl-CoA

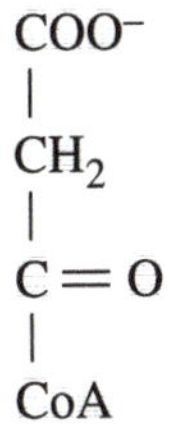

$$\text{COO}^-$$
$$|$$
$$\text{CH}_2$$
$$|$$
$$\text{C}=\text{O}$$
$$|$$
$$\text{CoA}$$

Malonyl-CoA

PATHWAYS

Synthesis and Deposition of Fatty Acids and Triacylglycerols

Synthesis of fatty acids starts with conversion of pyruvate to acetyl-CoA in mitochondria; the acetyl groups must first be moved into the cytosol (using citrate as the "carrier" molecule; see Figure 11·3, page 228) before they can be incorporated into fatty acids (Figure 12).

FIGURE 12

Synthesis of fatty acids

Fatty acid synthesis occurs mainly in the cytosol of the liver in humans. A fatty acid is built up, two carbons at a time, by an enzyme complex called fatty acid synthase. The growing chain is bound to the enzyme complex by acyl carrier protein (ACP). The cycle occurs seven times to make a 16-carbon fatty acid.

PATHWAYS	ENZYME	COMMENTS
$CH_3 - \overset{\displaystyle O}{\overset{\|}{C}} - CoA \ + \ H - ACP$		
CoA ↘	*Acetyl-CoA-ACP transferase*	Attach acetyl-CoA to acyl carrier protein (ACP)
$CH_3 - \overset{\displaystyle O}{\overset{\|}{C}} - ACP$		
Malonyl-CoA ↘ CO_2	*β-ketoacyl-ACP synthase*	Add malonyl-CoA
$CH_3 - \overset{\displaystyle O}{\overset{\|}{C}} - CH_2 - \overset{\displaystyle O}{\overset{\|}{C}} - ACP$		
NADPH + H⁺ ↘ NADP⁺	*β-ketoacyl-ACP reductase*	First reduction with NADPH + H⁺
$CH_3 - \overset{\displaystyle OH}{\underset{\displaystyle H}{\overset{\|}{\underset{\|}{C}}}} - CH_2 - \overset{\displaystyle O}{\overset{\|}{C}} - ACP$		
↘ H_2O	*β-ketoacyl-ACP dehydrase*	Remove water
$CH_3 - \overset{\displaystyle H}{\overset{\|}{\underset{\displaystyle H}{\underset{\|}{C}}}} = C - \overset{\displaystyle O}{\overset{\|}{C}} - ACP$		
NADPH + H⁺ ↘ NADP⁺	*Enol-ACP reductase*	Second reduction with NADPH + H⁺
$CH_3 - CH_2 - CH_2 - \overset{\displaystyle O}{\overset{\|}{C}} - CH_2 - ACP$		

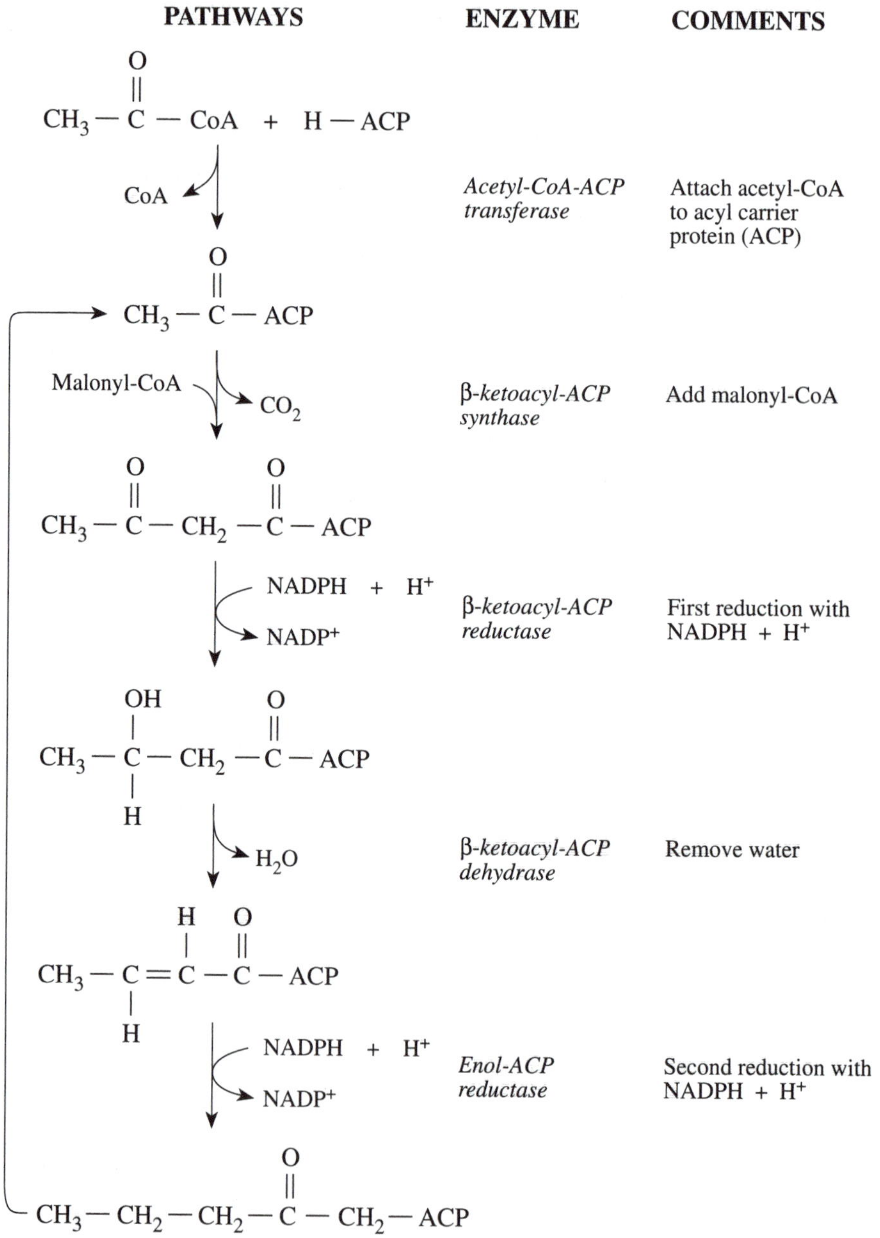

Triacylglycerols are synthesized in the liver using fatty acyl-CoA released from fatty acid synthesis; the deposition of triacylglycerols from the blood into adipose tissue uses essentially the same pathway (Figure 13).

FIGURE 13

Synthesis of triacylglycerols

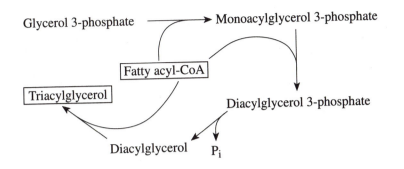

When fatty acyl-CoA is incorporated into triacylglycerols, α-glycerophosphate (derived from glycolysis) provides the glycerol backbone.

Breakdown of Triacylglycerols and Oxidation of Fatty Acids

Triacylglycerols are hydrolyzed from adipose tissue by hormone-sensitive lipase (see Chapter 11, pages 231–32). The fatty acids released are carried in the blood bound to serum albumin and are taken up by tissues as needed. Oxidation of fatty acids requires their activation to CoA derivatives in the cytosol, transfer into the mitochondria using the carnitine transport system (Figure 14), and oxidation by fatty acids to acetyl-CoA (Figure 15).

FIGURE 14

Carnitine system for the transport of fatty acyl groups into mitochondria

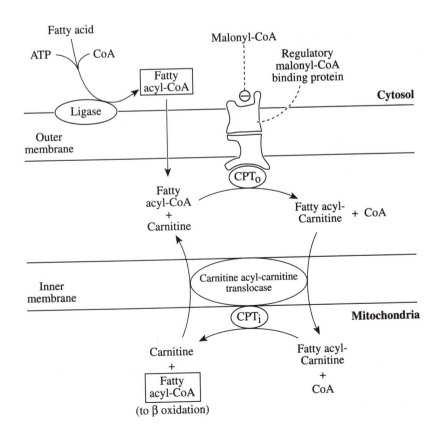

Malonyl-CoA, the major substrate for fatty acid synthesis, inhibits the transport of fatty acyl-CoA into mitochondria, thereby inhibiting the oxidation of fatty acids.

FIGURE 15

β Oxidation of fatty acids

Oxidation of fatty acids involves the sequential steps illustrated, resulting in the removal of two carbon fragments as acetyl-CoA with each spin of the cycle.

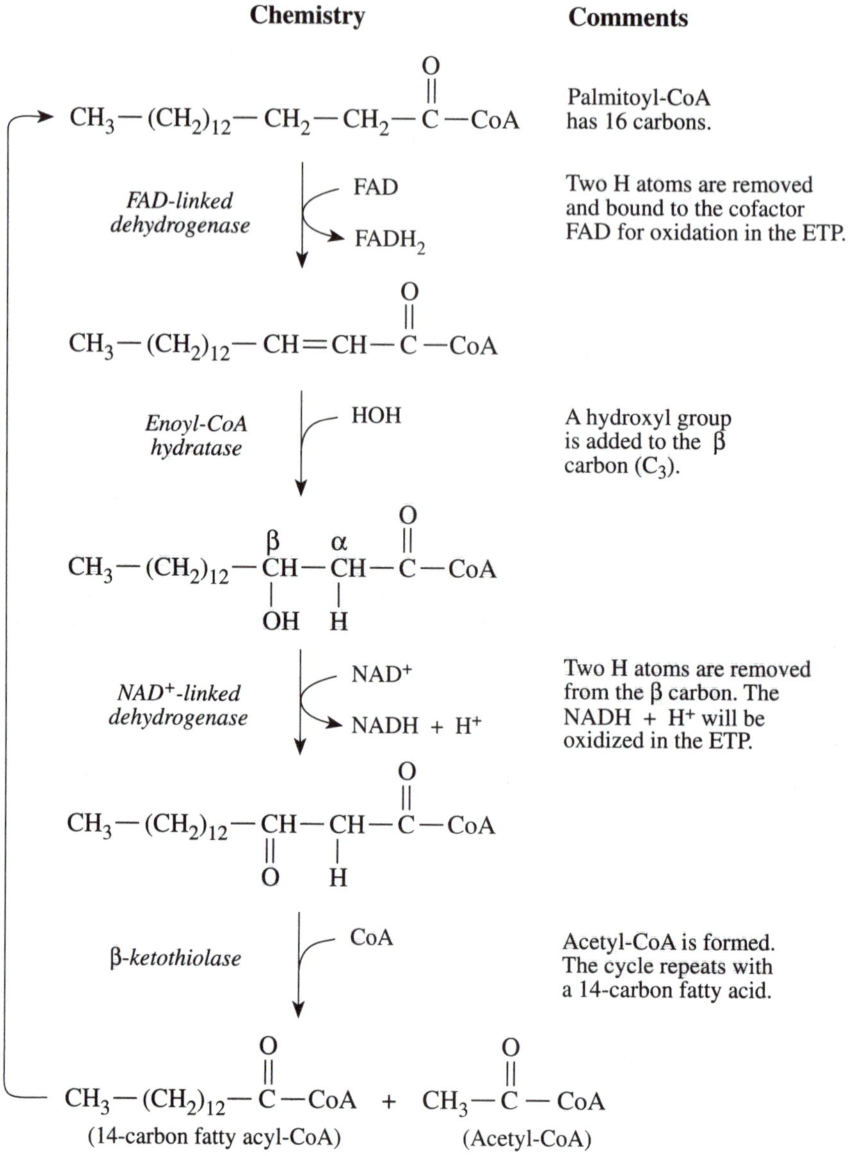

Chemistry   **Comments**

CH_3 — $(CH_2)_{12}$ — CH_2 — CH_2 — $\overset{\displaystyle O}{\overset{\|}{C}}$ — CoA — Palmitoyl-CoA has 16 carbons.

FAD-linked dehydrogenase   FAD → FADH$_2$ — Two H atoms are removed and bound to the cofactor FAD for oxidation in the ETP.

CH_3 — $(CH_2)_{12}$ — $CH=CH$ — $\overset{\displaystyle O}{\overset{\|}{C}}$ — CoA

Enoyl-CoA hydratase   HOH — A hydroxyl group is added to the β carbon (C_3).

CH_3 — $(CH_2)_{12}$ — $\overset{\beta}{\underset{OH}{C}}H$ — $\overset{\alpha}{\underset{H}{C}}H$ — $\overset{\displaystyle O}{\overset{\|}{C}}$ — CoA

NAD$^+$-linked dehydrogenase   NAD$^+$ → NADH + H$^+$ — Two H atoms are removed from the β carbon. The NADH + H$^+$ will be oxidized in the ETP.

CH_3 — $(CH_2)_{12}$ — $\underset{O}{C}$ — $\underset{H}{C}H$ — $\overset{\displaystyle O}{\overset{\|}{C}}$ — CoA

β-ketothiolase   CoA — Acetyl-CoA is formed. The cycle repeats with a 14-carbon fatty acid.

CH_3 — $(CH_2)_{12}$ — $\overset{\displaystyle O}{\overset{\|}{C}}$ — CoA   + CH_3 — $\overset{\displaystyle O}{\overset{\|}{C}}$ — CoA

(14-carbon fatty acyl-CoA)   (Acetyl-CoA)

Ketoacid Synthesis and Breakdown

Synthesis of ketoacids, shown in Figure 11·7, page 233, is active only in mitochondria of the liver during chronic fasting or ketoacidosis. Ketoacids are formed from the acetyl-CoA produced by rapid oxidation of fatty acids. Once the ketoacids are exported from the liver, they are delivered via the circulation to the brain and kidneys for oxidation.

Before a ketoacid such as β-hydroxybutyrate (β-HB) can be oxidized, it must first be converted to acetoacetate by β-HB dehydrogenase via a simple NAD$^+$-linked equilibrium reaction. Acetoacetate must then be activated to form its CoA derivative. This activation is linked to the TCA cycle; the CoA is transferred from succinyl-CoA, a TCA cycle intermediate, to acetoacetate (Figure 16).

FIGURE 16

Oxidation of ketoacids

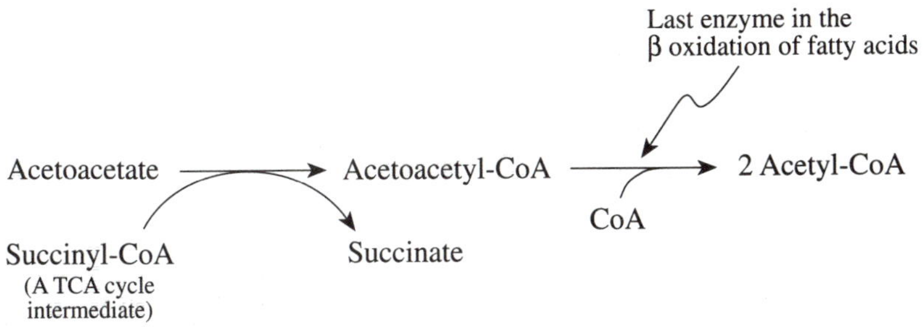

Last enzyme in the
β oxidation of fatty acids

Acetoacetate ⟶ Acetoacetyl-CoA ⟶ 2 Acetyl-CoA

CoA

Succinyl-CoA   Succinate
(A TCA cycle
intermediate)

THE PROTEIN SYSTEM

The protein system consists of the amino acids, proteins, and urea.

STRUCTURES

Circulating Fuels in the Protein System

The 20 amino acids circulate in the blood. They share a common general structure of
a carboxyl group and an amino group on the adjacent carbons. The side chains, R,
are different for the 20 amino acids (Figure 17).

FIGURE 17

Structure of an amino acid and of urea

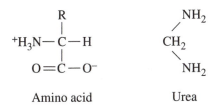

Amino acid   Urea

Unique Bond in the Protein System

Amino acids are held together in proteins by peptide bonds (Figure 18) that link the
carboxyl group of one amino acid to the amino group of the next amino acid.

FIGURE 18

The peptide bond

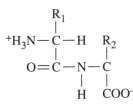

PATHWAYS

The protein synthetic pathway involves the transfer of genetic information from DNA to the precise sequencing of the amino acids in the growing protein chain; details of the mechanisms seem unlikely to help an understanding of energy metabolism.

The breakdown of proteins is simply hydrolysis of the peptide bonds that link the amino acids. Each amino acid has its own pathway for breakdown. A brief outline of some of these pathways has already been provided in Chapter 12.

TOXINS

The three toxins mentioned in this book are methanol (pages 56, 60), ethylene glycol (page 56) and toluene (pages 56, 71). Their structures and that of ethanol appear in Figure 19.

FIGURE 19

Structures of ethanol, methanol, ethylene glycol, and toluene

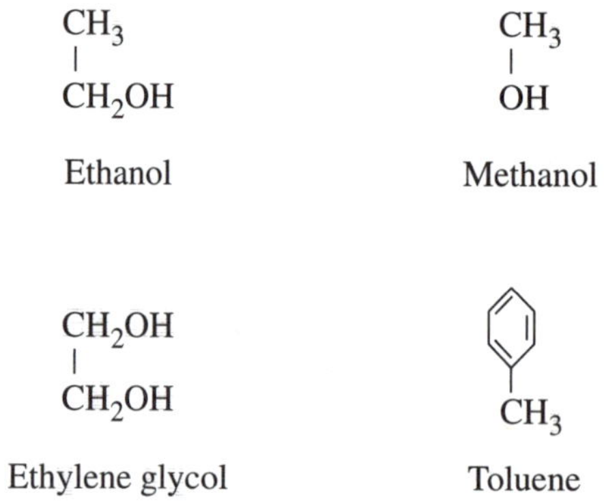

$$CH_3$$
$$|$$
$$CH_2OH$$

Ethanol

$$CH_3$$
$$|$$
$$OH$$

Methanol

$$CH_2OH$$
$$|$$
$$CH_2OH$$

Ethylene glycol

Toluene

SELECTED READINGS

CHAPTER 1

Fields, A., et al. 1982. Theoretical aspects of weight loss in patients with cancer: Possible importance of pyruvate dehydrogenase. *Cancer* 50:2183–88.

Linder, M. C., ed. 1991. *Nutritional biochemistry and metabolism with clinical applications.* 2d ed. Elsevier Science Publishing Company, Inc.

Owen, O. E., et al. 1983. Ketosis of starvation: A revisit and new perspectives. *Clin. Endocrinol. Metab.* 12:359–79.

Owen, O., et al. 1987. A reappraisal of the caloric requirements of men. *Am. J. Clin. Nutr.* 46:875–85.

Randle, P. J. 1986. Fuel selection in animals. *Biochem. Soc. Trans.* 14:799–806.

Rosen, O. M. 1987. After insulin binds. *Science* 237:1452–58.

CHAPTER 2

Carlisle, E., et al. 1991. Glue-sniffing and distal renal tubular acidosis: Sticking to the facts. *J. Amer. Soc. Nephrol.* 1:1019–27.

Elsner, R., and M. Daly de Burgh. 1988. Coping with asphyxia: lessons from seals. *NIPS* 3:65–69.

Halperin, M. L., and R. L. Jungas. 1983. Metabolic production and renal disposal of hydrogen ions. *Kidney Int.* 24:709–13.

Halperin, M. L., et al. 1987. Disposal of the daily acid load: An integrated function of the liver, lungs and kidneys. *Trends in Biol. Sci.* 12:197–99.

Halperin, M. L., et al. 1991. Biochemistry and physiology of ammonium excretion. In *The kidney: Physiology and pathophysiology*, ed. D. Seldin and G. Giebisch, 1471–89. New York: Raven Press.

Nelson, R. A. 1980. Protein and fat metabolism in hibernating bears. *Fed. Proc.* 39:2955–58.

Oh, M. S., and H. J. Carroll. 1979. The anion gap. *N. Engl. J. Med.* 297: 814–17.

Rahn, H. 1979. Acid-base and the "milieu interieur." In *Claude Bernard and the internal environment: A memorial symposium*, ed. E. Robin, 179–90. New York: Marcel Dekker, Inc.

Vasuvattakul, S., L. C. Warner, and M. L. Halperin. 1992. Quantitative role of the intracellular bicarbonate buffer system in response to an acute acid load. *Am. J. Physiol.* 262: R305–R309.

CHAPTER 3

Bayly, W. M., et al. 1989. Exercise-induced hypercapnia in the horse. *J. Appl. Physiol.* 67:1958–66.

Cheetham, M., et al. 1986. Human muscle metabolism during sprint running. *J. Appl. Physiol.* 61(1): 54–60.

Newsholme, E., and T. Leech. 1983. *The runner.* Fitness Books.

CHAPTER 4

Atchley, D. W., et al. 1933. On diabetic acidosis: A detailed study of electrolyte balances following withdrawal and reestablishment of insulin therapy. *J. Clin. Invest.* 12:297.

Brownlee, M. 1992. Glycation products and the pathogenesis of diabetic complications. *Diabetes Care* 15:1835–43.

Ellenberg, Rifkin. 1990. *Diabetes mellitus: Theory and practice.* 4th ed. Ed H. Rifkin and D. Porte, Jr. New York: Elsevier.

Greene, D. A., et al. 1992. Complications: Neuropathy, pathogenetic considerations. *Diabetes Care* 15:1902–25.

CHAPTER 5

Hale, P. J., and M. Natrass. 1989. Metabolic profiles in patients with insulinoma. *Clin. Endocrinol.* 30:29–38.

Jaspan, J. B. 1989. Hypoglycemia: Fact or fiction? *Hosp. Prac.* 11–14.

Unger, R. 1983. Insulin-glucagon relationships in the defense against hypoglycemia. *Diabetes* 32:575.

CHAPTER 6

Flatt, J. 1972. On the maximal possible rate of ketogenesis. *Diabetes* 21: 50–53.

Halperin, M. L., and S. Cheema-Dhadli. 1989. Renal and hepatic aspects of ketoacidosis: A quantitative analysis based on energy turnover. *Diabetes/Metabolism Rev.* 5:321–36.

Halperin, M. L., et al. 1983. Metabolic acidosis in the alcoholic: A pathophysiologic approach. *Metabolism* 32:308–15.

Hochachka, P., and T. Mommsen. 1983. Protons and anaerobiasis. *Science* 219:1391–97.

Lutz, P. L. 1992. Mechanisms for anoxic survival in the vertebrate brain. *Annu. Rev. Physiol.* 54:601–18.

Robinson, B. H. 1989. Lactic acidemia. In *Metabolic basis of inherited disease,* ed. C. Scriver, A. Beaudet, W. Sly, and D. Valle, 869–88. New York: McGraw-Hill.

CHAPTERS 7–12

We recommend the use of standard biochemical textbooks for details concerning the pathways of energy metabolism. In addition, we suggest the following articles:

Bender, D. A. 1985. *Amino acid metabolism.* Wiley.

Bosca, L., and C. Corredor. 1984. Is phosphofructokinase the rate limiting step in glycolysis? *TIBS* 9:372–73.

Cohen, P. P. 1981. The ornithine-urea cycle: Biosynthesis and regulation of carbamyl-phosphate synthetase I and ornithine transcarbamylase. *Curr Top. Cell. Reg.* 18:1–19.

Crabtree, B., and E. A. Newsholme. 1987. A systematic approach to describing and analyzing metabolic control systems. *TIBS* 12:4–12.

Denton, R. M., and A. P. Halestrap. 1979. Regulation of pyruvate metabolism in mammalian tissues. *Essays Biochem.* 15:37–77.

Denton, R. M., and J. G. McCormack. 1990. Ca^{2+} as a second messenger within mitochondria of the heart and other tissues. *Annu. Rev. Physiol.* 52:451–66.

Hatefi, Y. 1985. The mitochondrial electron transport system and oxidative phosphorylation system. *Annu. Rev. Biochem.* 54:1015–69.

Hers, H. G., and E. VanSchaftigan. 1982. Fructose 2,6-bisphosphate two years after its discovery. *Biochem. J.* 206:1–12.

Hue, L. 1981. The role of futile cycles in the regulation of carbohydrate metabolism in the liver. *Adv. Enzymol.* 52:247–330.

Jungas, R. L., M. L. Halperin, and J. T. Brosnan. 1992. Lessons learnt from a quantitative analysis of amino acid oxidation and related gluconeogenesis in man. *Physiol. Rev.* 72:419–48.

Kascer, H., and J. W. Portius. 1987. Control of metabolism: What do we measure? *TIBS* 12:5–14.

Katz, J., et al. 1986. The glucose paradox: New perspectives on hepatic carbohydrate metabolism. *TIBS* 11:136–40.

Krebs, H. A. 1970. The history of the tricarboxylate cycle. *Perspect. Biol. Med.* 14:154–70.

McGarry, J., et al. 1987. From dietary glucose to liver glycogen: The full circle round. *Annu. Rev. Nutr.* 7:51–73.

McGarry, J., et al. 1989. Regulation of ketogenesis and the renaissance of carnitine palmitoyl transferase. *Diab. Metab. Rev.* 5:271–84.

Mitchell, P. 1961. Coupling of phosphorylation to electron and hydrogen ion transfer by a chemiosmotic type of mechanism. *Nature* 191:144–48.

Nicholls, D. G. 1982. *Bioenergetics.* Academic Press.

Nicholls, D. G., and E. Rial. 1984. Brown fat mitochondria. *TIBS* 9:489–91.

Owen, O. E., et al. 1983. Ketosis of starvation: A revisit and new perspectives. *Clin. Endocrinol. Metab.* 12:359–79.

Pande, S., and M. Murthy. 1989. Carnitine: Vitamin for an insect, vital for man. *Biochem. Cell Biol.* 67:671–73.

Pilkus, S. J., M. L. Mahgrabi, and T. H. Claus. 1988. Hormonal regulation of hepatic gluconeogenesis and glycolysis. *Annu. Rev. Biochem.* 57:755–83.

Racker, E. 1980. From Pasteur to Mitchell: A hundred years of bioenergetics. *Fed. Proc.* 39:210–15.

Rolleston, F. S. 1972. A theoretical background to the use of measured concentrations of intermediates in study of the role of intermediary metabolism. *Current Topics in Cell Reg.* 5:47–75.

VanderLars, W. J., G. Elzinga, and R. C. Woledge. 1989. Energetics at the single cell level. *News in Physiol. Sci.* 4:91–93.

INDEX

Index

Phosphoenolpyruvate (PEP), 191F, 204, 205
Phosphoenolpyruvate carboxykinase, 200F, 202F
Phosphofructokinase-1
    control of, 195–197F
    fructose 2,6-*bis*phosphate inhibition of, 198
    glycolysis control and, 195F–197F
Phosphofructokinase-2, 198F
Phospholipids in membranes, 180F
Polydipsia, 15
Polyphagia, 15
Polyuria, 15, 70, 159
Potassium. *See* hyperkalemia; hypokalemia
Product inhibition, 16, 171–174
Prosthetic groups, 185
Protein, 5
    amino acids and, 29F, 238F
    consumption during fasting, 31T, 66F
    glucose from, 6, 66F, 119, 262, 265, 272
    disposal of dietary, 29F
    energy content, 5T, 6
    H^+ excretion and, 66F
    system, 25F, 26T, 237–246
Pulmonary reserve, 41
Pyruvate, 25F
    controls over, 212F–213F
    crossroads, 179F, 202F, 204
    structure, 279F
Pyruvate carboxylase, 200F, 202F
Pyruvate dehydrogenase. *See* PDH
Pyruvate kinase, 200F, 204–205F

Receptors, 134, 183F–184
Red blood cells, 27, 95
Renal failure. *See* kidney failure
Renal insufficiency. *See* kidney failure
Respiratory acidosis. *See* acidosis
Respiratory quotient, 86, 89

Seal, 70, 75–76
Second messengers, 183F–184
Sigmoid kinetics, 175F

Sodium acetate, 56, 255
Sorbitol, 115, 116F
Spleen, 76, 96
Sports anemia, 86
Sprint
    anaerobic glycolysis and, 82, 88
    ATP source for, 80, 82, 88
    hyperventilation and, 41, 50–51
Starvation, 24. *See also* fasting
Stomach function, 34F–35, 250
Substrate activation, 16, 171–174
Sucrose, 5, 194, 271
Sulfur, 55
Sulfuric acid, 55
Sweat, 7,123
Sympathetic nervous system, 123, 126

Thiamine, 130, 151
Thiazide, 106,112
Thyroid hormone, 135
Toluene
    metabolism, 56, 71F, 72
    structure, 292F
Transamination, 243–244
Transport, across membranes, 181–183, 184F
Trauma, 8,14F, 247
Triacylglycerols, 5
    adrenaline and, 36F
    breakdown, 230F–232, 289
    energy content, 5T, 6
    exercise and, 90
    fasting, consumption during, 31T, 230F
    fat system and, 225F–231
    formation from glucose, 26F
    structures, 286F
    synthesis, 229, 289F
    transport in lipoproteins, 12
Tricarboxylic acid cycle (citric acid cycle,
    Krebs cycle), 215–217F
Tumor
    fuel use and, 32, 33F, 267
    hypoglycemia and, 133, 138F

Ulcers, urease and, 67
Urea, 10, 63
    excretion of, 63, 66F, 241–245
    nitrogen source in ruminants, 67
    structure, 291F
    water reabsorption and, 64
Urea cycle, 244F, 245, 257
Urease, GI ulcers and, 67

VLDL, 12F, 17, 223, 226F
Volume of distribution, 112, 113T, 151
Vomiting
    acid-base imbalances and, 56, 153, 154,
      161, 164, 165–166
    alkalosis and, 162
    sodium loss from, 7

Weight control, 24–31. *See also* weight gain;
    weight loss
Weight gain, 3, 7F
Weight loss. *See also* cachexia
    considerations in, 7F
    fasting and, 31T, 61, 66
    IDDM and, 18F–19F
    NIDDM and, 106,112
    surgery and, 3
    trauma and, 8,14F
    triacylglycerol consumption and, 37
Wernicke-Korsakoff syndrome, 130, 151

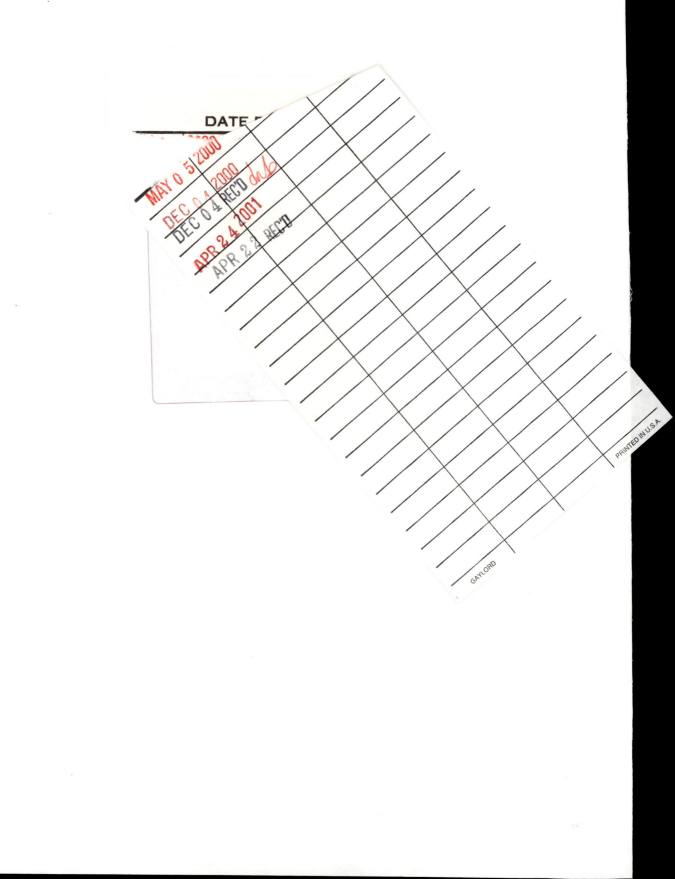

DATE

MAY 0 5 2000

DEC 0 4 2000

DEC 0 4 REC'D

APR 2 4 2001

APR 2 2 REC'D

GAYLORD

PRINTED IN U.S.A.